THE ENGINEER ENTREPRENEUR

Daniel T. Koenig, P.E.

New York　　　ASME Press　　　2003

© 2003 by The American Society of Mechanical Engineers
Three Park Avenue, New York, NY 10016

All rights reserved. Printed in the United States of America. Except as permitted under the United States Copyright Act of 1976, no part of this publication may be reproduced or distributed in any form or by any means, or stored in a database or retrieval system, without the prior written permission of the publisher.

INFORMATION CONTAINED IN THIS WORK HAS BEEN OBTAINED BY THE AMERICAN SOCIETY OF MECHANICAL ENGINEERS FROM SOURCES BELIEVED TO BE RELIABLE. HOWEVER, NEITHER ASME NOR ITS AUTHORS OR EDITORS GUARANTEE THE ACCURACY OR COMPLETENESS OF ANY INFORMATION PUBLISHED IN THIS WORK. NEITHER ASME NOR ITS AUTHORS AND EDITORS SHALL BE RESPONSIBLE FOR ANY ERRORS, OMISSIONS, OR DAMAGES ARISING OUT OF THE USE OF THIS INFORMATION. THE WORK IS PUBLISHED WITH THE UNDERSTANDING THAT ASME AND ITS AUTHORS AND EDITORS ARE SUPPLYING INFORMATION BUT ARE NOT ATTEMPTING TO RENDER ENGINEERING OR OTHER PROFESSIONAL SERVICES. IF SUCH ENGINEERING OR PROFESSIONAL SERVICES ARE REQUIRED, THE ASSISTANCE OF AN APPROPRIATE PROFESSIONAL SHOULD BE SOUGHT.

ASME *shall not be responsible for statements or opinions advanced in papers or . . . printed in its publications* (B7.1.3). Statement from the Bylaws.

For authorization to photocopy material for internal or personal use under those circumstances not falling within the fair use provisions of the Copyright Act, contact the Copyright Clearance Center (CCC), 222 Rosewood Drive, Danvers, MA 01923, tel: 978-750-8400, www.copyright.com.

Library of Congress Cataloging-in-Publication Data

Koenig, Daniel T.
  The engineer entrepreneur / Daniel T. Koenig.
    p. cm.
  ISBN 0-7918-0193-4
  1. Engineering firms—Management. 2. Entrepreneurship. I. Title.

TA190 .K63 2003
620'.0068—dc21   2002036668

To my wife Marilyn and my sons and daughter-in-law Michael, Alan, and Cindy for once more showing compassion and understanding and tolerance in supporting my quest to write this book. And to my granddaughter Alison who may some day read this book and profit from it. And finally to my late daughter-in-law Deborah, may she rest in peace. I love you all.

Contents

Introduction   ix

1. A Short History of the Organization of Engineering Work: Where We've Been and Where We're Likely to Go   1
    The Organization Man Concept 1
    The Change in the 1980s 4
    The Resulting New Role of the Engineer 7

2. The Engineering Education Needs for the Future   11
    The Skills Needs Assessment, Then and Now 11
    The Skills Philosophy Needed for the Successful Engineer
        Practicing in the Early 21st Century 14
    Current Training Regime for Engineers While in School,
        and Why It Is Inadequate 16
    How the Curriculum Must Change to Be an Effective Training
        Mechanism 18
    How to Introduce Entrepreneurship into the Curriculum 26
    Summary 32

3. The Entrepreneur's First Step: Learning About and Using the Art of Project Management   35
    Enlightened Self-Interest 36
    The Skills to Bridge the Engineer-Employee to the Engineer-
        Entrepreneur 39
    A Short Tutorial on Project Management 39
    Some Final Thoughts on Project Management 69

4. From a Business Team Member to a Business Owner   71
    Purpose of a Business 71
    Vision and Mission Statement 74
    Functions of a Business 78
    The Basic Organization of a Business 79
    Using the 7 Steps of the Manufacturing System 81
    The Intricacies of Organization in Applying the 7 Steps of
        the Manufacturing System, Using the Basic Manufacturing
        Resources Planning (MRP II) System 86
    Developing an Organization Compatible to the 7 Steps
        of the Manufacturing System 89

Differences Between a Small Business and a Business Team...............92
Company Support for Business Teams Compared
  with the Entrepreneurial Approach 94
The Continuing Road Toward Entrepreneurship.........................96

5. All Communications Are Good, Some Better Than Others   99
The Communications Process 100
Barriers to Effective Communications......................................103
Some Guidelines for Effective Communications......................106
Learning How to Listen 109
Communicating with the Customer 113
Summary..122

6. Going from Raw Emotions to a Polished Commercial Offering; The New Product Introduction Process   125
Using the Scientific Method to Introduce a New Product or Service.....125
Reason for a Business—A Great Idea that Can Be Commercialized......126
The New Product Introduction Process Philosophy 128
The New Product Introduction Process Technique..................130

7. Financial Potpourri   157
The Classic Ways of Measuring Financial Status.....................157
Constructing Budgets 166
Budget-Related Measurements..174
Some Useful Management Techniques for Maintaining Financial
  Viability...180
Keeping Focused on Controlling the Operation 190
Financial Measurements and the Business Plan 193
Summary..194

8. E-Commerce and the Virtual Corporation: Opportunities for the Entrepreneur   195
Imperative of Being Wanted 196
The Supply Chain Evolution..198
How to Join a Supply Chain 201
Introducing Your New Product to the World Using
  a Supply Chain 203
Being Able to Handle Transactions via E-Commerce...............204
Some Concerns About E-Commerce Order Entry Systems.....213
Summary..214

9. The Business Plan for the Entrepreneur   215
The Content of the Business Plan 216
Some Final Words on the Structure of a Business Plan 225
An Example of the Development of a Business Plan 227
Summary 268

10. Toward World-Class Performance   271
Introduction 271
Basic Truisms 272
Understanding the 7 Steps of the Manufacturing System 272
A Discussion of Each of the 7 Steps 275
Why the 7 Step Sequence Applies Even Though a Company
  May Be Ignorant of Its Existence 283
Why Understanding the 7 Steps and the Linkages Between Them
  Is Necessary for Developing "Communications Excellence" 285
Why the 7 Steps of the Manufacturing System Are Universal,
  No Matter What the Product Is 286
When a Step Is Not Readily Apparent: Why That's Dangerous
  and How to Find the Missing Step(s) 290
Why the Facts (Data) Need to Be the Same and Cascade from
  Step to Step, and What Happens When This Rule Is Disregarded 291
How to Evaluate Your Company to Determine How Well You
  Comply with the 7 Steps of the Manufacturing System 292
How to Compare Your Results Against Best Practices,
  A Gap Analysis 297
Defining the Opportunities for Improvements 301
Summary 304

Appendices   307
Appendix A: Investigation Points (product company) 309
Appendix B: Investigation Points (service company) 319

Glossary   329

Selected Related Readings   337

Index   339

Introduction

Many books are written about starting and managing small businesses, especially now, with Internet e-commerce opportunities beckoning. You have many excellent general-purpose references to chose from. This being the case, another startup book may seem redundant and not worth your time and resources to pursue. However, few books, if any, focus on how to start up and run an engineering-based company. If this is what you've been considering, this book is for you.

Why is a book on engineering company management needed? Simply because engineering-based companies focus on scientific principles for their products and services, as opposed to just providing for the wants and needs of their customers. A customer may have all sorts of needs, ranging from entertainment to legal representation to sustenance. Whatever those needs are, if they require compliance with the laws of science and mathematics, this puts a significant constraint on the provider. Engineering-based businesses always have that constraint placed on them. This means that running an engineering-based business includes elements beyond normal businesses. If these constraints are not understood, then failure looms.

An engineering-based company has a special focus: always complying with the laws of the applicable science. This is a constraint that the novice engineer entrepreneur must be aware of and understand how to manage. The successful engineering-based business is born not simply out of the desire to have one's own business, but are equally based on applying sound engineering principles in providing the product or service. The merging of these overriding laws of engineering-based business is what this book is about.

This book has 10 chapters; several present the various phases of setting up and running a business based on exploiting technology (perhaps better expressed as applying technical know-how to satisfy customer needs). Each chapter has illustrations to take you from theory to pragmatic "how to" scenarios. I will take you on a journey from contemplation through application to help you determine whether you should consider starting a business. I will show you the traits and prerequisites for success.

Several chapters are on the basic skills that are useful for the soon-to-be engineer entrepreneur. You will view scenarios of what business structures are evolving into (some say e-commerce and virtual corporations). The last portion of the book describes the cornerstone of planning—the business plan and discusses what it means to be "world class." All segments of the business plan are fully explained and annotated with examples. Being "world class" means performing at optimum level; I have explained in much detail what that means. I also include something near and dear to the heart of any engineer—many checklist summaries of all the things the entrepreneur ought to do to ensure a successful startup or continue the growth of an existing business. Although the book is designed to answer the question, "Should

I have my own business?" it will be equally useful for those already running a small engineering-based business who feel they need help to make it better.

Why am I qualified to write this book? First, I've run my own business for many years. Second, I consult on technical systems, business optimization, and organizations. Third, I learned my trade by being a turnaround manager within large, medium-sized, and small companies, not in the ordinary financial orientation, but from being involved with a combination of heavy operational corrections activities and implementing financial practicalities. Since I do a considerable amount of work along the lines of turnarounds—in other words I fix businesses that have failed or are in the process of failing—I've developed a good sense of the pragmatic do's and don'ts for actions to take for success. And since I operate my own technical-based consulting service company, I've had to practice what I advocate.

In addition, I've written 3 books on manufacturing management, on manufacturing engineering, computer integration of business operations, and shop operations work station dynamics. All of these books deal extensively on the interdependence of an integrated team working toward a goal to optimize profits. I've taught at the university level on all of these subjects plus turnaround management as an expert adjunct professor off and on since 1976. These topics make up the core competencies of running small businesses. I consider it my avocation to help fledgling businesses and those in trouble become optimum performers. So, in writing this book I am documenting how I currently earn my living.

I hope after reading the book you will have enough insights to make that all-important decision: Do I take the plunge? Do I become an engineer entrepreneur? Or do I continue the safer route of remaining an employee of some organization, large or small? Do I want security over all things? Or do I take my chances grabbing for the brass ring? The ultimate decision is yours. I would like to think that after reading my book, you will have the right information to make the decision that's best for you.

This book has taken almost 2 years to produce, since it has been a labor of commitment to our profession and has been done between and around business assignments I've taken on. It has meant time "mentally" away from my family, and I thank them for having patience with my pursuit and supporting and encouraging me every time I needed that push to continue with the preparation of the manuscript. Marilyn, Mike, Alan, and Cindy, I hope you find the outcome worth the disruptions.

Daniel T. Koenig, P.E.
Lake Worth, Florida

Chapter 1

A Short History of the Organization of Engineering Work: Where We've Been and Where We're Likely to Go

THE ORGANIZATION MAN CONCEPT

When I left the United States Coast Guard in 1966 for a position in industry, the business world was a much different place than it is today. International competition was more theory than practice. Engineering work transpired more in large organizations than in entrepreneurial smaller firms that are commonplace today. Engineers entering the work force were juniors in large bureaucratic organizations with very large resources potentially available to the technical worker. In fact, when I started in industry I felt very comfortable in my role. I was a junior officer in the service, and as an engineer in a large organization I felt as if I was once more a junior officer. In industry, the engineers and managers were the officer class, the technicians and shop foremen were the sergeants and higher-ranking enlisted personnel, and the rank and file factory workers were akin to the enlisted ranks. As I said, this was very familiar, and in truth it should have been because it used the same hierarchical organization structure from the "ultimate leader" down to the lowest-paid job in the company.

The way of work was very direct. Many engineers did the work, perhaps too many. Tasks were assigned in duplicate, often to see if the solutions proposed were complementary or at odds. Junior engineers virtually never were allowed to work on their own, but predictably with a more seasoned colleague. The type of work the junior engineers did would be considered similar to technician assignments today, if done at all. They did the "grunt number crunching," solving numeric problems, making graphs, setting up tests, and making suggestions rather than decisions about conclusions. Decisions were reserved for the senior engineers and managers. The junior engineer's work was routine problem-solving in a manner prescribed by the company's procedures for conducting engineering work. The excitement of working on technical problems was there but perhaps more from the viewpoint of an interested bit player and not as a critical contributor. The truth was, the only critical contributors were the leaders of this rather large "platoon" of engineers. He (and it was almost 100% male) had all of these resources to deploy at will. He also had a host of backup support in reserve. These were "expert" professionals more like current-day academic researchers, and they were essentially on call to answer questions as they came up. In addition, these

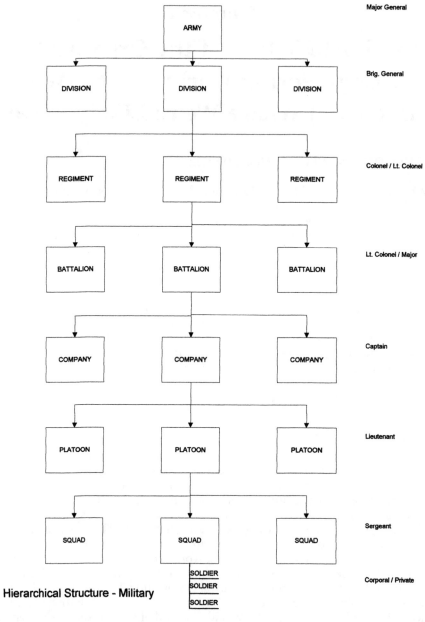

Figure 1-1a. Hierarchical organization structure—military compared to industry.

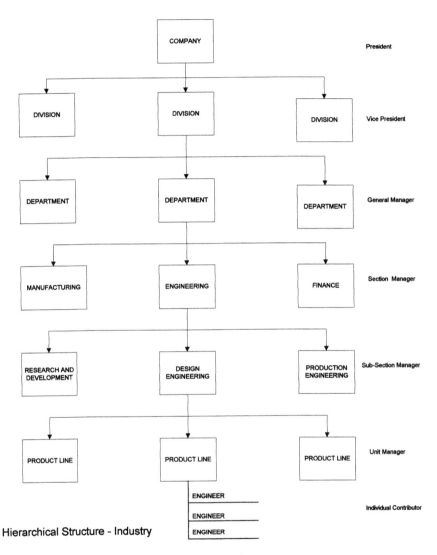

Figure 1-1b.

experts did basic research, wrote papers, and gave dissertations on their specialties, all with the purpose of supporting the company's commercial goals.

Another social structure the young engineer faced was that of paternalism. In the large organization, the underlying theme was that "father knew best." In this case, "father" was the large company itself. It had mired rules and regulations ranging from everything like dress codes to the size of office space dependent on rank within the organization. These large companies were governments unto themselves with all the executive and judicial authorities to control the activities of its employees, the "citizens" of the corporation. The creative engineer had to make sure his creativity fit within the administrative rules of the corporation; otherwise, no matter how good the idea, it would never see the light of day as actuality. In fact, lots of creative energy was by necessity transferred into how to get around the rules if the technical idea didn't fit. Companies acted like governments do today. Paternalism made for a sense of security for those who felt comfortable within its confines. The opposite was true for those who did not. Those who fit in had good secure jobs for life. Those who didn't were the rebel outcasts who did unusual things like starting entrepreneurial companies of their own at least 2 decades before economic necessity made it the thing to do.

An interesting aside: the modern entrepreneur startup of today probably owes its popularity to the initial downsizing of these monolithic vertically integrated companies of the 1960s and 1970s. The initial downsizing of these companies disbanded many of the support groups, the engineering think tanks. These people had to find work where opportunities were being curtailed; hence, they used their expert knowledge to launch new companies and new products. Also, these new products came into being sooner than would have been the case if the downsizing had not occurred and the monoliths maintained their total control of the technologies. If that had been the case, any new product would have had to fit the business strategy of the monolith to see the light of day.

Back to the situation of the 1960s and 1970s. The point to be made is that the engineer in industry had vast resources to draw from to get virtually any answer he needed to solve a technical problem. This was the glory years of large project undertakings being able to be commanded by single entities. This was a time when engineers could and did specialize in nitches and could expect to enjoy an entire career within a narrow specialty. All the support and resources were there; all one had to do was make use of them. It meant that an engineer in one discipline could be almost totally ignorant of the technologies managed by other disciplines. The lines could be drawn between mechanical, industrial, civil, and electrical, for example. Aside from knowing where that line was, the project manager of the time only needed to know when another discipline had to be brought in. Projects in that era needed lots of different engineers to complete because cross-discipline training, to say nothing of experience, was not a common commodity. And so it went until the era of globalization forced significant downsizing of industrial organizations. This trend started in the 1980s and continues today. It totally changed the nature of engineering work.

THE CHANGE IN THE 1980s

The oil shock of the mid to late 1970s led to the demise of the large monolithic industrial organization. Suddenly energy became more expensive and managing cash flow more intensive. We discovered that energy cost directly or indirectly affected everything a company does, from the cost of running the heating and air conditioning facilities to the cost of raw materials going up due to transportation cost increases. The price of everything went up and inflation became a household word in North America. This meant every opportunity had to be taken to reduce costs just to survive. Large corporations had a much more difficult chore to accomplish this because of their large intertwined structures. Many of the nonline activities fought ferociously to protect their turf. They instinctively reacted to defend their way of life and the very purpose of their existence. This made it very difficult for large corporations to reduce operating expenses enough to absorb the increases in energy costs. One thing was certain: fewer people would have to do more if the large entity was to survive.

On top of this and perhaps coincidentally we experienced new competition. International competition came to be a major factor. The age of "globalization" was upon us. At the end of World War II, international competition was severely depressed due to the destruction brought about by that conflict. For the next 25 to 30 years we lived in an artificial world where the majority of economic activity existed primarily in the United States and Canada (although to a much smaller extent because of Canada's population being only 10% of its southern neighbor). So we had markets where by all the players had virtually the same labor, materials, and distribution costs, and these costs could be set to any level we chose. We competed mostly within national boundaries and we all played by the same rules. The rules were very comfortable and didn't require "ungentlemanly" methods for us all to share a goodly piece of the pie.

At the same time we were involved with this internal game, our diplomats were striving to build up the devastated economies to keep them from joining the Soviet Union's sphere of influence. We were very successful at that. In the late 1970s and 1980s, the Japanese and Germans were ready and able to enter the world market once more, and since the truly significant market was North America, they were drawn here as strongly as a moth is attracted to a flame. The results were an economic Pearl Harbor. They came with lower prices and better quality. They came to compete successfully and did. The large corporations were caught flatfooted. The cozy national markets suddenly became global, where global meant that North American markets were now open to all comers who wanted to compete. Since these renewed international competitors learned their lessons well after World War II, they won virtually every battle for the market they engaged in with the large monoliths. To compete, the North American companies had to reduce their costs and improve their quality. This was a significant challenge and unfortunately, many had extreme difficulty coping with this new paradigm. This was another cause for the onset of downsizing.

With the advent of new and fiercer overseas competition coupled with higher cost of energy, there was a more urgent need to generate a substantial positive cash flow. The need for quick profits to meet demands of financial markets of the 1980s became intense. It became obvious that financial markets were not going to favor local, that is, national, companies over more efficient foreign corporations. The ability to support large networks of staff thus lowering profit margins was not acceptable. This became evident because of the paradigm shift. What used to be acceptable profit margins in the 1960s and 1970s was no longer so. The more efficient foreign-based competitors could show better results at lower prices; hence domestic competitors were hard-pressed to match that performance without doing things differently.

The obvious thing to do was to cut expenses every way possible without utterly destroying the company's ability to make its products or deliver its services. This meant paring down of resources—line predominates. Staff functions, since they didn't directly affect the "current" ability to make products or deliver services, were considered luxuries that these hard-pressed monoliths could no longer afford. Hence, the people and their material resources were disposed of as quickly as possible. In fact, a new term was introduced into business language—"reengineering." Consultants had a field day helping companies cut costs. The process employed focused not so much on becoming more efficient, but on lopping off activities that weren't by definition direct cause and effects of profit achieving. Obviously this meant that the superb staff support groups for the product groups were high on the list for corporate execution—and most were eliminated.

In addition, line functions were closely examined for excess capabilities. And just in case, reserve capabilities were looked at closely to ensure that only those needed were kept. The phrase "lean and mean" entered the lexicon and we still hear it to this day.

What this all reduced to was companies with less overhead, thus with lower costs, could begin to compete with the efficient foreign competitors. Coupled with all-out drives to lower energy, direct and indirect related costs made it possible for some of the monoliths to survive. However, as we know, many disappeared or simply became woeful shells of their former prowess. Companies did only what they had to do and used a minimalist philosophy in doing so. The concept of buy as needed versus having on the shelf for when needed became the name of the game. Research became applied research and had to be firmly wedded to the profit motive. Many engineers in industry pleaded that we were eating our "seed corn" by not putting funds into basic research. But the economic sense of the time precluded any significant change of heart.

The demise of the vertically oriented company became a reality to be replaced, with the discovery of the supply chain. Many companies became partial producers of a product they once produced in its entirety. Some even went so far as to dispose of manufacturing capabilities and just retained design and marketing functions for their product. Although many who tried this tactic soon found out that the manufacturing capability was the heart of the operation. And by teaching other companies how to manufacture their product was simply giving away technology that was the only true barrier to entry before. This was especially true for companies that went to Asia to get lower manufacturing costs by hiring other firms to do their manufacturing. They soon found that the Asian companies were reverse engineering their products and not only profiting from being the recipient of outsourcing, but by competing with those outsourced products. Since they already had a manufacturing cost advantage, many of the outsourced products simply disappeared from the marketplace.

The result of the great paradigm shift of the last 4 decades has been the demise of paternalism. The monoliths found that they simply couldn't afford to continue on that path. We know this has led to enormous disruptions in the orderliness that the engineering community enjoyed, orderliness that may never return. It has resulted in entirely new ways of educating engineers on how engineers work, and what I consider a good thing, the re-kindling of the entrepreneurial spirit among engineers.

THE RESULTING NEW ROLE OF THE ENGINEER

The engineer of the 21st century is a "gun for hire." She—and the engineer is very apt to be a she rather than a he—goes where the challenge is and doesn't put a high priority on job security, instead being interested in career security. Once these were one and the same, but not anymore. Jobs are assignments. A career is a build up of experiences that allows the practitioner to take more and more complex and challenging assignments as time goes by. This new breed of engineer is not looking for long-term security in the old paternalistic fashion, but rather opportunities to grow professionally and to be amply rewarded for his or her accomplishments in the service of the client. Notice the term "client." The new engineer considers the company he or she works for to be the engineer's client, and the engineer works hard to satisfy the needs of his or her client.

This is a paradigm shift in itself. The engineer conceiving of herself or himself as an individual contracting to a company, albeit for many years, is a psychological change. It may be a mere nuance but it sets the stage for self-preservation over company preservation, which is entirely different from the 1960s. Then, the engineer simply pledged allegiance to the company and assumed the company would do right by her or him. Now the engineer recognizes that one must look out for oneself and that the company is not responsible for the engineer's well being, simply to provide a mutually advantageous working relationship. This relationship may be long-term, medium-term, or short-term. It could last decades or be over with within days. The important part is that the implied contract is to perform a task for the company for a fee because the company needs it, and the engineer finds it professionally challenging and can do it. By performing the task, the engineer is fulfilling contractual terms, and that's all. No loyalty is implied, other than to honor the contract. Each party is free to go their different ways, once the implied contract is no longer valid or has been fulfilled.

Those wishing for the "good old days" say that loyalty is gone. But is it? What could be more honest than loyalty to oneself? In the good old days the company said be satisfied with slow seniority style, and promotions and the firm will reward you with a decent salary and good pension. But in many cases the company said that with fingers crossed and pleaded to the "deity" so that good fortune came their way so they could make good on those promises. Many didn't. They wanted to but couldn't because of circumstances beyond their control, the results being a broken promise and perhaps accusations of dishonesty. Contrast this with the new paradigm. The engineer expects to be paid for services rendered. Part of that pay may be 401(k) or pension benefits and part may be medical and other insurances. But the compensation is part of the implied contract. It is up to the engineer to negotiate what the engineer considers to be fair. Also, there is no implied "work for me forever and I will take care of you and yours forever." It is simply a contract for a defined or undefined time period offering compensation for services provided. Clean, to the point, objective. Allegiance is to self and secondarily to the employer as long as there is mutual advantage.

This is the consultant philosophy while employed. It has led to the concept of the engineer being a technical-based businessman or businesswoman. Let's explore this. The engineer can operate independently or as part of a team within a company charged with producing a product or service. The company likes the team concept because it focuses on products and services from which it derives its profits. Everyone who enjoys a long-term, ongoing relationship with the company is doing so within a team that is measured and rewarded on the basis of the success or failure of the product or service. All the people on the team have direct responsibility for success. They are line operators with responsibility for success of the product. That means the team needs to have full resources and be able to deploy them.

What does full resources mean by today's paradigm? By resources I mean engineering, manufacturing, sales/marketing, finance, and human resources. So we see that the engineer as a team member works across disciplines, no longer with only other engineers. The engineer is the technical resource for the team and as such needs to supply "all" the technologies needed for the project. This may be, and usually is, a significant challenge. All means all. In the good old days the engineer would work only on a small subset of a substantially larger task. If the engineer had any technical questions, the support staff was there to assist. Today, as part of the business team, the engineer has to take care of all aspect technical about the product. If the engineer doesn't have the answers, she or he needs to get them. In today's paradigm that means going outside the company for help. This could mean making simple calls to colleagues in other companies, getting help from various technical societies, or actually hiring an expert. But make no mistake about it, in today's downsized companies and startups, there are no hidden resources within the company waiting to be called upon.

In the team concept, all members are responsible for success. In fact, teams are like mini-companies, each having to fulfill the entire range of tasks a company needs to perform, but in this case, limited to the product or service. Also, for efficiency purposes, resources may be shared across the entire company. Figure 1-2 illustrates a typical matrix organization within which the engineer would participate. By being a team member, the engineer can't simply ignore the needs of manufacturing, or finance, or for that matter any other item that's necessary for product success. The engineer in the 21st cen-

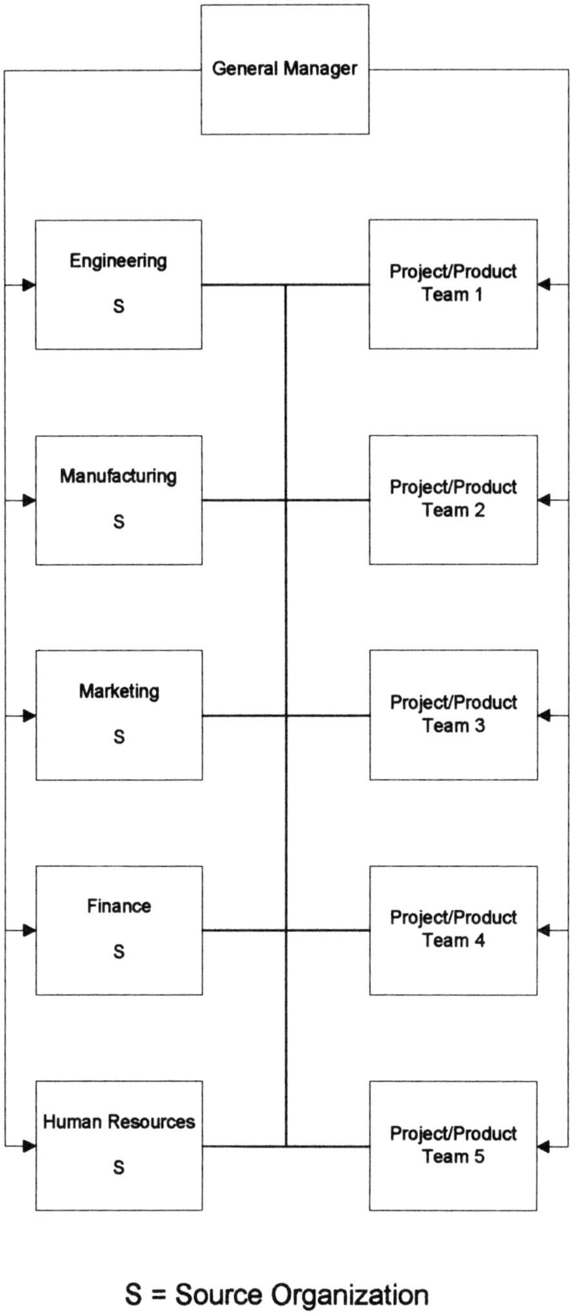

Figure 1-2. The new matrix organizational concept.

S = Source Organization

tury will find himself or herself being the technical advocate within the team, but also possess a vote and hence a responsibility for any activity the team needs to have accomplished. If sales are key, the engineer will have to learn selling processes and techniques to be a good team member. If it is fund raising that is important, then that skill will have to be learned. The point is that the engineer ceases being a technical being but rather a technical-based businessman or businesswoman who must do a lot more than simply engage in solving technical problems.

The need to be a technical-based businessperson is perhaps the single most important result of the years of change that engineering work has experienced in the last 40 years. It is a result of (1) engineers in industry having to fend for themselves with the virtual disappearance of paternalism and (2) downsized companies no longer having totally contained technical resources viable for the entire needs of the product. The engineer, from the first job out of school, is thrust into the role of technical manager and must find ways to cope with all the technical problems the product needs to have solved. Hence, these become laced with managerial problems on how to apply resources to get results. So the engineer quickly learns business skills, not previously thought of as within the domain of the profession.

As the engineer's career proceeds, the engineer find she or he is better suited to solve one type of problem verses others. This may lead to a desire to specialize. Unfortunately, being a team member makes that difficult to do. So here we see experts leaving to form their own consulting or specialty firms. This would have been rare indeed in the good old days. The engineer would simply have moved over to a staff position and probably at a higher salary. Now that same engineer opens her or his own business and once more needs to be a good technical-based businessperson to survive and prosper.

This book is about how to do just that—start a business based on a technical-based product or service and be profitable. It is important as we go through the process of setting up the technical-based business that we understand the environment we are operating in. Engineers form their own companies based on our culture change over the past 40 years; how we employ others in our firms will be based on those same experiences. History is a great teacher. We must take heed to apply the lessons learned correctly or be relegated to experience the same travail over again.

Chapter 2

The Engineering Education Needs for the Future

Presently, engineers coming out of school to their first job are really novice scientists, not pragmatic technical problem-solvers. In the future, just as now, we will need both. But we need many more of the latter than the former. For an engineer to establish that initial contract between engineer and employer, she or he needs to bring more than theoretical science knowledge to the enterprise. That may be good for the initial hiring because that's the present level available for companies to chose from, but it's not sufficient for future engagements with lean-staffed companies.

Being a technical problem-solver doesn't mean just solving technical problems. It means applying the "scientific method" to all forms of business-related problems. In truth, the entrepreneur needs to have an inherent sense of order to take a good idea to reality. The way engineers should be trained is to objectively analyze problems, articulate choices, and go with the one that is most likely to succeed. This is what is necessary in any startup or ongoing situation. The basic premise of the technical-based education aims to do this. Unfortunately the focus is too narrow. Young engineers are led to believe that it only applies to science-based problems. Nothing could be further from the truth. All business problems benefit from this approach, and the more we can do to instruct our young engineers to recognize this and practice it, the better they will fit into the company of tomorrow and the better they will be prepared to run their own enterprises. In this chapter, I will present a "non-educators" evaluation of current methods of training engineers and present a modification of the present approach that is more suitable for success in the 21st century business scenario, be it as an employee, contractor, or entrepreneur growing an idea into a business.

THE SKILLS NEEDS ASSESSMENT, THEN AND NOW

We have gone full circle in the skills package engineers need to have to be successful in the business world. A hundred years ago, virtually any trained professional in business was an engineer. Many others managing businesses may have had university degrees but they weren't considered professionals because what they learned in school wasn't applicable to their jobs. There was no such thing as business majors. Lawyers practiced law, not business management. Accountants, if they were called that, worked in and ran banks. The only university-trained people destined for the world of commerce were engineers. I think the same is basically true today but we seem to forget it. Most engineers today are destined for careers in commerce. The difference is that 100 years ago, the business professional, that is, the engineer, was trained for business and that's not true today. Unfortunately, experience shows that's what is needed today.

Looking back on history, businesses, circa 1900–1910, engaged in technologies of the time were frequently managed by disciples of the early mechanical and industrial engineers such as Gantt, Gilbreth, and Tailor, or by inventors who possessed significant engineering skills such as Ford, the Wrights, and Edison. The striking similarity was that they all shared a common trait: They were first and foremost businessmen who could apply technology. The successful ones had a sixth sense of how to manage resources to get the most out of the technology. In other words, they knew how to optimize the potential for technological success through proper husbanding of resources to get the best "bang for the buck" in competing through application of their technology.

Let's look at the skill sets a recent engineering graduate possesses, then and now. Figure 2-1 shows a comparison.

We can see from the figure that engineers had more hands-on courses than generally offered now. We see that mathematics and sciences have increased, reflecting the huge gains in knowledge made over the past century. Also the number of engineering courses has increased, again due to the expansion of the technological base. Humanities appears to be the same except for the general abandonment of foreign language studies. But this is deceptive. Most engineers now take watered-down courses in the humanities and the credit hours are much less than they were 100 years ago. In fact, engineers in 1900 took the same humanities courses as liberal arts majors but perhaps not as many. Similarly, they shared the same mathematics and science classes as their nonengineering major classmates. The truth is, engineers at the turn of the past century were more literate than those of today.

Note the types of course taken in 1900. There were more applications-oriented courses, such as properties of steam related to boilers. Iron and steel metallurgy courses reflected the desire to perfect and use more reliable materials for the vast industrial and transportation infrastructure expansion that was still in its formative years. Courses were aimed at fueling this expansion. Today our engineering courses seem to be focused on creations of new theory, which may or may not see its way into new and profitable commercial products. Today we are aimed more at discovery of new applied science facts rather than direct feeding of the economic engine to increase the worldwide standard of living. This is a very telling truth since most schools of engineering do not bother to offer engineering economics as a required course. This is a very interesting dichotomy because the professional engineering state licensing boards have a large section of their examinations dedicated to engineering economics. Why aren't young engineers taught that in school? Perhaps there is a significant difference of opinion between the professional engineering state boards and the engineering education community as to what engineering really is. And perhaps this is the

Typical Mechanical Engineering Curriculum

Circa 1900	Circa 2000
Mathematics	Mathematics
Algebra	Calculus
Trigonometry	Differential equations
Geometry - plane and solid	Linear and Boolean algebra
calculus	Set theory
Differential equations	Matrices
Engineering economics	Probability and statistics
Science	Science
Inorganic chemistry	Inorganic chemistry
Newtonian physics	Organic chemistry
	Newtonian physics
Engineering	Relativistic physics
Statics - Newtonian	Atomic physics
Dynamics - Newtonian	Quantum mechanics
Electricity - DC	
Thermodynamics	Engineering
Properties of steam	Statics - Newtonian
Iron and steel metallurgy	Dynamics - Newtonian
Strength of materials - elasticity	Electricity - circuits
Surveying	Electricity - power
Machine shop	Electronic circuits
Engineering drawing	Computer design theory
Soil mechanics	Control theory
Machine design	Systems design and modeling
	Thermodynamics
Humanities	Heat transfer
Foreign language	Strength of materials - elasticity/plasticity
History	Metallurgy
Composition	Nonmetallic materials
Literature	Machine design
Philosophy	Manufacturing engineering systems-CAD/CAM
Ethics	NanoTechnology
	Humanities
	History
	Composition
	Literature
	Philosophy
	Ethics

Figure 2-1. Skills of recent engineering graduates—circa 1900 vs circa 2000.

problem that needs to be resolved for engineers to have a good chance at being effective entrepreneurs. It is my opinion that many engineering educators subconsciously think of themselves as applied scientists rather than technology-based businessmen and businesswomen. I think this is a natural bias due to most professors never gaining "real" business experience since their exposure to industrial jobs is very limited.

THE SKILLS PHILOSOPHY NEEDED FOR THE SUCCESSFUL ENGINEER PRACTICING IN THE EARLY 21ST CENTURY

As described in Chapter 1, the nature of the workplace is changing to become more pragmatic and more dedicated to the profit motive. This is fertile ground for the engineer entrepreneur. Many young engineers quickly become immersed in the pragmatic applications mantra mandated by their work environment. Currently, most young engineers must develop skills on becoming pragmatic profit-oriented businessmen and businesswomen. It is better to change the way we train engineers so they have a more realistic expectation of what they will be doing and have the skills to do. What are the skills they need to have? Let's take a look.

We can say that we should go back to pragmatic application courses. A sort of back to basics. But what are the basics? Surely, we would all agree that engineers have to be well-versed in the physical sciences and mathematics, as well as economics and communications skills. But is that sufficient as the starting point? We must have a definition of what engineering is now and will continue to evolve into in the 21st century. Also, to be of any use, the definition has to match the prevailing industrial environment. If the term engineer entrepreneur, which is a subset of entrepreneur, is to be meaningful, the prime definition of engineer has to be compatible with the desired end results expected of entrepreneurs. So for a definition of engineering I would propose the following:

> Engineering is the application of scientific principles to solve business problems.

Notice that my definition doesn't say "technical problems." I use the broader term "business problems," and this is where I differ from many current definitions of engineering. I believe "technical problems" limits the scope of the engineer to that narrow role of being a gadgeteer guru instead of being a full and equal member of the firm's team. Besides, in today's environment, most business problems have significant technical components, be they hard technical processes based on scientific principles or soft processes that need an integrated systems approach to solve. Even more important is that the technical content is usually interwoven with the nontechnical, and it is impossible to solve the problem without considering the mutually dependent relationships.

If we can accept my definition of engineering, then "back to basics" means the curriculum must contain subject matter that has useful applications for solving business problems. The results means we need to teach interrelationships between technology and nontechnical aspects of applications with much more emphasis than is currently being considered, if it is being considered at all.

The range of these problems can go far beyond the science of making the product work. It can relate to understanding the dynamics of systems to control processes, as well as finding the most optimal use of all resources, such as money, materials, machinery, applications of human resources, conservation of energy, flux density in a nuclear reactor, or wing profiles for optimum lift. The one thing the curriculum for an engineering degree cannot do is simply teach science for science's sake. Engineering study does not have the luxury of "wasting time" to simply understand science if no foreseeable application is in the relatively near-term future. Capitalism does not permit that. Engineering study has to know its limits and those limits are the near-term practical application of technology. Anything that is not, is the domain of the science curriculum, not the engineering curriculum.

Quite frankly, I believe many engineering schools today are mixed up on this point. They confuse science with engineering. They fail to differentiate. The point where engineering ends and science begins is admittedly blurred. But if we can answer yes to the following simple question about a course or its contents, it can be considered an engineering course: Is there relatively near-term practical application possible with this technology? If we simply don't know the answer, then it is a science course.

"Back to basics" means the top priority in the education of engineers must consist of focused courses that teach science for the sake of practical application. Those who are not interested in practical application need to become science majors. They will not thrive in the current industrial paradigm. Obviously the line between what is or is not practical is not always known, so engineers will always find themselves at or near the border. This is good because it is more likely to be the pragmatic engineer entrepreneur who will drive technology to useful application, not the inquisitive scientist.

Not all engineering students will be well-served by this back-to-basics approach. A significant minority of engineers, perhaps as high as 10%, will always work in the "blurred" area between basic science and application-oriented engineering. These are the people who take longer-term applicable science and bring it down to near-term applicable science. They mine the scientific depths for technology that may someday metamorphose into profitable technologies. These are the research and development (R&D) engineers. They are a very important hybrid between pragmatic entrepreneurial-oriented engineers and pure scientists. The back-to-basics curriculum doesn't suit them well. The back-to-basics engineering degree isn't enough training for these engineers; they need more preparation in much narrower, focused areas of their choice so they can pursue these interests. But the modern industrial company does not depend directly on these people for success. There is an indirect link with basic research companies that have taken the place of the virtually extinct large corporation research lab. And the link is usually through mutual interest trade societies, engineering technical societies, and university outreach programs. The link occurs when engineers in industry spend a few days to a few weeks a year maintaining currency in their respective fields and "mine" the new ideas for potential technology for evolution to near-term commercial products. The most prevalent example of this is the electronics industry's various trade shows and deal making that goes on between research companies selling technology to industrial companies to bring forth new products.

The engineer who is a member of a business team is rarely an R&D engineer. He or she needs to be trained to be an effective engineering performer to make the team successful; and that training needs to be based on the back-to-basics philosophy. In the succeeding sections of this chapter, we will look at the experiences that engineering students undergo now, why and how the curriculum needs to be changed, and how to go about doing that. If engineers are to be a significant force in entrepreneurial developments, their education process must mirror the work they will perform. Later I will give real examples of techniques that can be taught in engineering schools to prepare engineers to be more entrepreneurial, but are now mostly learned on the job. These successful techniques would be in more widespread use today if they were part of the core curriculum used to train today's engineering students.

CURRENT TRAINING REGIME FOR ENGINEERS WHILE IN SCHOOL, AND WHY IT IS INADEQUATE

Today's engineering curriculum is producing pseudoscientists. They are far too divorced from practical courses for them to really know how to do an industrial-based job when they arrive at that first position. Engineering courses are taught by lectures out of textbooks using problems that do not reflect real situations. The model is the scientific experiment where all variables can be controlled. Virtually all problems are designed to reinforce the theory of having 1 correct answer with little interpretation of assumptions. There are virtually no follow-up problems that bring "real" situations into the learning process. The idea is to teach theory with the proposition that once theory is known and mastered, the student can then solve "real" problems. This, of course, is false. If students aren't used to handling "real" situations we can't expect them to miraculously transition from pretty textbook solutions to messy real-world situations where the data first has to be validated, and expect them to be successful. Some current curricula try to resolve this situation by offering "capstone" design courses in the senior year to teach real-world situations. But often this is too little too late.

The method of instruction described earlier is fine for basic science courses. Here it is possible to reinforce theory with controlled laboratory experiments that bring home the learning points. Real-world engineering is not like that. It is full of truths, half-truths, and totally false assumptions. It requires digging to get all the facts and then even more digging to verify that what has been discovered is relevant to solving the problem at hand. As I said, real-world engineering is messy and disorganized, and this is the antithesis of the domain of the scientist working with a well-conceived and controlled experiment. The fact that it is messy is often overlooked in engineering classes to the ultimate distress of the graduate of the curriculum. We need to get down and dirty in the instructional methods. We must understand that solving a "real" problem is based on the best data we can find within the time constraint and cost constraint and act to the best of our ability to that set of stimuli. We need to teach pragmatism. Let me illustrate:

Example: We want to determine the size of an anchor bolt for a pedestal used to support a road sign that is subject to hurricane wind force. We will never know the true load the device will experience, either static or dynamic. We will have to select a load based on history and then add a factor of safety on that load to compensate for not knowing the direction of the wind. In addition, we will have to make an educated guess as to what kind of load will be imposed. Is it tension, compression, bending, torsion? By the hothouse textbook solution, we would be told what to assume and solve for that. In the real world, we simply don't know so we need to solve for the worst condition. And since we can't even be sure what the worst condition is, we would have to put a margin of safety on top of what we calculated. Then we would size the bolt and feel with a high degree of confidence that the structure would survive, but never know for sure.

We've used pragmatic engineering judgment to make the call and solve the problem situation the real world presented to us. Is it the absolute correct answer? Have we used too much material, that is, is the bolt oversize? We really don't know, but if we surmise the worst-case scenario with enough conservatism (but not too much), we are confident that the probability of success is very high in our favor.

Unfortunately, the typical engineering student isn't taught how to do this simple design problem making assumptions and deciding what has to be solved for. Typically he or she would opt to solve for every conceivable thing, such as bolt compression as the potential cause of failure even though it is highly unlikely. Even more disturbing, the student may simply give up because all the data isn't available and wouldn't attempt to suppose some data just to see what happens. The student may also be frustrated because of an inability to set up tests to get data that models a hurricane situation. The point is, the way we teach today requires tightly controlled experiments and we don't expect the student to "rough it," to create an acceptable solution. We want the student to find the answer given in the instructor's answer manual. Another point lost in this method of instruction is that we infer through repetition and drill that all problems have 1 correct answer. It comes as a shock to young engineers when they find there is no such thing as a single correct answer; there is only an array of acceptable answers, each with its pluses and minuses with respect to the problem at hand. One of the most valuable methodologies I ever learned was that in order to solve an engineering problem, first gather all the known facts and determine what I didn't know, then make assumptions as what those missing facts may be, and finally, test those assumptions and see if they made sense. My mentors used to say, "Does it pass the eyeball test? Does it hold water?"

Another thing we find in inexperienced engineers is paralysis by analysis. Their scientific instead of engineering training doesn't allow them to stop analyzing until they have all the facts. And they don't know how to determine what is relevant or irrelevant. There comes a time in any problem situation when an action needs to be taken to try a solution. As an aside, an allegory comes to mind. In the famous Steve McQueen movie *The Sand Pebbles*, about a U.S. Navy gunboat on the Yangtze River in China, the hero and his crew are in a desperate situation, bound only to get worse through inaction. Compounding the problem, the crew doesn't have all the facts to effect a certain correct course of action. So the hero utters a now famous line to his commanding officer, "Just do something!" This is often the case with real problems requiring a course of action. Especially where no action is worse than a partially satisfactory action. In real engineering problems, partially satisfactory solutions are more the norm than we care to admit. And those solutions are acceptable as long as they meet the "must" of the situation, even though not all or perhaps not any of the "wants" are achieved. Again, pragmatism rules. The entrepreneurial-trained engineer will always ask, "What kind of answer do you want? A 5-minute answer? A 5-hour answer? A 5-day answer? The amount of analysis I do before taking action depends on your answer." And that is true engineering: pragmatic solutions based on known facts, gathered within a given time constraint, supported by assumptions based on personal experience or previously known situations, to do the best job possible for the given situation, that passes the eyeball test. Now all we have to do is get engineering schools to teach this philosophy from first-year courses through capstone design seminars.

HOW THE CURRICULUM MUST CHANGE TO BE AN EFFECTIVE TRAINING MECHANISM

I have no doubt that the engineering curriculum has to favor pragmatic application to remain relevant to current and future business needs. I think a good curriculum would have 2 phases:

Phase I. Foundation
Phase II. Application

No doubt current university engineering department chairs would say this is what they're currently doing and what the Accreditation Board for Engineering and Technology (ABET) requires. But is it, and is the ABET requirement sufficient? Let's take a look at what I mean about Phases I and II, then answer the question.

Phase I. Foundation

The foundation is the core science, mathematics, and humanities courses from which the engineer will draw knowledge to solve future "real" problems. These include physics, chemistry, and mathematics through whatever level is currently in vogue for pragmatic engineering design success. They also include communications skills gained through studying the humanities such as literature, creative writing, histories, public speaking, and perhaps foreign languages. Finally, the core list of foundation courses ought to encompass economics and the basics of contract law. This list creates an educated person who can understand the modern world we live in. Now on top of that we can superimpose the special knowledge the pragmatic engineer needs to know to become an effective practitioner.

I'm not going to dwell on the basic sciences and mathematics requirements. We all agree this is absolutely core to being a competent engineer. But I think it's necessary to champion the need for the humanities. Every aspect of business deals with communications. In my book *Computer Integrated Manufacturing, Theory and Practice*, I said it mattered little what type of computer information system existed in a company, as long as the system communicated directions, responses, and ideas accurately and efficiently (fast and completely). I said a good computer database system needs to be an excellent communicator between modules for computer integrated manufacturing (CIM) to work. In a broader sense this is also very true for interactions between people within an organization and between organizations. The number 1 prerequisite for success is that we be able to communicate our thoughts to each other and to organize our capabilities to achieve common goals. Without such abilities, organized pursuit of common objectives is impossible. Unfortunately, the ability to work well within a group is not necessarily an inborn instinctive ability of our species. Yes, we are social animals, and yes, some of us can socialize better than others. But how to optimize group behavior for the common good is a science (perhaps quantified as a soft science; theorems are less defined) that can be taught and learned. And a large part of being successful is being able to express oneself clearly without ambiguity and to do it in a way that doesn't disturb our companions to the point of tuning us out.

Engineers have been given the tag of not being "gifted" in group behavior, socializing, and communications skills. This may or may not be true; however, these social skills are necessary in business applications. Indeed, they are necessary in all human endeavor. An individual may be a brilliant engineer who has solved a very perplexing problem. That solution may be a boon to humankind, but if that engineer cannot articulate the solution, it is not implemented and the problem remains unsolved for all practical purposes. Therefore, it is urgently necessary for engineers to be good social engineers capable of communicating and working well within groups. They need to master the psychology of group dynamics and then be leaders of the group when mastery of the hard sciences is required. For this reason, engineers need to excel in fields of study that sharpen these skills. They need to be able to formulate and express ideas beyond postulating science applications. They need to tie their knowledge of science and technology to the human side of the equation, and this is done through merging culture with technology. It requires a sense of understanding of the human psyche as defined by the great philosophers, to understand what drives people to do what they do.

They need to communicate ideas via the written and spoken word. If they are successful at this, they can surely tie technology to their social discourses as the situation requires. In fact, by being well grounded in the humanities as well as the sciences makes the engineer going through this educational process the only truly educated people of the 21st century.

Now that I've expressed the idea of combining sciences and humanities as the foundation of the engineer's education, we should consider how to evolve it into the second phase of the preparation to become an effective engineer. Let's look at phase II, Applications.

Phase II. Application

Applications is the job-specific instructions part of the engineer's education. Once the student is well grounded in the fundamentals of the sciences, mathematics, and humanities, the nature of the education process must change to represent the work environment. We must present all course material as close as possible to how it will be received in the workplace. This means theory has to be couched in terms of real situations, not laboratory experiments. This is a complex and perhaps difficult task to accomplish, but it must be. We need to represent theory in such a way that it immediately becomes apparent as to how it can be applied. I know many of my professor friends would say I'm downgrading the pursuit of knowledge and trying to make education simply a "how to do it" experience. But isn't it? Don't surgeons learn by doing, but backed up with sound science principles? I think yes. Why can't we do the same for engineers? Why can't we teach advanced techniques by applications-dominated processes instead of relying on abstract absorption of theory with no experience factors? We can and we must. For example, fuzzy logic is a mathematics-intense control logic for a subset of artificial intelligence computer programs. It can be very abstract and intimidating to many people who are not experienced computer science engineers. In fact, reading the current texts can leave the engineer baffled and confused because there are very few applications examples to help in the learning process. Having spoken with those who understand it and asking for a demonstration based on a step-by-step explanation of the theory, I can begin to understand it enough to use it. Think of how much more proficient a student engineer would be if the student were taught fuzzy logic by example laced with theory instead of simply via the traditional lecture method with lots of abstract symbolism.

Sometimes I think professors are really in the business of intimidating students with their narrowly focused knowledge. They use advanced mathematics peculiar to their specialty and make assumptions that the students fully understand it, while in truth the students don't understand and are afraid to ask. The professor knows this and either doesn't care or cannot be bothered to teach it. It's as if the professor is challenging students to ask so he can destroy the students' egos to build up his own. I think those who have worked in industry know this isn't the way it happens in real life when a new technology needs to be learned. Often a business team will need to become proficient on a new technology. Perhaps one member of the team has more knowledge of the subject than others. It is then that person's responsibility to bring the rest of the team up to his level. In industry this would be done by on-the-job learning by doing, guided by the person who already has the level of knowledge required for the need. The instruction is done by demonstration, experimentation, learning by doing, perhaps backed up by some classroom-type lectures. The key is that those learning the new technology are being exposed to theory and direct application to solve a task at the same time. There is no reason this model cannot be emulated in the university. The results would be a more committed student who gets a clearer understanding of the subject matter in a manner that is conducive to learning. This replaces the old model of finding a way to glean enough from the lecture notes just to "ride the curve" and get by. By the current model, the student learns only when and if she or he needs to have that knowledge to succeed in a so-called capstone design project.

Perhaps the problem is in the academia model itself. Are professors hired to teach the applications of engineering or to do basic research? The official answer is both. But which one takes precedence? My impression is that professors think of themselves as researchers first and teachers second, and in my opinion this is a mistake. The old university model of apprentice, journeyman, master dressed up with appropriate titles (such as undergraduate student, graduate student, learned professor) is a throwback to the Middle Ages and mirrors the cloistered centers of slow-paced deliberative learning. This has no counterpart in industry. In fact, only the engineering education seems to be saddled with this antiquity. We find ourselves in a position where entry-level engineers are being trained by nonpracticing engineers. Our professor class is really R&D engineers who are certainly not mainstream engineers in the modern context. Their methodology of teaching may be okay for foundation-level courses, but not applications-level courses. They teach in the parlance of R&D engineers; therefore, their model does not represent the modern team-oriented enterprises that most 21st century engineers will work within. The Middle Ages origin method is no longer sufficient for teaching applications or applied technology courses such as thermodynamics, heat transfer, strength of materials, etc. What I will describe next is my opinion on how the applications content of the engineer's education should be managed and taught.

Applications courses need to emulate real-life situations faced by engineers. Many differences exist between an academic setting and one encountered on the job. Besides the obvious profit motive of jobs vs. school, probably the most important one is that a large part of the work done by the engineer is done in a team mode. This means engineers and other colleagues consult about problems and solutions. They do this in a manner that reinforces good ideas and discards bad ones. Technical problem discussions are based on technologies, and learning takes place by one engineer adding his or her strengths to another. This way it is common for individuals to grow their skills and focus on those segments of technology that are important to the business. We can emulate this process for applications courses. Here's how.

Applications of technology most often occur in group settings within industry. Therefore, teaching applications should emulate group settings as much as possible. Recall that engineers in industry are often members of business teams. In solving problems, team member piggyback off each other's ideas until the best solution that meet all the constraints is arrived at. This doesn't mean it's the most optimum solution, only that it is the best for the current situation. This means that engineers learn from each other and have sufficient feedback as to the relevancy of their solution proposals, and in essence, the solution is endorsed by all members of the team. This way we are assured that the best technology is used to solve the problem. In the case where the team is not satisfied with its solution, it would probably resort to a group research search where a team member would be appointed to get as much information as possible on the problem, then report back to the group. Here we see groups increasing their knowledge base at the same time they are solving the current problem. There is no reason, except that it's different than the current model, that this cannot be emulated in the classroom.

In Figure 2-2 we see an outline of a lesson plan that can be used to teach applications courses that models the process found in business teams procedures. We start with an introduction to the specific subset of the technology in a lecture manner as with Phase I-type subjects. We quickly introduce real-world applications, first with "hothouse" textbook problems, then with "dirty" problems. A "dirty" problem is one that contains data that would be found in practice, such as data that contains gems of truth and irrelevancies, as well as data that isn't correct. The instructor would show how to sift through this information, make assumptions where necessary, and then solve the problem. The instructor would then make sure the class understood the validity of the answer, the ways validity can be examined, and the ways the answer could be made more robust so it or its evolved state could be used. This would introduce such pragmatic features such as relevant codes and factors of safety, the so-called allowables. It would also allow the introduction of probabilistic approaches to ensure robustness of the usable answer. Finally, the students would be assigned teams and asked to solve and explain practice "dirty" problems. Here students are given the opportunity to practice applications of technology against the background of a very "neutral" real world where things are not being set up for ease of solutions for grading purposes.

This model of a lesson plan requires much more thought and attention to "what ifs" by the professor. It's harder to develop, and it takes real interest and knowledge of the psychology of teaching than is required by the current crop of professors. It would mean that our professors would need real-world applications experience to teach the course, not just book theory. It probably means we would need to change the paradigm of university teaching of engineering. Theoreticians would be fine for the foundation courses, but practicing engineers would be required for applications courses. In my opinion, this would get us more in line with the other professions. I think you would agree that it would be appalling if surgery were not taught by practicing surgeons. The same is true for engineering. We must get away

- Instructor introduces a segment of the applied technology in the lecture.
  o Theory
    - Root scientific basis
    - How derived
    - Applications
  o Simple application examples
    - "hot house" examples
    - Relevancy of "hot house" to real world
    - Methodology of applications in the real world
- Instructor introduces the theory to a real problem where data is "dirty"—must be
  o Evaluated for accuracy
  o Tested for validity—relevancy for the problem at hand
  o Assumptions made for missing data
- Instructor does a real-world problem
  o Evaluation of the accuracy of the answer arrived at
  o How the answer should be modified before applied as a solution (or step toward the solution) to the problem
    - Code requirements that may affect the answer
    - Factors of safety relevant and customarily used
    - Where probabilistic methods are usable for justifying use as a modifier of the answer
- Students are assigned to groups to solve a real-world problem using the applications technology introduced in the lecture
    - Each group is asked to propose a solution to the problem relevant to the given scenario
      - The group is expected to define the problem
      - Evaluate the data
      - Make assumptions based on the nature of the problem
- Each group has to prepare a written report and
  o State their approach to the solution of the problem
  o Show their calculations with all the assumptions defined and explain technologies used
  o State why they think their solution is adequate
  o Determine the degree of confidence they have with their solution
- Each group, based on their written report, verbally presents and defends their solution to the instructor and the other groups

Figure 2-2. Applications teaching generic lesson plan.

from an education system whereby young engineers are forced to learn their profession only after they have their degree.

We cannot conclude a discussion of teaching applications without mentioning teams. It is a fact of business life that most engineers will work in some sort of team. It may be a product development team, a team devoted to improving manufacturing effectiveness, one involved in marketing, or more likely a unified business team as described in Chapter 1. Whatever it is, the engineer will almost certainly not be working as a sole contributor outside the context of the modern business structure. However, we teach engineers as if they will be doing just that by the way we measure their performance as a student. We give exams in ways that create an artificial mechanism of students regurgitating what they have learned back to their professor. We teach so they can pass exams as a singular entity. Therefore, the exams are designed to test the ability to memorize "stuff" so the student can show that she or he can recall information about which the instructor lectured.

Exams do not reflect the way the business world works. In the business world, we debate and research solutions. We look at the pros and cons and struggle with a solution that is at least acceptable, recognizing that there are hosts of other solutions available but not selected for various reasons. We either satisfy the customer or not, depending on the choices we made. We may be totally correct, but still fail because of facts not known to us or because the customer's selection process is flawed. What I'm getting at here is that the answers we give in solving a business problem are never 100% clear cut for a variety of reasons. Contrast this with a typical thermodynamic quiz. It asks students to solve particular problems and gives a set of data that can lead to only 1 correct answer; there is rarely, if ever, any constraint that makes it impossible to select the best solution. If we had a problem to solve for the best heat rate for a heat exchanger we could do that. But what would the answer be if several of the tubes were clogged? And to get the best results we would have to spend time and labor (which means cost) to repair the tubes. What if the cost to repair is greater than the gain to run with a lesser heat rate? What then is the best answer? Is the engineering student trained to do this? Can he or she consult with another student or do some research during an exam to get a better solution? Probably not, because then the professor doesn't know how to dispense grades. The tragedy is the students are taught to not seek advice or do additional research because the reward system doesn't exist for it; in industry just the opposite prevails.

Teamwork must be taught in upper-level engineering courses because that's how the world really works. How do you do this? Make the reward high for doing so. Grade everyone on how well they perform in a team setting. This means a structure for evaluation has to be set up that requires participation by all students. For example, a project grade must contain the name of every student, and the work done on a project team's report must indicate each member's participation. This gives the instructor an idea as to who's doing what. Make sure that all team members' grades are strongly influenced by how the team organized to do its assignments. Exams as we know them, that is, stand alone individual tests, need to be minimized and in fact de-emphasized. In their place, the grading emphasis should be placed on how well the team performs on a project that highlights the subject matter of the particular course.

There is no doubt that grading based on team performance is harder than grading via traditional exams. It is more indirect than traditional exams, and it requires closer observations of students than is traditional within universities. It would mean that a professor's emphasis would be placed on teaching rather than research. This could be the catalyst for a paradigm shift in engineering education, which I believe is needed. Right now, our best professors are rated as being best because of their ability to do research and bring in research money to the university. This is fine for those wishing to pursue a science research career. But all too often the professor has to make choices and ends up favoring research and has to find time to teach. Many well-meaning professors rationalize that their graduate student assistants can teach the undergraduates. After all, it is basic stuff that their graduate students know; therefore, they can stand in for the professor. This leads to inferior education. The person who is the expert is not translating that knowledge directly to the client (the student). It is being filtered by an assistant who may or may not have the skills to teach. This brings up another question. Is the professor skilled in teaching? If not, all the knowledge in the world will not help in transferring information to the student. The paradigm shift I mentioned must lead to the primacy of teaching skills and the desire to do so over the quest for research grants and the thrill of experimentation. This is the dilemma we face today. Teaching is not considered as important as research. How do we fix it?

Again, the reward system has to be changed. Universities, which are well-versed in avoiding the truth, must recognize that the traditional master–journeyman–apprentice method of the Middle Ages is not appropriate for today's industrial needs. We need to have dedicated teachers capable of teaching novitiates the true methodologies of their future work. We must reward professors for teaching well, as we currently reward them for their research prowess. In fact, probably professors should be rewarded more for their teaching skills than for their ability to run a good experiment. Herein lies another problem. How many engineering professors are good engineers? By that I mean how many meet the standards necessary to thrive in today's teamlike atmosphere and highly competitive situations of commerce? Probably very few. Most are removed from that type of work environment and are not interested in it. Therefore, we have engineering students being taught by a cadre of teachers who have no experience or interest in the world their students will enter. Do we have the right people teaching engineering students? For the most part I don't think so. To teach engineering you need to know how to do engineering work, and as I've demonstrated, that is not the same as science work.

So again, how do we change the reward system? We need to get more instructors who are engineers in practice to teach at the universities. The reward has to be financial and prestigious. Engineers from industry should be recruited with zeal by universities. Their real-life experiences ought to be incorporated into the applications curriculum. They need to be the role models for the engineering students. Also, we need to send professors into industry to relearn applications-oriented engineering. This can't be just for a year-long sabbatical on an offline job everyone in industry recognizes as a semi-vacation for the professor. It has to be in a line function contributing to the profit-gaining needs of the company. I would also recommend that no one be allowed to teach engineering or qualify for a tenured engineering professor position until he or she has had a minimum of 5 years, 10 being better, of real industrial line engineering experience. This way our professors can represent the profession to the students and not simply represent an academic study of the profession as an interested observer.

I am reminded of a tale that a young engineer told me 5 years after his graduation. He was asked what he thought of his alma mater on graduation day by the mechanical engineering department head. The department head was proud of the young man; he had the third highest grade point average in his class.

The graduating student said, "Not much."

Startled, the department head asked "What do you mean?"

The young man replied, "I came to the university after working as a technician for 2 years. I had to earn enough money to pay for school and support myself. I wanted to learn how to be an engineer. The science and math I learned was good and I appreciate the effort the professors made in teaching it to me. But then I took junior- and senior-level courses—engineering courses, and nobody could teach me how to engineer. Everything the professors did was scripted to give 1 correct answer. And I know that's not the way it was when I worked with engineers before I came here. I wanted to learn how engineers made judgments on what to do to solve what they were working on. And nobody here could teach me that. Now I have an engineering job and I have to figure that out myself."

I don't think we want this to be the end result of an engineering education. We want engineers to know how to engineer when they come to their first job.

HOW TO INTRODUCE ENTREPRENEURSHIP INTO THE CURRICULUM

The modern business team comprises engineers and other professionals who represent all functions of the organization. Their function is to make their company profitable through proper entrepreneurial management of their product(s). The business plan they work to consists of optimizing the 7 steps of the manufacturing system (Figure 2-3) of which steps 1, 2, 5, and 6 have large engineering contents. Hence we have engineers participating as equal members of the business team.

The purpose of the business team is to make money for the company. Therefore, all members need to be cognizant of the methodologies of accounting and how to evaluate ideas for their potential for improving profits. In other word, all team members need to be businessmen and businesswomen first, before specialists in their particular profession.

Engineers coming out of school with a mastery of the current curriculum are not equipped to handle anything but pure technology. This is not satisfactory. How do we train the engineer to be an equal partner on the team when a large part of that responsibility is entrepreneurship? How do we train student engineers to be entrepreneurs? This is the topic of this book and obviously my answer would be to read and master the techniques

The 7 Steps of the Manufacturing System

(1) Obtain product specification

(2) Design a method for producing the product, including the design and purchase of equipment and processes to produce, if required.

(3) Schedule to produce.

(4) Purchase raw material in accordance with the schedule.

(5) Produce in the factory.

(6) Monitor results for technical compliance and cost control.

(7) Ship the completed product to the customer.

The 2 "Knows" of Manufacturing

(1) Know how to make the product.

(2) Know how long it should take.

Figure 2-3. Steps of the manufacturing system.

presented within. However, there is more to it than that. We need to expose our student engineers to certain concepts and philosophies so they understand entrepreneurship. Let's start out with a definition of entrepreneur.

Entrepreneur: One who organizes, manages, and assumes the risk of a business or enterprise.
(from: *Webster's New Collegiate Dictionary,* G & C Merriam Co., Springfield, MA, 1976)

The definition points out 3 characteristics that would have to be explored for us to say a student has been exposed to entrepreneurship. Lets look at those 3 topics.

The first characteristic is "organizes." This means the student needs to understand organization vs. disorganization. Some forms of mathematics approach this concept in an abstract manner, such as the chaos theory, and the student should be exposed to that. Also the student should understand the basics of the social groups that we human beings have been forming for eons. The student should understand why we are social animals and congregate in groups as do lions vs. isolated individuals like leopards and tigers. This leads to an understanding of group dynamics and how we work together. So soft science courses are required, namely, psychology. In some format, this has to be added to the curriculum.

The second characteristic is "manages." This is a natural outgrowth of "organizes." If we know the psychic deeply hidden reason we do what we do, then we can postulate ways to create ways to organize for effective work outcomes. This leads to a whole host of management stratagems, which in turn leads to a cadre of measurement strategies that tell us how well we're doing in running the organization vis-à-vis a set of goals. This requires management strategies courses.

The third characteristic is "risk taking." An entrepreneur takes risks to achieve goals. It is as simple as that. But entrepreneurs do not take blind risks. They calculate the odds and do everything they can to mitigate adverse results. This is relegated to technique by thoroughly understanding the mathematics of probability and statistics. We couple this with training in goal setting and how to evaluate risks. Then for each risk we also determine what to do if the "bad" thing happens. We learn how to develop workable contingency plans and understand the trigger mechanisms to set them off. In modern parlance, we teach these principles beyond probability and statistics in risk management courses, most often coupled with project management techniques. Probability, statistics, and risk management are taught as part of the Master of Business Administration curriculum. They should be taught to engineers as part of the applications segment of their education at the undergraduate level.

From a practical viewpoint we should also teach Quality Functional Deployment (QFD) at the undergraduate level to coalesce the entrepreneurial risk-taking philosophies. Quality Functional Deployment is a methodology used to determine whether a company or an organization has the capability to perform tasks and make products that a potential customer desires. It's a way to determine whether a company would be successful in fulfilling a customer contract. This a very pragmatic way for a company to assess risk.

I've found a technique to be successful in making QFD a practical tool for business plans, which I provide to my clients. These same principles can be taught easily at the undergraduate level. By doing so we can either introduce or reinforce the business nature of engineering practice and how engineers cooperate with others as members of business teams.

This 3-phase approach, used to define an opportunity for a sale and evaluate a company's ability to successfully compete for it, is paramount for understanding the nature of risk-taking. It's obvious that the better a match exists between a client's needs and a company's capabilities, the less of a risk is involved. Being able to have a "feel" for the amount of risk associated with any project is a skill that can be taught and ought to be taught at the undergraduate level. The model as shown in Figure 2-4, while originally conceived

The Engineering Education Needs for the Future • 29

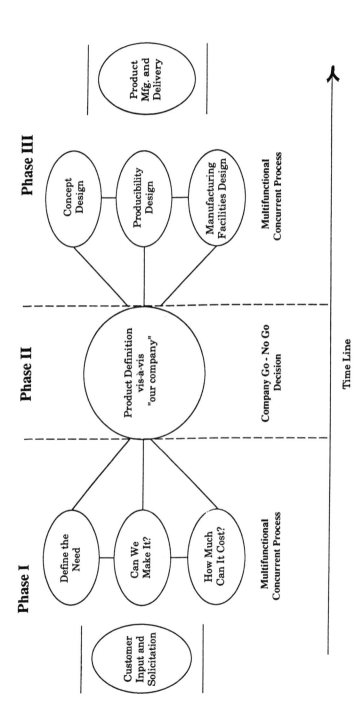

Figure 2-4. The concurrent engineering schematic.

- Identify customers' desires by customer visitations and discussions.
- Have customers rank these desires.
- Rank the business team's capabilities using "ask the experts" techniques.
- Match the business team's highest capabilities with the highest customer desires.
- By subjective evaluation, determine if the match is high enough to warrant a concurrent engineering approach study. If yes, gain customer concurrence and proceed.

Figure 2-5. Practical aspects of applying quality functional deployment.

as a marketing tool, is equally correct for any engineering problem. When a solution to a problem is proposed, a variant of QFD can be used as a quick check to evaluate the risk of the solution. In other words, how good a fit is it to the problem and what is the confidence level that it will be satisfactory? Let's see how that is used and could be applied in the applications portion of the engineering curriculum.

The process of pragmatic QFD is a 5-step approach, as shown in Figure 2-5.

We see that the steps go through the business team process of validating their ability to service a customer's needs. They are effecting a risk-assessment process, the third step of the entrepreneurial definition. We can make this an effective teaching tool for Phase II subjects by prefacing problems whereby the student teams are asked to do a QFD on the assigned problem. This would force them to determine which solution technique has the best chance of succeeding. Let's look at a simple example of applying QFD to a design problem in a Phase II strength of materials course.

Example: The problem assigned to the team is to design foundation bolts to hold up a 10 foot high by 15 foot wide by 1 foot thick rotating sign on a 15-foot-high column, 1 foot in diameter. The sign is aluminum, the pedestal is steel with a concrete base that can be as wide as 4 foot by 4 foot, depth as deep as required. The sign has to withstand hurricane-force winds up to Class 5. The design has to be producible in a typical job shop with lathes, arc welding, and milling equipment and within a budget of $5000.

Using the QFD approach, the team would assess its risk to be able to satisfy the customer's requirements. They are told that if they cannot do the job, they are to decline to quote. They would demonstrate how they used the QFD process to assess risk and how they reached their conclusion. For grading purposes, if they decline they must specify why and then determine what they have to do to give their team the ability to quote with low risk for future jobs. This would include learning the technologies they do not have and report on it to the class. If they chose to quote they must again demonstrate how they used the QFD process to assess risk. Then they are to act as if they won the bid and have the job. They would then proceed to do the design work and be graded on that, as well as their QFD analysis. Here we see a model of true engineering application in industry: Not all requests for quotes become quotes and not all quotes become jobs. Let's go through QFD process:

Step 1. Identify customer's desires:
   We have a customer requiring a rotating pedestal sign to be bolted to the ground and to be able to withstand the force of a Class 5 hurricane.

Step 2. Have the customer rank these desires:
   In the class, setting this would come from discussions with the instructor acting as the agent of the sign owner. In this case it would be a foregone conclusion that being able to withstand a Class 5 hurricane is the most severe requirement and be number one. Withstanding a tornado is not required.

Step 3. Rank the company's ability to satisfy the requirements:
   This is the assessment of what the students know about designing the bolts and what they don't. In this case, they probably understand how to solve for bending stresses and moment, tension and shear load, and perhaps thread form loadings. They may not know much, if anything, about concrete strength and how to anchor bolts into concrete.

Step 4. Match the company's capabilities with the customer's needs:
   Here the student team members need to assess what they can and cannot do and see if they know enough to bridge the differences. In other words, can they obtain the necessary additional information with sufficient capability to proceed to a successful outcome. This means they are making a risk assessment. In fact, they may want to learn some of the probabilistic methods used in risk assessment and try them. In this problem, their lack of knowledge may be concrete design, especially pullout forces cones and distance of bolt separations required in concrete.

Step 5. By subjective evaluation, determine if they should bid on the project:
   Here the student team would evaluate the risks and make a decision. And most important, the team would outline what the risks are and what the contingencies are if the risk is greater than thought, in other words, how to cope with the unexpected.

Here we see a simple example of how pragmatic real-world engineering can be introduced into a classroom situation and give students an appreciation of how theory can be applied to real situations. This exercise would demonstrate the range of ancillary problems associated with any one subset of technology. The students would see for themselves that problems related to one technology cannot be isolated from other influences, no matter how hard we try or wish it to be. This introduces risk assessment into the equation in that the engineer needs to know how well she or he can define the boundary conditions in solving for the client's needs. Entrepreneurship is solving for needs by organizing, managing, and taking risks with low probabilities that they will occur. We can teach this and do it at the same time we teach engineering technical applications.

SUMMARY

Engineering education is not the same as a science curriculum. Engineers are mostly technically trained businessmen and businesswomen who apply their science-based skills in solving commerce-related problems. Engineers are expected to be integral members of their respective company teams who, for the most part, are focused on making a profit.

Since most products have some technical or science basis, the skills an engineer brings to the party are often critical toward the success of the venture. In fact, most business failures occur because the technical issues are violated, that is, the processes for making the product or the design of the product is not done in accordance with science facts. Think about what happens when a salesman promises a product at a particular calendar date and the company misses, thereby setting itself up for penalties or simply not being paid at all. The technical question of design adequacy, capacity, and/or capability were not addressed properly. An engineer on the product team could have, or possibly should have, prevented such errors from occurring, provided, of course, the engineer was a fully participating member of the team. Sadly, if an engineer isn't trained to think as a businessperson as well as a science-based professional, he or she wouldn't be asked or even think of making inputs to alleviate the situation. In many situations, the engineer may not have even been on the team because of a perceived reluctance to become involved. This leads to problems for the company and is the antithesis of the modern business model.

Engineering education needs to be biased toward satisfying the team-based business model. The puts and takes into the curriculum, that is, the need for engineering economics and other business courses to be put into the core curriculum, will necessitate deleting some science courses. It's also evident that extensive skills-building in report writing and presentations are required. What needs to happen is to work out, based on the model, what's worth keeping and what isn't, and why.

How to overcome this inertia within the academic community resisting change is the problem facing the entire engineering community today. We seem to have a split personality in the profession. The 75+% of the engineering community involved in industry is moving very quickly to the 21st century business model, whereas the remainder, the academic-dominated portion, appears almost oblivious to this paradigm shift. Or perhaps they think it doesn't apply to them. They are wrong. Their customer, industry, will put up with unprepared professionals coming into their domain only so long. Then they will fix the problem to their advantage, possibly through very results-oriented on-the-job training, lacking in intellectual vigor. In essence, they will degrade the need for the academic degree, and eventually our intellectual capital will dwindle to the point we can no longer create new products and processes as well as other societies. When that happens, our standard of living declines, much to the detriment of us all.

How do we fix this problem? First, the academic community has to do a QFD on their ability to meet their customers' needs. If they're honest about the assessment, I believe they will come to share my viewpoint. When that happens the battle is won and it's only a matter of time before the engineering undergraduate curriculum begins to realign itself with the needs of industry. Certainly the engineering professional societies, through their affiliation with the Accreditation Board for Engineering and Technology, can act as a catalyst to speed the process. But perhaps another paradigm shift has to occur. We need to split teaching engineering from research in engineering. And both have to be of equal status in the academic community. Teaching has to be pragmatic and cross-functional. Teaching 21st century engineers is not like teaching 17th century philosopher scientists. Teaching has to be a reward for being an excellent practitioner. But even that is not totally sufficient. A teacher has to be, above all else, an excellent purveyor of knowledge to his or her students and have a sense of mission in that pursuit.

Chapter 3

The Entrepreneur's First Step: Learning About and Using the Art of Project Management

So far I've presented a picture of how the engineer's work environment has changed from a specialist applying his or her talent to a specific project to one that is engaged in a product team environment. In that environment, the engineer is expected to be a generalist technical expert and at the same time be a businessman or businesswoman very directly engaged in the process of making a profit. I've also devoted time to explain the education needs for the engineer to be successful in that environment. Now I'd like to pursue these 2 compatible issues, work environment and education needs to succeed in that work environment, and show how that is really the basis for the first step toward becoming an entrepreneur. This is the initial step for skills development toward establishing an independent business. The skill I will introduce now is project management. But first we need to understand why project management proficiency is necessary to evolve into a successful entrepreneurship.

Let's make sure we understand that entrepreneurship does not necessarily lead to the engineer abandoning the business team and striking off on her or his own venture. It is a method of working that is a necessary condition to be an independent businessperson, but not totally sufficient for it to happen. Being an engineer exhibiting the desirable traits of entrepreneurship is just as important for the individual who stays on the company's business team as it is for the person desiring to own his or her own business. Let us just say that the first step in establishing an independent business is learning how to be an entrepreneur. Whether the engineer ever does start an independent business depends on a host of other factors. Some of these are the right product or service development, financial situations, and opportunities within the current job.

We will start with an explanation of how the current environment the engineer dwells in invariably drives "aware" toward the entrepreneur model. In this environment, it is easy to see why learning entrepreneurship skills becomes necessary simply to survive. If the engineer is aware of these forces driving career success, then he or she will probably do a better job absorbing the skills of entrepreneurship and in the bargain, also learn quite a bit about being a company owner. The engineer would also learn that being the owner is something very special and liberating and is not alien to a successful engineering career. Learning how to be an independent businessperson is actually a good thing. It teaches self-reliance and an ability to stretch one's sense of capabilities. But perhaps more important, it teaches a sense of realism, a healthy degree of pragmatic philosophy to mitigate unbridled idealism. It makes the individual aware that great things can be done but that there is a need to carefully plan for success, including looking at situations as they really are and not viewing them through rose-colored glasses. It also means being prepared mentally to shift gears and cut losses if what appeared to be a good idea turns out not to be. It means always having contingencies to turn to, to enable the entrepreneur to go on to new opportunities.

ENLIGHTENED SELF-INTEREST

In today's environment, and probably well into the future, the engineer no longer can count on a firm retaining his or her services out of loyalty. Firms are much smaller than the megaliths of the past and no longer have the cash flows that can smooth over a lack of revenue returns from employees with good potential but not contributing to the bottom line. Many firm don't even have the ability to carry individuals with "outstanding" potential. Everyone has to contribute to the profits, and if they aren't contributing they are no longer associated with the firm.

Economics of business today doesn't allow much slack for maintaining a human capital reserve. We find that smaller firms need to treat human capital as if it were physical capital from a payback point of view. We all concur that capital has to be used wisely. We want to gain optimal return for our capital. The same is true for the application of human capital. We need to pick and choose the individuals that can bring the most profit to the company. If the firm cannot afford to carry people who are not contributing, it is because the payback for those individuals is beyond the acceptable hurdle rate for the capital investment. Therefore, the investment is unacceptable and needs to be rejected, that is, the purported noncontributors are fired.

Hurdle rates in finance are the amount of time a company can afford before an investment has been paid back and the returns are positive, that is, generating profits. Hurdle rates are different for different companies. Obviously, a company that has large cash reserves can wait longer for an investment to be recouped and profits to flow from it, than one with lesser cash reserves. The modern company, being smaller, generally has less cash reserves; therefore, its hurdle rate would be shorter than the norm experienced for the larger and usually better capitalized firms that dominated industry in years past. Equating this principle to people, we see that the tolerance levels for noncontribution to profit shrinks in proportion to the reserves. Does that mean that loyalty between the company and the employee has disappeared? No, it's probably the same as its always been, except that the margin between "wait and see" and "the need to take corrective action" has shrunk. Simply stated, loyalty has never been a determining issue. It's always been profit generation and nothing more.

Understanding this principle is paramount to understanding business in any age. The engineer working in industry must be part of a productive team whose output needs to generate profit for the company. If the specific team cannot, it is draining profits away from those that do, and if allowed to continue could jeopardize the survivability of the entire company.

We can take this thought down to the individual within the team itself. Each team member has to carry a fair share of the profit-generating load in order to remain on the team. Those that don't are draining the profitability of the team and presenting an overhead cost that cannot be borne. "Fair share" means doing the assignment the individual team member has in a manner that fits within the team's schedule and in a way that the results are usable by other team members. For example, the engineer member of the team needs to have the design work done in a manner that is interpretable by the shop operations member and materials member so those people can do their jobs without undue delays caused by incomplete or erroneous information. In addition, fair share means pitching in and helping when an unforeseen incident occurs outside the scripted plan. The best way of dividing work up into tasks based on required skills and overcoming unforeseen incidents is project-management techniques, as we shall see.

To stay on a team, a team member needs to perform up to expectations. As much as we say that engineers now work in teams, they are in many ways actually competing to stay on the team. But who are they competing with? In a football team, the quarterback is not competing with the left tackle. The same is true in a product team. The design engineer is not competing with the finance representative. Then where is the competition the team member is competing against? It is competition against a set standard, the goal the team has had set for it or by it.

Each team member is competing against the goal. In essence, the team member needs to achieve the goal. Keep in mind that every goal has many components associated with it. Some are group-oriented and they tend to be overall goals. An example of a group goal is achieving the profit rate based on sales offset by costs. But most goals are singular or individually oriented. For example, the design engineer would have a goal to produce a workable design accepted by the team by a specific date. Here the primary responsibility for making the goal can be measured and applied to the design engineer. In fact, singular goals are desirable because they allow for direct measurements to a larger degree. We can see if a team member is doing her or his fair share much easier than we can with a group goal.

By measuring, we can see if a team member is cutting the mustard, so to speak. This all boils down to each team member needing to be cognizant of the requirement to perform to the set standard, and doing so. This has to be done while working in a group. If the standard is easy to achieve, then the individual being measured enjoys high praise. If the goal is extremely tough, then missing it becomes more likely and the member can suffer ridicule, which in extremis would probably lead to being dismissed from the team.

Notice I haven't said anything about competency of the individual. In an ideal world, the competent would always flourish and the incompetent would be weeded out. Unfortunately that is not always the case. Many times an impossible task is given to a team, either on purpose or in error (we would hope mostly in error and not in some devious way to cause failure for the purpose of fostering dissatisfaction with the members). So it is important to understand that measurements aren't always objective in intent. This is a lesson that team members need to learn. They must understand the nature of the goal and whether or not it is a reasonable goal for them to pursue. Each individual on the team needs to understand his or her role in achieving the goal. Can it be done, and what are the risks? The ability to carry a fair share will be determined by the severity of the goal being used to set the fair share.

With this being the case, we can define a very important lesson to be learned by the engineer entrepreneur in the making: Make sure that the risk being undertaken to achieve a set goal is reasonable.

Each individual assigned to the team needs to do this, not only the engineer. I mention engineer because we don't often think of engineers having to accept business risks. We expect them to accept technical development risks and the others accept the profit and loss risks. But that's no longer true. On a team, the engineer has a primary responsibility for technical risks, but also a shared responsibility for business risks. To logically accept risks, there has to be a reward for successfully outwitting those risks and achieving the goal. In business this is associated with the payment fees a company receives for doing a task successfully. Translating this to the team, we would expect the team to share in the rewards, such as bonuses and salary increases. It is not uncommon, nowadays, for a team to negotiate a bonus prior to starting a project. Obviously, the higher the risk, the bigger the bonus becomes.

So we see a team becoming a small business unto its own. It has a contract to do a specific project with accepted milestone measurements of performance and with set payments for services performed and products produced. It accepts the risk of failure: the team could go out of business for a total failure, which means the members lose their jobs. But it also knows that if each team member pulls his and her fair shares, they will reap multiple benefits.

This leads to the individual. Each person has to make sure he or she understands the risk and is willing to participate. If not, they should seek reassignment or perhaps leave the company. Using the sports analogy once again, it is like being the shortstop and negotiating a contract. His worth is dependent on what he can, in theory or previously demonstrated, do for the team. The shortstop can stay with the team, that is, he believes the risk is worthwhile and manageable, or go to a competitor with a better chance of success, for example, of winning the World Series. The analogy to the business team is quite close. The engineer is in reality an individual practitioner who offers services to a firm for a fee (his or her salary). And as long as he or she gives value to the firm and the fee is sufficient, we have a business contract.

Many engineers find themselves in the role of being an individual business enterprise in concept, even if not legally, rather than an employee. This leads to the primacy of entrepreneurism as being the philosophy the individual uses to justify his or her actions. It's every person for himself or herself and company loyalty is an obsolete concept, if it ever was a concept at all. What's needed is team loyalty for members who need to support each other so each can make the required contributions as defined within the team's implied or real contract with the firm. And the individual's implied or real contract has to be compatible with the team's commitment. It is only a short step from the feeling of being an individual contractor within a company to actually becoming one in fact.

THE SKILLS TO BRIDGE THE ENGINEER-EMPLOYEE TO THE ENGINEER-ENTREPRENEUR

The theme of the individual having to fend for herself or himself, contrasted with the thought that the company she or he presently performs services for will take care of the individual, is the driving force to learn new entrepreneurial skills. I will elaborate on the theme of individuals who fend for themselves contrasted with the thought of the company for whom they perform services taking care of them. There are significant ramifications of this change in philosophy regarding what skills, in addition to technical skills, an engineer needs to survive. We will see that those skills are strikingly similar to those exhibited by men and women running their own firms. It is only a small step from being a good performer in today's team-based firms to that of being the entrepreneur, running and working on the team they really own.

What are the skill sets required of the entrepreneur? We've already discussed the education requirements for today's engineers, which is a reasonable basis for going forward. The key skill we haven't discussed is good project-management skills, complemented with good negotiating skills focused on contracts. Both of these skills are what I consider amalgamation skills. They come about by learning many subsets, and then becoming adept at putting it together in a comprehensive compatible manner. Let's look at what the subsets of project management are, then discuss how they are learned. Later I will address negotiating skills.

A SHORT TUTORIAL ON PROJECT MANAGEMENT

All successful entrepreneurs exhibit excellent project-management skills. This is definitely a skill set that can be learn, but I strongly suspect that many people who willingly take risks to achieve success have an instinctive understanding of project management, even though they may not be able to define what they do and why. Obviously it is better to know why you do certain things, so we'll proceed on the premise that given a blank mind, we can imprint project-management skills into it in such a way that it will perform the task admirably well, much like putting a well-proven program into a computer.

Project management is described as follows:

> Project management: Formalizing a task into achievable and measurable steps to reach a set goal.

As soon as an engineer becomes involved with "real-world" work, he or she realizes that success depends on effectively working and communicating with others. Obviously this is more than simply being cooperative and willing to work on assigned tasks. It takes more than a desire to pull one's weight. It takes a certain learned skill set to be able to join in the workflow in an effective manner. Perhaps a better working definition of project management is as follows:

> Project management: Formalizing tasks necessary to track complex activities effectively and communicating interim and final results to all those who have an interest in the activity.

Another way of understanding the project-management process is to think of it as an effective code for, and the related process of, organizing group work. New engineers rarely learn these skills during the course of their formal education. In fact, most engineers are not exposed to project management until they're well into their first jobs in industry. Unfortunately, most of this exposure is as a "worker bee," not fully understanding the process. However, to migrate toward an entrepreneurial mind set, it is necessary to have a virtually instinctive understanding of the project-management process. This comes about by learning the project-management skill set, then looking at each activity the engineer is assigned to as a project and applying project-management skills to them.

The process is straightforward but appears complicated by the division of complexity issue. Projects are divided into 2 categories: simple and complex. Simple tasks use a 4-step approach, whereas more complex activity requires a 6-step model.

It is interesting to note the universality of the models. We can use either model for virtually any activity and be confident that by doing so we have the project under control. The ability to visualize all jobs within the domain of the 2 models is a necessary condition for entrepreneurs. People who habitually take calculated risks must be able to size up the steps, hence the resources, needed to reach a successful conclusion. Project-management skills provide a sense of security to the entrepreneur that his or her activity is under control. In effect, project-management skills instill the ability to assess situations, then organize an approach to reach desired conclusions. To operate in the entrepreneurial mode, one needs to learn these skills and apply them as if this is an integral part of the person's subconscious response to stimuli.

Both models shown in Figure 3-1 have *organizing, assessing,* and *executing* phases. We typically use words like concept, definition, and planning to

Simple Project Steps	Project Phase
1. Concept	Organizing
2. Planning	Organizing/assessing
3. Execution	Executing
4. Closeout	Executing

Complex Project Steps	Project Phase
1. Concept	Organizing
2. Definition	Organizing
3. Design	Assessing
4. Development or construction	Assessing
5. Application	Executing
6. Post-completion	Executing

Figure 3-1. Two project management models.

define the organizing phase; words like design, development, or construction imply the assessing phase. And the words application, post-completion, and closeout refer to the executing phase.

Project management goes 1 step further by introducing more definitions for the 3 phases (I promise this is the end of the definitions maze). This path is necessary to gain an understanding of what we're doing and why the structure is necessary.

The organizing phase is expanded into what is called the Work Breakdown Structure (WBS). The assessing phase is commonly known as Risk Analysis or Risk Management. Finally, the actual carrying out of the tasks to compete a project denoted as the executing phase is known as the Action Steps. The reason I bring out these convoluted paths for naming and renaming the 3 linked activities of project management is to point out that we are using a language of words to define a synergistic linking of ideas. Sometimes we feel a need to express ideas with many adjectives, even though they really mean approximately the same thing. So I will get off the Roget's Thesaurus explanation of language and back to project management. I will use what I consider to be the most common and accepted format to describe project management.

- Work Breakdown Structure
- Risk Analysis and Management
- Action Steps

What follows will be a summary of project management theory and application.

Work Breakdown Structure

The ultimate end result of project planning will be a series of action steps designed to achieve the given or self-imposed goal of the project. The action steps, if defined properly, will be a series of investigations or probes for proper applications of resources followed by actual applications of those resources to achieve the goal. The Work Breakdown Structure is the organizing step of a project that structures the action steps. Logically, before we can do anything we need to define what that thing is and what we are being asked to do. So the Work Breakdown Structure starts with 4 questions:

1. What has to be done and in what order?
2. Who will be doing the specific tasks?
3. How long should it take to do each specific task?
4. What's the cost, both financial and in people resources?

These 4 questions force us to deal in specifics instead of generalities. We have to define the tasks to the point that there is no doubt within any team member's mind as to what has to be done. We take it down to dotting the i's and crossing the t's. For every step in the project that we think we need to do, we have to assess what we know and don't know about doing the task. What we don't know we have to learn about to the point that we now know or "feel" we know (this becomes part of risk analysis). Also, when we eliminate doubt by creating "knows" out of "don't knows," we have to be able to apply this new knowledge.

The output of the Work Breakdown Structure gives us a list of tasks in the approximate order they will be accomplished, along with a succinct evaluation as to whether or not we possess the information and resources necessary to do the task. This allows us to review for completeness and correctness of the project before we attempt to do it. A properly done WBS gives the project manager a macro- and micro-look at a project. By doing so, the project manager can determine if the steps make sense. Let me illustrate this with a simple example, shown in Figure 3-2.

The purpose of this project is to enable our batter to have the best chance of hitting the ball with all the power he possesses. We see the end result is to swing at selected pitches after the batter has properly prepared to do so. This project is simple because I'm assuming we're all familiar with the game of baseball and, to a certain degree, the process of batting. But for our English colleagues who may be reading this book (after all, it is their native language, too), it may be a mystery to them. They have to turn every "don't know" into a "know" and then determine if the sequence of steps is proper. The North American knows it is important to observe the pitcher's motion before getting into the batter's box. However, it's conceivable that our English colleague would have that step out of sequence. So we see that the WBS shows where each step falls in sequence after each step has been carefully analyzed; it also shows whether any apparent voids are in the listed steps. For example, the batting project would be incomplete if the step of taking practice swings were missing.

The successful at-bat project:

- Select a bat.
- Go to the on-deck circle.
- Dry hands with resin.
- Apply pine tar to handle of bat.
- Take practice swings to limber up.
- Observe pitcher's motions.
- Take place in batter's box.
- Swing at selected pitches.

Figure 3-2. The successful at bat project

A good project design ensures that the steps flow without any logical breaks. The WBS, by its nature, allows us to spot discontinuities. Since most of us are familiar with baseball, we can easily review the steps to make sure there are no discontinuities. This review is critical for gauging completeness of the potential project. Most projects that are unsuccessful fail because of incomplete planning. Usually this means there was no logical bridging between steps. The WBS process is very useful in minimizing this failure.

The 4 questions we must answer create a situation whereby importing bridging steps are needed to fill the gaps. For example, let's say we had a project that required delivery of perishable products to a customer. The logical sequence would be to stage the product and then ship it. But there are lots of steps in between. Step 1 of the WBS requires that we understand what has to be done and in what order. The following logical sequence may be too macro: Stage the product and then ship it. In fact it is. We haven't differentiated between any product and a perishable product. We need to know the characteristics of this perishable product. So a better logical sequence would be the following: Understand the nature of the perishable characteristics of the product, establish procedures to control it, maintain the integrity of the product while being staged, and then ship it. We've added an entirely new subtask to the project, which is entirely necessary, that allows us to bridge between staging and shipping. Without this task, there is a high risk that we would be unsuccessful in performing the goal of the project: to ship products to customers.

A successful listing of all steps sets up solutions to Step 2: the assignment stage of who will be doing the task. Without addressing the perishability control need of the sequence, we may not have assigned the proper skills to the tasks. If we didn't think about the perishable products needs as being part of the WBS, then perhaps we would have assigned a shipping clerk to gather the materials and ship it. But with "perishable" now a word in the planning of the project, we would learn what needs to be done to care for the product during staging and transportation. In this example, it is conceivable that we would also assign the appropriate technical support that is familiar with the way this perishable product decays and be charged with setting up procedures to prevent it. Obviously, if we're talking about a food product for which refrigeration minimizes or delays decay, then the need for proper equipment would be addressed as part of the project. If we didn't consider the perishable nature of the product, we probably would have spoilage at an unsatisfactory level.

This thought process filters down intact to Steps 3 and 4. Certainly knowing about the true level of complexity will impact the cycle time to do the project. Without proper recognition of the perishable nature of the example product, we'd have a grossly underestimated cycle time. The exact same thing can be said about cost. If we don't know the full story of what has to be done, we can't determine the true cost. As an aside, most projects that are over budget are so for this reason.

The experienced entrepreneur will review the steps of a project with a very critical eye to see if it is complete. As I said before, good entrepreneurs have excellent skills in project management. Some people say it is an instinctive flair for the process. I believe it is something they have learned, either formally or through trial and error. Whatever the case, it is apparent that good entrepreneurs make sure that all bases are covered before they proceed. This is what the Work Breakdown Structure requires the project manager to do.

With a WBS completed, we have a road map for creating the project prose and presentation. Some projects are shown as Gantt charts, others as outlines with the proper indentations showing the relationships between steps. Another way to record the plan (which is what a project is) is as an essay narrative with specific instruction, and at the other end of the spectrum would be a PERT chart or critical path diagram. Whatever the methodology of presenting the project document, it will be an exposition of the 4 steps of the WBS for getting from the beginning to the conclusion without any discontinuities in the process.

Risk Analysis and Management

The Work Breakdown Structure defines what must be done in the sequence that it needs to be accomplished. We saw in the WBS discussion that we had to convert "don't knows" to "knows" in order to get the project done. The question then becomes, how well did we do that? If we follow steps A, B, C, etc. to the end, will we reach a successful conclusion? We would hope so. However, the phrase, "the only certainty is death and taxes," is a truism, no matter how whimsical it appears, and one that must be contended with. Uncertainty is a fact of life. No matter how well we plan we can be assured we could have planned better. Risk analysis is the process of defining the level of uncertainty, that is, the probability that an unplanned event will happen, and then developing contingencies to deal with unplanned events. Projects, especially complex ones, have to deal with unplanned events because we can never be sure our planning is 100% complete and correct. Many techniques have been developed to minimize risk and we will look at the best ones as they relate to project management.

I should point out that risk analysis is like a physical exam. We try via risk analysis to define, in a quantitative manner, what the likelihood is of a project going astray. By knowing whether there is a high or low probability of an unplanned event happening, we can select appropriate standby corrective measures to take if it does happen. This infers that when we apply risk analysis depends on the degree of uncertainty the project exhibits. This is basically what we want to find out. So the first item on the agenda is to quantify the probability of something going wrong with a project. The less of a probability that a project can go wrong, the less the need for risk analysis and contingency planning. To quantify, we need to understand what the causes are for a project going wrong. If we can gain that understanding we have a good chance of mitigating them. Entrepreneurs do this all the time. They are constantly trying to visualize what could go wrong with a venture so they can bolster themselves against that potential.

There are basically only 2 major categories of what causes projects to go wrong:

1. A poorly planned project, one that exhibits a high degree of discontinuities in the sequences of project step events
2. A change in plans for whatever reason

Category 1 is the domain of the Work Breakdown Structure. If it is done properly, discontinuities are rarities. The risk assessment of a project with lots of discontinuities is always high compared with a project with few undiscovered discontinuities. If discontinuities are affecting the potential for success, we would need to employ some corrective actions. However, when we're talking about risk analysis, we normally consider only well-structured projects where discontinuities in the step sequences have already been attended to, therefore are not important factors.

Category 2 is the normal domain for risk analysis. From Category 1 we know that careful planning is the hallmark of mistake minimization. We also know that change enhances risk because as we change, we rarely have the opportunity to do a full cause-and-effect planning routine before the change. Hence the ability to carefully plan decreases. Time to plan is compressed and it's less likely for the change plan to be thoroughly evaluated. And the less thorough the planning is, the more likely a deviation from the expected will occur.

Another factor increasing the probability of failure due to change is the dynamic nature of change itself. It is not static standstill, which means we are less likely to have enough facts to react with a new best course of action. For these reasons, we have another factor to look at and understand. It is important in a controlling sense to understand what the nature of change is and how to control it.

Change and Change Control

Let's look at change and change control as it affects analysis of risk, and then get to the more encompassing topic of Risks and Contingency Planning.

If we look at reasons for considering making changes in goal-oriented activities, we find that they results from 3 major causes:

- Changes in the environment necessitating a change in the goals of a project
- Erroneous assumptions made during the planning of the project
- Personnel changes requiring re-planning to suit different skill sets

When we talk about environment, we mean the situation the project team finds itself in with respect toward progress in meeting their goals. This recognizes that few if any projects exist in isolation. And it is the effects on the task by outside influences that constitutes the environment. For example, early or late completions of supporting projects can greatly effect the team's activity. In the "preparing to bat" example discussed previously, suppose there are 2 outs and you are the batter in the on-deck circle (you bat next after the current batter completes her or his turn). However, unfortunately for your team, the batter strikes out and ends the inning. This means you're first up in the next inning. The change in the project then is that you will not be able to observe the pitcher's style against a batter before you bat. You'll only be able to observe warmup pitches that may not indicate how the pitcher behaves in a "live" situation. The pitcher you're facing may be canny enough to not show too much of his motions during warmup. After all, warmup is just that: limbering the muscles to give best performance when it counts. This is change and it reduces your ability to create effective contingencies.

Another factor that will affect the environment and will do so greatly is a change in direction of overriding strategies. Again, going back to the batter project. What would happen if the other team changes pitchers, going from a left hander to a right hander? You are a right-handed batter, and your team's manager knows you've been in a slump against right-handed pitchers. Suppose, also, your team is losing by 1 run and you're the first batter scheduled for the ninth inning. A change in direction of overriding strategies could be the manager lifting you from the lineup for a pinch hitter—one who has demonstrated better recent success batting against right handers.

A third environmental factor is an internal one. It is the corrections you must make because of a discovered error your team made. This is the most devastating environmental factor because it is self-inflicted. Errors are frequently caused by taking action before the facts of the situation are fully known. Inexperienced project managers often jump the gun and try to do too much too soon. Most good project managers have learned that it is foolhardy to proceed unless all the pertinent facts are known. They use the skilled carpenter's rule: Measure twice, cut once.

Or another way of expressing it: Aim, aim, aim, then shoot.

Most experienced entrepreneurs, while outwardly exhibiting a person of supreme confidence and action and perhaps a little flamboyance, are in truth finding ways to become confident about their project and not willing to expend real capital on it until they are sure it is the correct thing to do. They are constantly balancing the risks against the rewards and will not initiate an action until the risks of doing so have been minimized. When we say minimized, we mean to the point that we can cover the risk of losing and still have capital left over to try again, only this time with an improved plan.

Perhaps you've already deduced what risk analysis is. It is trying to find all the facts pertaining to the problem and only launching an action when we're very (and I emphasize very) confident we will succeed. Risk analysis in business is not a wild Las Vegas–style spin of the roulette wheel.

Now let's see how we can put logic into evaluating whether a proposed change should be made and if so, do it with the lowest possible risk.

Before any change in a project can be made, it is prudent to determine its potential impact on the stated goal. The only exception to this bit of advice is if the proposed change is due to a change in the goal. If that's the case, then the change to the project to realign with the goal is mandated. This is true because a change in the goal automatically changes the scope of the project. Here's an example to illustrate this point.

Example: A project is set up to transport the U.S. Olympic Women's Soccer Team to 5 sites throughout the United States to play exhibition games against 5 other national teams, one team at each site. The entire trip is being sponsored by the U.S. State Department. A change is made that changes the goal of the project. Instead of playing 5 other teams, each at a different site, the State Department has requested the team to play only the team from China at all 5 sites. Also, the teams will travel and live together. This means the transportation project has at least doubled in size, with all the added complexity this creates. There is no need to make an evaluation of the change and its worthiness before going ahead with it because the sponsor has changed the goal. Since we can assume that the transportation can still be arranged, the change will be made.

The example points out that what would have been a set of milestones to move 1 team, now has become a set of milestones to move 2 teams and to do it in a way that leaves the game on the field, not in the airplane being shared by the teams. This is a mandated change because the goal the project supports has changed. But most changes are not as a result of mandated requirements. There is another category of change known as "non–goal-mandated changes." These changes occur only after the proposal for the change has been evaluated and found to be beneficial to the ability of the project team to achieve the goal. Here are the steps to evaluating whether or not to implement the proposed non–goal-oriented changes.

Non-goal change evaluation steps:

1. How will the change affect the project?
   a. Scope
   b. Cost
   c. Schedule
   d. Capital plan
2. Will the change affect adjacent, predecessor, or successor projects?
3. Will the change affect the financial justifications for doing the project? If so, are they still positive for doing the project?
4. What is the justification for the change?
   a. This implies a test for validity of the request.
   b. It is proper only if the organization requesting the change is equal to or lower than the organization structure.
   c. It is good management practice to tell subordinates the reasons for changes.

The answer for all of the evaluation questions should be positive. If not, there is a lesser likelihood for the change to be beneficial to the end results of the project. Remember, change increases risk. So the change should be more beneficial than the original plan to justify the additional risk. It is also good practice to document change requests and to maintain an audit path of all investigatory activities. Since most projects have many interrelated steps, it becomes necessary to carefully document changes to prevent confusion and perhaps even missing steps that may have been impacted by the change.

How to Calculate Risk Factors

We understand that change increases the risk that a project can fail. You now also know how to evaluate whether a change should be allowed. But are there ways to calibrate the level of a risk? This is certainly a question an engineer trained to work with data and understanding probability and statistics should ask. The answer is yes and here's how. We can divide the term "risk" into 2 words and we often do. The word "risk" is often used interchangeably with uncertainty. But as you would surmise from an engineering usage, they have 2 different meanings. I favor the definitions proposed by N. Barash and S. Kaplan in their book *Economic Analysis for Engineering and Managerial Decision Making*, 2nd Edition, McGraw-Hill, NY, NY 1978.

Risk:
A variance from average (expected) value, which occurs in random chance patterns. The larger the variance, the larger the risk.

Uncertainty:
The degree of unknown caused by errors in forecasting 1 or more of the factors significant in determining the future values of the variables making up the variances.

In mathematical terms, variance is the standard deviation used in probability and statistics calculations. The most widely used expression for variance in risk analysis work is:

$$\sigma^2 = \int_{-\infty}^{\infty} (x - u)^2 f(x) dx$$

If we can find the value of the variances, we can equate it to a risk value with an experimentally determined constant. Unfortunately, the constant would not be the same for all situations. Therefore, using the expression for variance is not an easy task, so for practical purposes is not used. However, it is possible to use mathematics to determine a probability of risk associated with a decision. Let's look at an example.

Example of an Analytic Method for Evaluating Risk

Suppose you are an entrepreneurial owner of a startup company with a newly patented process for applying oxidation-prevention coatings on metals. Part of that process requires a precision grinding step that would normally be done with a computer numerical control (CNC) grinder. You can either buy a CNC grinder or subcontract out the work. What is the risk in the choice you have to make?

- The first set of evaluations we would make would be a judgment factor based on previous experiences, either your own or those of valued advisors. If the judgment says the decision is a good one and should be pursued, we would want to hold for a short time and first do a risk assessment. Now judgment and risk assessment are blurred entities of almost the same thing. Good judgment almost always contains some form of instinctive risk assessment. If you needed funds to buy the CNC grinder and there was a bank with an unguarded cash drawer it would be tempting to help yourself. But good judgment would deter you from doing so because you know that the probability of getting caught is high and it is immoral to behave in this manner. So here judgment and risk assessment are one and the same. But in most instances, this type of evaluation of risk is not sufficient. We have situations where judgment calls cannot be made because of the abstract nature of the situation where our experiences and those of our counselors are not sufficient to allow us to make a decision. We need to do something more. This leads to the second approach, a more analytical process.
- The second approach is a financial evaluation of the proposed decision that can be used to help assess risk. Financial evaluations allow us to put dollar numbers on the various aspects of the decision and theoretically

show us which way gives the best value for the money spent, the best value thereby purporting to be the best risk vs. benefits choice.

Oh if it was so easy! So straightforward!

But a financial evaluation is not sufficient by itself to decide if a decision is valid. All business decisions need to pass some sort of hurdle rate that tries to define whether or not implementing the decision is a good use of the company's resources. Some would say that the only factor to consider is if the decision, if implemented, would be the best use of the company's money. This is too simplistic an approach because most decisions have many intangibles, such as items of benefit that cannot be easily measured, if they can be measured at all. Unfortunately, intangibles are real and cannot be ignored. We can't say since they are difficult to measure, we will ignore them by giving them a zero value in the overall benefits rating. So, financial evaluations are important, but not sufficient to allow a decision to be implemented or not. We need to do more.

- The third set of information that must be generated is a determination of how likely to be valid the assumptions are that are used to make the financial evaluations. We do this in a probabilistic sense, as we shall see. When we get to the third set we have facts and specific assumptions that we play against 2 questions: 'How real is it? Will it happen as we think it will?'

And we answer in percent probability. If we say we are 100% certain, then without hesitation we say "implement." If we are more than than 50% certain it will happen as predicted, we weigh the consequences of being wrong. And if we can survive being wrong, we would also go ahead with the decision. Surviving means we have contingency plans that we feel very confident will work if the original decision turned out to be wrong. Finally, if we are less than 50% certain the decision will work to our favor, we should forget the whole thing and go back to the drawing board.

Using the CNC grinder example, we can trace the evaluation of risk through the 3 phases.

Phase 1: Judgment

Buy or subcontract. If we buy a CNC grinder, we have complete control of the process within the company. However, this means allocating cash to purchase or lease the equipment, dedicating manufacturing space for the operation, and hiring trained personnel to operate the machine and all the other subsystems including maintenance to support the machine. On the other hand, to subcontract means we eliminate all of the infrastructure necessary to support the grinding operating with all of its supporting costs. There is definitely less of an administrative obligation we would have to undertake. But we would be vulnerable to vendors for quality and timeliness of deliveries. We would also have to set up a more complex purchasing strategy to mitigate the possibility of rising costs due to vendor inefficiency and/or greed and be concerned with being locked into 1 or possibly 2 vendors. We would need to establish a supply-chain strategy, more so than with raw material vendors because the process is so much closer to the ship time to our customer.

We could use judgment to determine the risk in both choices, but only if we had access to experience within either choice. The experiences would also have to closely parallel this current scenario to be valid. If trusted people on staff or advisors to the company are familiar with running a machining operation and are available, they could be used to do a preliminary evaluation of the grinding in-house option. Likewise, if the company is strong in purchasing a variety of commodities and special purchase items, it could scope the subcontract option. On the basis of the report of the 2 groups, you could make a decision. Unfortunately, many times in this type of scenario, the advocate for the position who articulates best would be favored even if wrong. So we see that judgment is mostly subjective and we wouldn't really know with a high level of certainty which way to go. So that leads to the next phase.

Phase 2: Financial Evaluation

After we have made a judgment evaluation, we probably have a much better feel for the decision to be made and what the risks will be. Since we can't really evaluate those risks at this point, we need to gather some hard financial facts about the project and merge them with what we know about the 2 choices: make or buy.

Financial evaluations come in 3 primary methodologies, ranging from quick and least accurate to more time-consuming and theoretically very accurate. They are in order of ease to do:

- Payback period method
- Return on investment method
- Present worth method

For most capital projects such as purchasing a CNC grinder, all 3 methods would be used. Let's see what they tell us.

Payback period method: The CNC grinder selected for technical capability reasons costs $125,000. Is this a good use of company funds from a financial perspective? Or could the money be applied elsewhere more effectively? The payback period method is a tool to help make that decision. The equation is as follows:

$$\text{Payback period} = \frac{\text{Cost of project}}{\text{Incremental project savings/yr}}$$

Analyzing the situation we have:

- Cost = $125,000
- Estimated savings data
  - Subcontract cost is quoted at $110,000 per year for the volume forecasted.

- Cost of running the CNC grinder:
  - Operators: 2 @ $10/hr +25% benefits for 2000 h/yr = $50,000/yr
  - Support costs: = est. $10,000/yr
  - Total cost of CNC operation = $60,000/yr
- Savings equal subcontract cost minus make cost = $50,000/yr

$$\text{Payback period} = \frac{\text{Cost of project}}{\text{Incremental project savings/yr}} = \frac{\$125,000}{\$50,000/\text{yr}}$$
$$= 2.5 \text{ yr}$$

Based on the company's policy for payback period, the project would be acceptable or not. Let's assume the policy is a 3-year payback for capital expenditure. In that case, the project is acceptable.

Return on investment method (ROI): This is a variation of the payback method in that the life of the investment is factored in. In this case, we are concerned with 2 things: The life of the technology—will it still be viable after 2.5 years? Also, will the CNC grinder last without requiring large repair funding beyond the payback period of 2.5 years up to the expected useful life of the equipment? We want to make sure the investment will last and be viable through at least the payback period as a minimum, but actually through the expected life of the equipment.

The equation gives us a percentage value from which to compare other investments, similar to a comparison to interest rates of bonds or other financial instruments. The value shows an incremental return on the investment. This value is then judged against other investments to see whether the return is sufficient, or passes an arbitrary hurdle rate, for the company to have made the investment. As a general guideline, an ROI greater than 15% is considered to have passed the hurdle rate, which is usually 2 to 3 times the normal bank interest rates for large sums of money. Keep in mind this is only a guideline. If capital is earning significantly more than that, then obviously the guideline is moot. The formula is shown below.

$$\text{ROI} = \frac{\text{Ave savings/yr} - \text{Cost of project/yrs of life}}{\text{Cost of project}}$$

For the CNC grinder example, the first item of business is to decide how many years of useful life we should get from the equipment. This information would come from the vendors of this type of equipment, from our previous experiences, or through benchmarking with other manufacturers that use CNC grinders. Let's say after we investigated, we concluded the equipment is expected to last 5 years. Now before we plug this number into the equation we should contemplate one more scenario. What is the likelihood that we will still be using this process 5 years from now? In other words, what's the expected business cycle for the process? If we expect that we will still be viable in the marketplace 5 years after product launch, then 5 is the number for the equation. If, on the other hand, we expect to have a superseding process that doesn't require the use of the CNC grinder and that product would sunset the current product in 3 years, then that value should be used instead.

This brings up some interesting alternatives to consider. If the payback is 2.5 years and the product is expected to be viable for only 3 years, can we make enough money in 3 years to warrant the investment? This shows why ROIs are calculated in addition to payback periods. The payback period shows we would earn money, not lose. But it doesn't give a clear-cut view of the magnitude of gain. ROI does.

Let's plug the numbers into the equation and see what it tells us.

For an expected life (or product cycle) of 5 years:

$$\text{ROI} = \frac{\$50{,}000 - \$125{,}000/5}{\$125{,}000} = 0.20, \text{ or } 20\%$$

For an expected life (or product cycle) of 3 years:

$$\text{ROI} = \frac{\$50{,}000 - \$125{,}000/3}{\$125{,}000} = 0.067, \text{ or } 6.7\%$$

We can see from this scenario that the product cycle of 5 years corresponding to a useful life of the equipment of 5 years the ROI exceeds the general guideline and is acceptable. However, when we replace the expected life of the equipment with the hypothetical expected useful commercial sales life of the product, the ROI becomes an unacceptable 6.7%. If the latter were the case, based on financial evaluation alone we should reject in-house manufacturing and subcontract out the grinding operation.

Present worth method: The present worth method is also known as the discounted cash flow (DCF), internal rate of return (IRR), and the net present value (NPV) methods. The present worth method takes an entirely different viewpoint. It says let's compare the project against not doing it at all and simply investing the money in the financial market. This is analogous to becoming a bank instead of an entrepreneur. Granted, this is the antithesis of the topic of this book, but as I've said, good entrepreneurs evaluate risks before engaging. So we ought to know what the alternative is to taking our money and putting it into equities and savings. Of course most budding entrepreneurs are trying to raise money to get their ideas launched and do not have the luxury of investing it for other than their prized dream; so this method, while illuminating, is not the prime method of evaluating the choice mix. But it will give a good indication of how potential investors in your venture will judge your idea. So let's take a look and see how it works.

We will compare the return of the investment to buy the CNC grinder against investing that same amount of money in financial markets. Now you might wonder, how you can do that with any degree of accuracy and you would be absolutely correct. What we do is select a rate of return we can expect to achieve, for example, a certificate of deposit for "X" percent return. We know we can lock that in for a sufficient period of years. We then use the time value of money formula, the typical interest rate calculation, to see what the value of the investment will be at the end of the period. We select the period to be coincidental with the payback period or expected duration (money-generating life) of the project (or product). If the gain from the investment is better than the gain from the project, then the choice is simple: Don't do the project. If it is the opposite, do the project. Let's demonstrate the method using the CNC grinder example.

The formula for the present worth method is the traditional time value of money equation:

$$P = F/(1 + I)^n$$

Where:

P = present value of money
F = future value of money after n interest periods
n = interest periods, usually years
I = interest rate

For the CNC grinder example, we expect the grinder to last 5 years; this means in 5 years we will have saved $250,000 excluding any interest gained if we were able to bank the money. In this evaluation we assume the money is not banked but used for other purposes. The $250,000 is the value for F in the time value of money equation. The interest period, n, is 5 years which is the expected life of the grinder and the project. The interest rate is an arbitrary selection based on a best guess as to what we can earn with the money invested in financial instruments. Usually a value is selected that represents currently available rates that can be locked in for the entire time period. Let's say we can get a rate of 10%. Now inputting these values into the equation we solve for P.

$$P = \$250,000/(1 + 0.1)^5 = \$\,155,230$$

This means a sum equal to $155,230 invested for 5 years at an annual interest rate of 10% would yield $250,000. But our original assumption, based on the facts we presumed to be correct, said that our grinder would save us $50,000 a year, $250,000 in 5 years for an investment of $125,000 (the cost of the machine). By mathematics alone we would be $30,230 richer at the end of 5 years by going the grinder purchase route. Will that really happen?

What happens if the grinder lasts only 3 years? We can calculate a P value for that based on 3 times $50,000 annual savings and an interest period of 3 years. Here the answer comes out to be a P value of $96,240. That amount would gain us $150,000 in 3 years. Compare this with the constant of $125,000 needed to invest to purchase the grinder. Here we see the evaluation becomes negative. We are $28,760 worse off for having purchased the grinder rather than becoming a banker.

So we have our financial numbers and we find we still can't make a real decision. We still need to determine the risk. This leads to Phase 3, a probabilistic approach to assessing the level of risk.

Phase 3: Probability Assessment

We need to know how good our assumptions are in reference to the numbers we are calculating. If we are going to invest in a project, we need to be able to define how risky it is. Obviously investing in real estate in the financial district of New York City is less risky than investing in land in a small south sea island that could develop into a tropical tourist paradise. Or is it? I bring up these questions and assumptions to point out that unless we have hard cast-in-concrete facts, we are forced to objectivize subjective reality. And we will do that using the best analytical tools we possibly can.

So a probability assessment comes out to be a set of subjective assessments that we can rationalize to give us comfort with the decision we make. It does one thing: It gives us reasons for doing what we chose to do that can be constantly revisited and fine-tuned. It gives us a base line to measure against. But by no means does a probability assessment give us a certainty that the decision is correct. It is probabilistic in that, say to a 90% confidence level, it tells us that 9 out of 10 times we will have made the right choice. But by no means does it ensure that at any specific time, the answer will have been correct. It is not a guarantee of success.

Let's look at this objectification of a subjective process once more using the CNC grinder example.

In the grinder example, we see that there are 2 major conclusions we need to validate: to make a decision to buy the machine tool or to purchase finished parts from a vendor. They are, in essence, the risk factors that must be dealt with, and they turn out to be:

- Will the savings be realized?
- Will the CNC grinder perform properly for 5 years?

The first risk factor is easier to deal with. If the CNC grinder is implemented, the extra people to do manual labor will not be required; hence, the expense will not be accrued. So the savings will be realized. The second risk factor will require more data. We saw from the examples used in the financial analysis that the CNC machine must operate for the full 5 years in order to make this a viable choice.

How do we know if the CNC grinder will run properly for 5 years? Again there are 2 things to consider. First, will it be necessary to fulfill the product life cycle for that period of time? If the selling life of the product is significantly less than 5 years, that is, the product becomes obsolete before the machine tool's life is expired, then the answer is no and the risk factor is too high to buy the machine. For the purpose of argument, let's say that's not the case. The product's selling life cycle is greater than 5 years. Then we would have to look at the second consideration, which is to find a way to quantify the probability of the machine being viable for that period of time. We can do that by investigating the CNC grinder model or similar models' performance histories. This again becomes a task that has a large subjective component to it. How will we get data? How good will the data be? How complete will the data be?

Getting data on machine tools is a classic detective story situation. We would have to find how many such machines or similar ones exist. Then we would have to know the severity of the applications they're applied to. This applications data would need to be compared with what we will do with our proposed machine so we do not end up comparing apples with oranges. We would also want to talk to other users to find out what they did to keep their CNC grinders operational and how difficult it was (or still is) to do that. We would also have to assemble a sufficient amount of data to ensure that we have a meaningful sample. Let's assume we've been successful in this task and have data as presented in Figure 3-3 below.

We see from Figure 3-3 that the life span of grinders, the same as or similar to the one we contemplate purchasing, is plotted. The graph shows us that no machine lasted more than 6.9 years. However, 5 machines (accounting for 8%) lasted that long. The data also tells us that 20 machines lasted up to 5.9 years, or 31% of the universe of the survey. We are interested in knowing what the probably would be of the machine we intend to purchase lasting 5 years. To get this answer from this data, we would determine the number of machines that lasted less than 5 years, calculate that percentage, and subtract the value from 100%. That would tell us the percent probability of our machine lasting 5 years. Mathematically we could express it as follows:

$$\% \text{ Probability} > 5 \text{ yrs} = 100\% - \text{Sum}\% (0\text{–}0.9, 1\text{–}1.9, 2\text{–}2.9, 3\text{–}3.9, 4\text{–}4.9)$$
$$= 100 - (0 + 7 + 14 + 18 + 22) = 39$$

From this calculation we see that the probability of the CNC grinder lasting 5 years or longer is 39%. We can see that the odds are not in our favor. For a baseball batting average it would be outstanding. But for a business proposition, it's hardly enticing. In this case, even though the financial numbers could be portrayed to be very favorable, the risk or uncertainty factor is too high to be a viable project. The entrepreneur should go ahead and send out the grinding operation to a specialty facility equipped to do the job, in this case, a machine shop with multiple similar grinders so that if one grinder is down for major overhaul, others are available to keep up with production needs.

Uncertainty Factors in Assessing Risks

The earlier example shows how we can use data to determine the extent of risk a certain decision will entail. But not all decisions encompass situations where data is available to set the probability level of success. Many entrepreneurs say they make those decisions "by the seat of their pants." They intimate that they have a feel for the decision being up, down, or sideways with respect to the situation. In other words, they have a way of assessing risk that makes sense to them. Most successful entrepreneurs will tell you that they always figure the risk percentage through an analytical approach as shown earlier or through some quasi-calculated approach they use to quantify a subjective decision. To do this, we need to be able to measure a decision choice against a ranking of situation facts. This is done through introducing an uncertainty rating on these facts. I use the term "facts" loosely because they are feelings and hunches and known conditions all lumped together; and in total they make

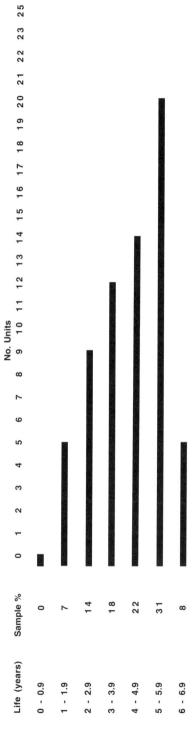

Figure 3-3. Life yield for 65 CNC grinders.

up the whole universe of the situation that the decision-maker is faced with. Let's see how we put some logic and order into this process to end up with a probability of success to use in making a business decision.

The concept of uncertainty looks at the degree of unknown in calculating the variance to what you think will happen vs. what actually does happen. This yields a risk factor. In the grinder example, the uncertainty was low because we calculated the risk using a sample of actual life data for 65 similar machine tools. Now, to illustrate uncertainty factor reasoning, let's say the entrepreneur needed to use a newer design in grinding because the work he would be doing would cause significantly more wear on standard machines. The designer of the machine tool says they would need to change the steel bed of the machine to incorporate protection from the more corrosive environment, and this involves new designs to channel the fluids back into a recirculating tank and different alloys for better wear and corrosive resistance. I list these details because this is change of a complex nature. It means that life data of similar machines may no longer be valid.

Change is the initiator of uncertainty. Now we have to investigate the magnitude of the change with respect to the previous condition. The change can range from minor to an entire paradigm shift. Investigating uncertainty virtually always is a pragmatic assessment based on circumstances. It is possible to mathematically calculate uncertainty, but the technique is cumbersome, depends on knowing specific facts generally not known to the degree of accuracy required, and is infrequently done.

The pragmatic approach is to put an arbitrary factor onto the risk factor to account for uncertainty. If there is no change in the environment, for example, the CNC grinder similar to the ones used to calculate the risk of failure, perhaps we should say probability of success (39%), then the uncertainty factor is 1.0. If there is change in the environment, then a value of less than 1.0 is used. Some practical suggestions:

Condition	Uncertainty factor
No change	1.0
Minimal change	0.9
Paradigm shift	0.1

Everything else is a judgment in-between. There are many tables and graphs we could find purporting to give values for intervals between paradigm shift and minimal change. They are all subjective therefore of little practical value. Instead of using someone else's subjective evaluation, it is better to use your own, based on what you feel the situation is with respect to your pending decision. This can be done by

- Listing what is known and unknown.
- Make value assessments that meet the specifics of the facts of the situation.
- Use this assessment to make a best-guess estimate of the uncertainty value.

When an uncertainty value is selected, we use it to modify the risk probability in the following equation:

Pragmatic risk probability = (Calculated risk probability) (Uncertainty factor)

The pragmatic risk is the probability of occurrence of the risk the decision choice entails. For the CNC grinder example, the calculated risk probability is 39%. Let's say that the change is minimal. The uncertainty factor is 0.9; note the following equation:

$$\text{Pragmatic risk probability} = (0.39)(0.9) = 0.351$$

Interpretation of this result tells us that there is a probability of 35.1% that the grinder would last the lifetime of the project, that is, 5 years. Or the risk of failure is 64.9%.

Severity of a Risk

We've modified the risk factor for uncertainty. Now let's look at another factor. How important is this particular risk to the success of the overall project? Knowing this will set the priority for developing a contingency plan. If the risk is only on the periphery of the main thrust of the project, then the severity of the event happening is low. For example, color selection for a new automobile inauguration is significantly less important than engine performance features. So, not having the first selected paint choice for the model is not very important and "what to do if" plans wouldn't need to be very elaborate if any existed at all. However, if a needed performance test was delayed and couldn't be performed before the scheduled unveiling, a significant contingency plan would have to be put in place to get the theoretical performance attributes known without revealing the lack of supporting test data. Since this is very important for the commercial success of the automobile, considerable time would have to be allocated in the project schedule to develop that plan.

The determination of how important a risk coming into reality is to the ultimate success of the project is called testing for severity. Let me reinforce this concept with another example and then let's see how we can quantify severity.

Once more let's look at our entrepreneur's CNC grinder and see how he could have evaluated the severity of a risk. Suppose the entrepreneur needed to have a well-running and tidy manufacturing operation to create an impression of a functioning company for both attracting customers and investors. As a matter of fact, the type of abrasives proposed for use with the machine (a combination of the most effective and lowest costs) also does an abrasive job on the machine itself. It makes it difficult to keep it fresh looking and in fact keep the paint on. This is a risk to contend with and it has a high priority of happening. What is the severity of this risk? If it came to be, how bad would it be for the entrepreneur and should he or she spend any time in pre-planning a mitigating strategy? The following questions test the severity of the risk:

- Is it very important to consider for operating effectiveness? No, not for the working of the machine. It is designed to work with the corrosive fluid, and paint on the machine is purely cosmetic. However, it could give the impression that the factory is sloppy in its housekeeping.
- Is it capable of giving a negative perception to customers and investors? Yes, for customers who are not cognizant of machining operations and the concept of a clean or dirty shop. This is probably not much of a factor for investors who are interested in manufacturing investment opportunities.
- How important is a contingency plan to mitigate this potential risk? It is higher than a non-problem, but certainly not a show stopper. It would probably be prudent to make sure that customers and investors are pre-briefed as to how grinding works and how it affects the machinery involved before letting them see the operation. This preemptive action, while not a true contingency plan, would go a long way to minimizing any downside effect associated with the visits. If this wasn't being done and was only contemplated being done after the new machine no longer looked new (this doesn't take very long with a grinder), then it would qualify as a contingency plan.

Here we see a severity test being done in the manner that is the universal method:

- Identify the risk factor.
- Measure the risk factor against the ability to successfully carry out the task.
- Assign a severity value to the risk factor.
- Determine the priority of the need to develop a contingency plan to be implemented if the risk is realized.

Basically this means how bad the risk could be if realized, and if it happens, what can I do to minimize its impact? How bad it could be for the most benign could result in nothing more than a shoulder shrug, which means I don't have to do anything. At the other end of the scale, it could mean that if it happens, we are in such bad trouble that it is indescribable and we need to do very heroic and risky things to fix it. The former is like discovering your tire is flat and calling the tow service, whereas the latter is like an airline pilot losing 3 of 4 engines on his 747 and frantically trying to get at least 1 restarted. Severity has a wide range; much subjectivity is needed to determine its importance. But we must try to put some sort of value to it. I like to use a scale of 1 to 10, with 10 being the most severe, most important case to mitigate, down to 1 being extremely benign. To give you a feel for the range, the need to mitigate investors' potential negative reaction to a dirty shop would likely be a 4.0 on a scale of 10.0. Of course this is totally subjective and open to other opinions based on others' experiences with such things.

Detectability of a Risk

Risks that cannot be detected pose more problems than those that can. This is so intuitively obvious that it is often missed entirely. Knowing something is about to go wrong allows us to fix it before it happens, thus preserving equipment, keeping a project on track, completing a journey, etc. These are examples of saves that can happen if we know something bad can happen and we know it is about to happen. We've seen that we can determine a probability of a risk occurring, we can evaluate uncertainty, and we can to some degree rationalize severity of impact if it happens. We have to add to this a methodology of detecting a risk that's about to become a reality just prior to it happening or very quickly after it does, before it creates significant damage—like using a fire extinguisher to suppress a flame before it gets so large we need the fire department to deal with it.

Again, like most of these risk modifiers, we are objectivizing a subjective value. I like to use the 1 to 10 range. For detectability, a 1 means it's easily detectable, hence unlikely to create a nasty surprise for the entrepreneur. Conversely, a 10 means it's hard to detect and could cause severe problems when it happens. Notice I'm hedging on what the results of going undetected would be. I'm doing this simply because I can't use a detectability measurement as if it were a severity measurement. Detectability is the ability to find a flaw. Severity tells us how dangerous it is. For the CNC grinder, how difficult will it be to determine whether a dirty machine tool will turn off potential investors? So let's give detectability a score of 6.0. Perhaps if your skills in reading body language and facial expressions were significantly advanced, the detectability score would be lower.

Risk Probability Number

For ease of making relative ratings between different risks, we can combine all the various elements and create a ranking value to compare a specific risk against other risks. This is called a risk probability number. The relationship is as follows:

Risk probability number = (Pragmatic risk probability) (Severity factor) (Detection factor)

Let's look at the relationship's utility. For the CNC grinder example we calculated the pragmatic risk probability of the machine lasting 5 years to be 0.351. Also recall that the severity factor is only 4.0. Finally, the detection factor is 6.0. Putting these values into the equation, we have:

Risk probability number = (0.351) (4.0) (6.0) = 8.424

This is a number with no particular meaning except to be able to compare various risks by ranking them against one another. Each risk can be evaluated against an objective ranking system. In fact, we can even determine what the highest risk probability number is and then compare our test number against it. The highest possible score is 100. This occurs with a probability of occurrence being 1.0—a certainty, and the other 2 factors being 10.0—the highest they can possibly be. Using 100 as the highest value, where the probability of the risk becoming a reality is virtually certain, we can compare that number with the one for a specific activity; the CNC grinder purchase or non-purchase. A value of 8.424 would indicate that the potential for problems occurring is relatively low.

The ability to take mostly subjective data and objectivize it to create a decision rule tree is a skill set that the successful entrepreneur has. We've seen how hunches and guesses can be quantified and made into a series of rules by which to guide decisions. Is it the best way to play the game? There's no answer for that. But the one thing it does do is give the practitioner a base line for comparisons, and sometimes that's the little extra that makes a project successful. And perhaps it does that because it forces the entrepreneur to examine all facets of the decision before it's made.

Mitigating Risks

Learning that an event may cause an adverse risk is not good enough. We need to know how to deal with it to minimize adverse effects on the process. Planning what to do if the risk comes to pass is call risk mitigation. There are ways to mitigate and even eliminate risk. They all start with an internal assessment of the entrepreneur's reference to its SWOT: strengths, weaknesses, opportunities, threats. With respect to risk probabilities, etc., it would be helpful to understand what SWOT mean.

Strength:
What assets can be employed to achieve the entrepreneur's goals? For example, in purchasing the CNC grinder, if the entrepreneur or an associate is experienced in manufacturing engineering processes involved in machine-tool purchases, that would be an asset. Employing that asset for negotiating with a vendor and setting the performance criteria for the purchase would be a plus for the project.

Weaknesses:
What are the lack of capabilities that we need to shield from being used? Referring to the grinder purchase, perhaps a weakness is that the entrepreneur has never entered into a contract for capital equipment and is afraid of being taken advantage of. This risk has to be mitigated. Maybe a way to do it is to hire an agent to negotiate the contract and or use the entrepreneur's associate who has manufacturing engineering experience as the prime negotiator.

Opportunities:
What are the areas of the project plan that can be exploited with the entrepreneur's and his or her associates' strengths to optimize reaching

their goal? Referring to the grinder example, perhaps the manufacturing engineer can find a cheaper alternative and at the same time mentor the purchasing agent on this aspect of capital equipment acquisition.

Threats:
What are the areas of the project plan that might require using weaker project team resources to achieve, and how can they be avoided? When the time comes for the entrepreneur to purchase the grinder, she or he may find that the associate with the best knowledge on how to proceed may not be available. This is a threat to the successful conclusion of this phase of the project and perhaps the only way to avoid it is to postpone the acquisition until the associate is available.

Going through a mental SWOT checklist for a major project decision step causes the entrepreneur to focus on selecting the best risk-mitigation strategy. The 3 generally accepted risk-mitigation strategies are as follows:

- Risk avoidance
  - Used when SWOT weakness is the dominant match up with the risk.
  - Tactic: Avoid taking action until the weakness is corrected.
- Risk minimization
  - Used when SWOT strength is the dominant match up with the risk.
  - Tactic: Use the highest-strength process the company possesses against the risk. This gives the best chance of blunting the risk before it becomes a reality.
- Risk transfer
  - Used when no tactic is viable to mitigate the risk.
  - Tactic: Try to find another way of accomplishing the goal whereby the company's strengths are used against the risk to minimize the likelihood that the risk will be realized.

Risk mitigation implementation is based on the common-sense strategy of not going out of our way to find a problem, and coupling this with a firm knowledge of the organization's limitations. It requires the entrepreneur to be honest with himself or herself and not view the world through rose-colored glasses. He or she must keep the planning for the endeavor within personal and associates' capabilities and never try to stretch those capabilities without first testing to see if it feasible. If an attractive opportunity is presented to the entrepreneur that would require skills beyond current abilities, then acquiring those abilities should be the first step in the plan. Even before that, evaluate if it is even possible to get them in a manner that is affordable and timely.

Another mistake that happens too frequently must also be avoided when trying to mitigate a risk: the error of not putting forth the best effort in performing each step of a project. Keep in mind that every step of a project is equally important. If it weren't, it wouldn't need to be done in the first place.

Then there is insurance, both literal and in the form of support. We can insure against risks coming to pass. For example, we can purchase business-interruption insurance to guard against the CNC grinder having a premature failure. We can have another form of insurance called contingency planning. If avoidance, minimization, and transfer don't work, we bring in "Plan B"— the contingency plan.

Contingency Planning

The first step an entrepreneur must take is to avoid unacceptable risks. However, all risks are not avoidable whether they're acceptable or unacceptable. Businesses make large profits from strategies that accept controllable risks, for example, building a new factory based on a projection of market growth. The projection is not a sure thing. It is the best possible analysis of a given situation. There is always the probability that the projection will not occur as forecasted. In fact, forecasting the future, which all businesses have to do (we call this business planning or sometimes strategic planning), is never a certainty. How much of the uncertainty we are willing to risk is the subjective nature of defining acceptable vs. unacceptable risks.

Businesses have to take some risks, but they are controlled. We control risks with contingency planning. Contingency planning answers the "what if?" question. What do we do if the CNC grinder fails after 1 year? A contingency plan might be to initiate the profit recovery allowed under the business-interruption insurance. Or it might be to have sufficient spare parts on hand to get the machine operational within an acceptable period of time. Theoretically we create a contingency plan for every identified risk. The Pentagon may be able to do this but most companies lack the resources to cover all risks. They have to pick and choose what appears to be the most frightful risks and make sure they're covered for these. We need to have a methodology for selecting those risks to protect against in a manner that gives the entrepreneur an adequate safety margin for the project. Let's see how that's done.

The process is another in a long line of objectifying a subjective process. This means much is based on previous experiences and hunches. Here's the process:

- Rank the calculated risk probability numbers from highest to lowest.
- Establish a cutoff level that represents the highest probability of a risk the company can tolerate without a contingency plan.
- With the remaining ranked risk probability numbers, group for commonalities. If some exist, this will make the list to contend with smaller.
- Establish auxiliary projects to work on that will cover as many of the risks on the list as possible.
- Every time an auxiliary project is completed, have a contingency plan for a risk or a set of risks.

The amount of contingency plans developed depends on the resources available to do them. Pragmatically, entrepreneurs do not have the resources to do much in the way of contingency planning. But they must recognize the risk potentials and realize they have a relatively low threshold before some sort of contingency actions need to be initiated. Therefore, most successful entrepreneurs have "unofficial" or "undocumented" contingency plans in their heads ready to go on a moment's notice. They may not be fully fleshed out and they may contain other new risks, but the "owner" is at least aware of a need to react to certain risk stimuli and will. This is a trait the engineer entrepreneur has to cultivate and be comfortable with. Risk is the nature of new ventures, and knowing when to try a different tactic or even pull the plug is what makes the difference between a successful entrepreneur or a bankrupt one.

Action Steps

As you can see, there is a great deal to consider in setting up a project and defining risks as defined by the project-management technique. How much work from the Work Breakdown Structure and Risk Analysis gets done depends on the level of complexity and anticipated cost believed necessary to complete the project. If we were planning to send a man to the moon, these first 2 steps would be done very thoroughly and in intricate redundant detail. Why? Because we couldn't afford to be wrong. If at the other extreme we're planning the office picnic, we can afford a more cavalier approach to the WBS and certainly drastically minimize the Risk Analysis. The amount done on these 2 steps is a decision made by subjective analysis and depends on the complexity of the project. However, the last phase of a project, the Action Step, always needs to be meticulously complete.

Action Steps are where "the rubber meets the road." Everything that needs to be done to do the project's work must be done in the order planned without compromise. If we plan how to swing the bat or deliver perishable foods, we can most likely optimize how we do it through minimizing the "don't knows" of the WBS and identifying potential troublesome risks in the Risk Analysis step. But we're not successful until we actually do the work of the project. We have to execute and do it—no more theory and contemplating what could go wrong. By the time we get to the Action Step, the only thing left to do is to initiate the action and hope the process plan employed is the correct one.

The salient points of the Action Steps is the procedure of defining the actions, doing them, and measuring to see if they have been done correctly. The most successful and yet still fundamental way of doing this is through Objectives and Goals Management. A more complete treatise of the subject is in my book *Manufacturing Engineering, Principles for Optimization*, 2nd edition, published by Taylor & Francis. Let's summarize and briefly review that process.

Objectives and Goals Management is a 3-phase system where we identify what we want to do and set out to do it with feedback of progress and needed corrective actions along the way. We start with the most generalized intentions and plan our way to meticulous details. We always check that the meticulous details are still focused only on the broad-based intents. Here is the triad of generalized to specific:

- Objectives
  - Broad-based generalized statements of intents
- Goals
  - Measurable statements of specific intent bounded by a specific period of time
- Projects
  - Specific plans with measurable steps that leads to an accomplishment of a goal. Several projects may be required to be completed to achieve a goal.

The triad goes from broad-based to very practical steps programmed to accomplish the need the entrepreneur has set out to achieve.

In project management parlance, a project is a goal. It has a major specific end point to accomplish a task and is very measurable, both in quantifiable terms and in time. So when we create a Work Breakdown Structure, then examine the "don't know" risks and build in contingency steps, we are in essence creating projects to achieve the desired goal of the activity. Hence we call it project management.

Action Steps are the project-specific plans with measurable steps leading to achieving the goal. Each action step is a critical cog in the machinery to reach the goal. If we go back to the simple "at bat" project we can illustrate this.

We see in Figure 3-4 that the project steps for the "at bat" project, goal in Objectives and Goals terminology* are listed in the left-hand column. This in essence is only the Work Breakdown Structure. The WBS tells us what we have to do; it doesn't deal with how we know we're successful. In the Action Steps portion of project management, since we must define what success is, we add the 2 measurements columns as shown in Figure 3-4. The 3 columns correspond to the definition of "Projects" in the Objectives and Goals Management process. This constitutes the Action Steps phase of project management and is the difference between it and the WBS.

In most projects, it's critical to have all predecessor steps done correctly and in order for the macro-work to be accomplished as required. The example shown in Figure 3-4 illustrates this. Notice that the order of the at bat is scripted step by step and needs to be followed that way. Also notice that every step results in a measurement of the action and a measurement of the time to do the step. This is done so we can evaluate the step for completeness and understand if the planning resulted in the results we really wanted.

---

* (Objectives and Goals terminology is slightly different than that used in project management; but the intent is the same. It's only semantics of theory and you shouldn't be concerned with nuances of difference. These definitions help to determine the extent of the measurements required. Goals contain projects, therefore have more measurements. The only difference is complexity. Keep in mind that both goals and projects are specifically measurable and bounded by time)

The Entrepreneur's First Step • 67

Project step	Success measurement—action	Success measurement—time
1. Select a bat.	1. Picking a bat out of the bat rack	1. Having selection done while predecessor batter is in "on deck circle"
2. Go to the "on deck circle."	2. Arrive in the vicinity without causing rebuke from the umpire	2. Being in the "on deck circle" as the predecessor batter enters the "batter's box"
3. Dry hands with resin.	3. Hands are dry.	3. Done before preparing bat with pine tar
4. Apply pine tar to bat handle.	4. Sufficient pine tar is on bat handle to get a firm grip.	4. Done before taking practice swings
5. Take practice swings to limber up.	5. Swings indicate no obstructions.	5. Swings performed at a time that tracks the pitcher's motions against the predecessor's at bat
6. Observe pitcher's motions	6. Actual motions checked against "scouting report."	6. Observing pitchers motions befor entering "batter's box"
7. Take place in "batter's box."	7. Feet and body are in correct position for an effective swing matched to expected pitching motion.	7. Being set and ready before pitcher starts his motion
8. Swing at selected pitches.	8. Good swing results in a ball hit with maximum energy transferred from the bat to the ball.	8. Making contact with the ball at the optimum time to achieve maximum energy transfer from bat to ball

Figure 3-4. The "At Bat" project.

Action Step management is really nothing more than taking the plan, applying resources to do the required steps, then measuring the results. If the results are what are required by the plan, we go on to the next step. If not, we analyze the measurements to understand why the plan didn't result in the correct outcome. We see if the plan was wrong or if execution of the plan was wrong. If the plan was wrong, we correct it. If the execution was wrong we learn why, then make it easier to do. This sounds very straightforward, and in theory, it really is. Where it falls down is when we're not properly prepared to execute or when we didn't plan correctly.

A look at the "at bat" project demonstrates how difficult it is to react to measurements in a real-time dynamic mode. Can you imagine the difficulty of correcting the batter's swing if he is not executing properly? This brings up another interesting point. We see the connectivity between the 3 phases of project management. If the people doing the Action Step phase need to implant some contingency planning conceived during the Risk Analysis Step, how can they do it in real time? This means the WBS didn't convert all of the "don't knows" to "knows." I'm not smart enough to learn how to make corrective actions effectively with the batter while he's in the batter's box. But perhaps if the WBS identified this as a step beforehand, the Risk Analysis contingency plan may have required the batter to wear a radio earpiece. Then as a minimal try at corrective action, the batting coach could tell him what he's doing wrong. Is this a good plan? Well it's hypothetical. Maybe the batter would feel better knowing why he is striking out. Of course this is a whimsical example but it illustrates the point I want to make: In successful project management, the Action Step is where it happens, but it can't be successful unless the Work Breakdown Structure and the Risk Analysis Steps are done properly.

Another look at the "at bat" example illustrates the difficulty in taking theory to perfection in practice. The batter could have done everything in accordance with the plan and still strike out. Why? There are multiple possibilities. Perhaps the batter has poor hand–eye coordination and no matter how well he observes the pitcher's motion, he will never be able to react fast enough to hit the ball delivered by this pitcher. Perhaps he didn't apply the pine tar adequately and couldn't get a correct grip on the bat. We could go on, but these examples point out the difficulty in executing a plan. Maybe the former is not fixable and the latter is. My purpose for bringing up this sour note is to point out once more from a different perspective that proper execution of the Action Steps is just as important as doing the WBS and the Risk Analysis portions of Project Management properly. We need to execute to be successful. In fact, the difference between successful entrepreneurs and dreamers is the former makes it happen; the latter wishes it would happen. The Action Steps is where it happens.

Meticulous attention to detail needs to be an attribute of the engineer entrepreneur. Thinking that entrepreneurs are big-picture people and not doers is false. They may not do everything themselves, but they most assuredly ensure that all details are attended to. If we look at all 3 subsets of project management, the commonality between them is complete understanding of all the pertinent facts to make the project successful. From this we can properly infer that entrepreneurs, who by definition must excel in project-management application, are meticulously detail-oriented. They may not write it down or have it neatly arranged, but they know how the plan goes together with all the checks and balances and they react quickly to variances of the plan.

SOME FINAL THOUGHTS ON PROJECT MANAGEMENT

Entrepreneurs understand the methodology of project management. They may not practice it in the textbook fashion but they surely subscribe to its philosophy and use it to fashion their actions. Project management is all about managing risk. It's not the type of engineering subject one learns by rote memorization, but rather by following the logic of the process and practicing it. This is definitely a case of "practice makes perfect." I suggest playing some mind games to learn the techniques. Any endeavor where there is a goal to achieve can be put into project-management format. This is probably a bother for simple tasks, but do it anyway. It's a way of learning and mastering the technique.

Let's look at a silly goal: buy lunch. List the steps needed to do so, list the measurements of successful achievement, list the risks involved in completing those steps, and list the contingency steps you would take to mitigate the risks. I won't list every conceivable step necessary to buy lunch but here are a few:

Select a food category choice.
Select an amount you're willing to spend.
Select a suitable vendor.
Go to the vendor and make your purchase.

Here are the simplistic measurements of task achievement:

Food category: You've chosen Italian.
$3.00 maximum: You can stay within goal with this category
Vendor selection: a pizza kiosk at the mall-within spending category
Purchased a slice of pizza and a diet coke (to mollify your conscience): $2.98

Now what are the risks? Here are some:

The food category choice is not available.
You have no transportation to get to the vendor.
You forgot your wallet and can't pay for the food.

So what are the contingencies available?

Select a different food category choice, perhaps a hot dog.
Borrow some money—probably need only $3.00.
Arrange for a ride with a friend.

The game can go on virtually forever. But the point is that by doing this you're practicing project management to the point that you automatically think about how to plan and do a task, how to measure, what the risks are, and what the contingencies are to mitigate the risks. These are skills entrepreneurs possess and they are gained solely by practice. Keep in mind that project management skills mastery is vital for the engineer entrepreneur, whether in your own business or working as a member of a product team. This is a skill set necessary for the successful practice of engineering in the 21st century.

Chapter 4

From a Business Team Member to a Business Owner

Now we will begin the journey of becoming a business owner rather than a team player within a company. There are lots of things about being an owner of a business that are similar to being a member of a business product team. Likewise, there are significant differences. It's important to understand what is the same and what's different because it's those differences that the engineer entrepreneur must understand and be comfortable with to make a successful transition. Being "comfortable with" sounds like a warm and fuzzy phrase; that's not my purpose. I want to make sure that the potential entrepreneur is forewarned about these differences and understands and can successfully cope with what he or she needs to learn.

I will explain the differences and point out skill sets that must be mastered to be successful in running your own business. We'll do this in a format that introduces all the basic business knowledge the entrepreneur needs in order to enjoy a positive experience with his or her company. This introduction will form the basis for greater depth examinations that appear in the latter chapters of the book.

PURPOSE OF A BUSINESS

The purpose of a business is to make a profit for its owners. This is the purpose that most people in the Western world would subscribe to. I agree with this principle, but it's not the only reason for a business to exist. There are others. What about the "not for profit" companies? The Ford Foundation, Harvard University, the American Civil Liberties Union, and the National Rifle Association are in every way, shape, and definition private corporations that compete against other corporations. But these companies, although voracious competitors, profess not to be in business to make a profit, at least not a profit as we generally define profits. So we must understand that there are two types of companies. Those that strive to make profits and those that do not. Immediately you think "well what's the reason for being of these not for profit companies and why do they try so hard to compete?" The answer is there is more than one purpose for a company to exist. The profit motive is not the only reason, as we shall see.

Before I get into other purposes for companies to exist, let's look at profits philosophically. We think of profits as being sums of money that the owner of the company puts in his or her pocket after all the bills are paid. This is definitely true for a sole proprietorship or for a partnership where the profits are shared among the partners. How about a corporation? Profits for a corporation go directly to the corporation. A corporation has legal status as an individual, so the General Electric Company, a previous employer of mine, is a legal entity and all the profits it reaps goes to its pockets. No individuals have a legal claim to the profits. There is no "divine right" for individuals to be paid shares of profits from corporations they own. A corporation doesn't have to pay dividends. Many do because another legal quirk says that corporations are owned by people (or by other corporations, which in turn are owned by people, and the people may be the collective we, for example, government entities). The people who own the corporation can decide whether or not they want to share the corporation's profits. So profits may or may not be dispensed by the corporation to the people who own it. To further muddle the distribution of profits for corporations, we have tax laws. Corporations that want to pay profits to their owners are taxed differently than those that choose not to. And even this can vary. Some corporations elect not to pay profits to owners as a hard and fast contractual rule among its owners. Some elect not to but can change their minds at any time. Microsoft Corporation comes to mind as an example of the latter. Harvard University would be an example of the former.

Does the fact that a corporation doesn't pay out its profits to its owners make it a not-for-profit company? No. To be a not-for-profit company, a company must state it has no profit motive, that all of its annual gains are used to further the cause it professes to champion.

Another way to look at profits is mathematical. A company sells its products or services for $X. It costs $Y to produce the product or service. If we subtract $Y from $X the difference is $Z, the profit (or loss if the $Z is negative). This is the classic business profit formula, the basis of all economics from the beginning of time.

$$\text{Selling price} - \text{Cost to produce} = \text{Profit}$$

This is true for the so-called "not-for-profit corporations" as well as for the "for-profit corporations." Having said that the not-for-profit corporations don't pay profits and the for-profit corporations do, then how do the not-for-profit corporations get around the business profit formula? They see to it that the business equation always comes out to equal zero. They take the money and immediately apply it to their reserves for future use (a so-called cost to produce—but in the future), or they pay it out in salaries to those who work for them at a very handsome level, or they donate it to other worthy causes; so the equation always equals zero.

Now that we see how profits are dealt with within the 2 styles of companies, we need to ask why the not-for-profit corporations even bother to exist at all, if all they're doing is disguising profits and calling it something else? This leads us back to the other reasons for a company to exist. I'm sure you have heard the phrase "Man cannot live by bread alone." We all have other driving forces that make us want to do certain things. I've always felt that one of those driving forces is to leave a lasting mark in the world after your time has passed, the so-called "do good " factor. This is something most companies strive to do, some more so than others. This is a very powerful reason for a company to exist, along with the profit motive.

So we see companies have reasons for being, and I'm convinced that it's more than making a profit. Psychological drives also have to be answered. People want to feel their efforts are for some noble good and are important. Maybe it's a way of giving thanks for the blessings they've received through high profits that have allowed them to enjoy comfortable standards of living. In certain areas this is easy to achieve. The medical profession comes to mind. They cure our ills and drive away our fears, but if they charge too much for their services, we begin to believe they're greedy and living off the suffering of others. This implies a need for a balanced reason for being that companies strive for: to make a profit (even in the convoluted ways of the not-for-profit companies) and to give charity to others as a tribute for having done well. As an entrepreneur, you need to be mindful of these driving forces because they will greatly affect how you will react in specific situations. And since it is a balance, you must do the abstract portion, "feel good," in order to achieve the real portion, satisfactory profits.

How do we define the "feel good" portions and do they help the real portion? The feel good stuff is not simply fluff. It is a message put out that states the core belief of the company that the staff of the company can subscribe to and by doing so will result in the real portion being highly satisfactory. So, yes it does help the real portion. Let's look at an example and see how.

Suppose we have a company in the business of making streetcars (we now know this form of transportation as surface rail or light rail). The company desires to have an after-tax profit of 10%. To do so, it needs to convince the municipalities that its costs and fees and resultant profits are a reasonable return for its good work. After all, by building the streetcars and the infrastructure, which provide cheap and efficient transportation, the company is improving the quality of life of the general public.

Therefore, to convince its customer of the worthiness of its project, hence the company, it needs to successfully convince the customer that company has a business reason for existing other than simply profit. It must be convincing in portraying the view that its objective is to provide quality transportation systems for a deserving public. If the company does this successfully, then the "feel good" motive will help the company achieve its real motive: achieving the 10% profit level.

The example shows how the profit motive and feel good motive work hand in hand to support the profitability of the company. Let me say one more thing about the business equation. We can see that there must be money coming in to keep the company viable. There must be a net cash gain, whether we call it profit or not, for the company to survive.

The feel good motive, which we now know is tied to profits, needs to be structured and focused for the effort to be successful. This is done through an exercise called defining the vision and mission of the company. The vision and mission, in turn, need to be compatible with the objectives and goals (see Chapter 3) for everything to come together and profits to be achieved. Sometimes vision, mission, objectives, and goals are looked at as a synergistic package called a business plan. I say "sometimes" because all too often, some would-be entrepreneurs do not take the time to tie all of this together in a cohesive plan and they pay the price later. Business planning is always the proper way to proceed. We will continue with that premise in mind and discuss the first 2 steps: vision and mission. The entirety of the business plan process will be introduced in a later chapter.

VISION AND MISSION STATEMENT

To protect real profits, as we've seen, it is necessary to excel in the other reasons for a business to exist. I've referred to them as the "feel good stuff," but please do not take this as a putdown of their necessity. It is absolutely vital that a business be well accepted by its customers and vendors. We've probably all had experiences with companies that simply turn us off. I can think of a retail electronics giant whose advertisements I won't even read anymore because of a bad experience I had with them supplying equipment to their retail outlets. That company lost a customer and I do my best to persuade others not to deal with them. I'm only a microscopic irritant to them, but if their attitude in dealing with customers and vendors is as it was with me, they are in trouble. Visions and mission statements are intended to put the spotlight on how a company should behave in pursuit of profit so that such things as mentioned above do not happen.

It's not necessary to be syrupy nice in public pronouncements, but it is required that the company be perceived to be a good corporate citizen. The vendor and customer community need to feel that the company is not out to take unfair advantage of them and will play by the rules in all matters pertaining to competition and delivery of goods and services. Vision and mission statements set the company in the direction to be compatible with that need. They codify the company's culture in how it will behave in all situations.

The vision and mission statement also forms the foundation of the company's business plan. Let me define the business plan (sometimes referred to as strategic plan).

Business Plan:
A business plan is a unifying document that coordinates and communicates all functions' responsibilities for carrying out the stated purpose of the business. It outlines the goals to be achieved over a nominal 3-year time frame primarily measured in profits to be achieved and spells out how those profits will be gained.

There is a hierarchy of ideas to be expressed in the business plan. We go from the most broad-based to the most specific. We also must make sure there are direct communications links between the levels so that the entire plan is cohesive and compatible.

Figure 4-1 demonstrates that linkage.

It is important for the entrepreneur to grasp the full picture of the business plan that the linkage portrays. To do less is like trying to navigate a ship through a fog bank without radar. It can be done but is much more difficult and considerably slower. An entrepreneur trying to start a new business is much like the captain of the ship. He or she needs all the help possible to navigate to success. By being able to get from broad-based to specifics, we can create a map for all the actions that need to be done to start the successful new company and to be able to manage it effectively when it has been established. Every action we do needs to be measured to see if it is the root step, that is, the most basic; therefore, there is no predecessor step. Or is it some intermediate step that requires a predecessor and successor action. By creating the linkage, we can see where a step should be placed. Let me give you an example:

Suppose our company has obtained a contract to make mailboxes, the ones we see at the head of every driveway in suburbia. I say to you, "Purchase aluminum sheet to form the basic shape." Can you do it? Not really. This is an intermediate step somewhere between preparing a design drawing and forming metal to shape. On the other hand, I say "Obtain the design criteria color choices from the customer." Can you do that? The answer is yes. We have a

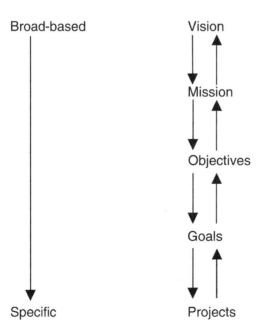

Figure 4-1. Hierarchy of strategic planning.

contract signed by the customer and we can ask for the basics to proceed. Color would not have been an issue for pricing the contract proposal.

This example demonstrates that work needs to be organized in a linked way. The vision, mission, objectives, goals, and projects hierarchy makes us do that on a macro basis for the entire company and makes us aware that what happens at one plateau is a result of a previous one and will affect the succeeding one. In the case of business planning, we start off with an idea, a vision, and constantly refine it by adding detail until we have a very specific and measurable project to accomplish the intent of the vision.

From Figure 4-1 we see that the vision is the broadest of the statements of intent, whereas the projects are the most specific, hence measurable, statements of intent. Objectives, goals, and projects were described in Chapter 3. But for clarity purposes I will carry out the integrated examples to include objectives, goals, and projects. Let's first see how the vision and mission statement fit with them at the broad-based end of the spectrum. I'll use a definition and an example to explain each.

Vision:
A concept or formulation of a desirable end point.

Example: Our vision is to create a database to service the knowledge reference needs of mechanical engineers.

The vision states what the company is going to try to do and why. In this case, the company is focusing its energies to create a database and they're doing it for their aimed market clientele—mechanical engineers. There is no bound on the size of the database, nor its structure. Also there is no time frame of when it will be completed. In fact, there is no end point at all since the database development can go on and on and on as long as the company wishes to tinker with it, that is, offer new revisions for sale.

Mission:
A clearly stated purpose showing how tasks will be organized to make the vision a reality.

Example: We will create an organization whose purpose will be to create a useful knowledge reference database for mechanical engineers.

The mission focuses on how the purpose will be accomplished. It takes the vision 1 step down the ladder to being more specific. In this case, the company is going to set up an organization to achieve the formulation of the database. The vision didn't state how the company would do the job, but the mission statement does. In this aspect, it's quite specific. We can tell if the mission is accomplished by the actions of the company. Here we see that to accomplish the mission, an organization must be in place to develop databases that are judged useful for mechanical engineers.

Now let's continue down the trail to specificity by continuing the example to show the linkages of vision and mission to objectives, goals, and finally projects.

Objective:
A singular or multiple sets of broad-based generalized statements of intent that define how the mission will be implemented. Usually ongoing for as long as the organization exists.

Example: Develop a series of databases for the various mechanical engineering disciplines and subdisciplines containing definitions and examples useful for the inquirers work application.

The purpose has some specificity to it now. We have fully charged the organization to implement the mission statement. The organization is required to set up a protocol outline of what kind of work will be done and how the user will make use of it. But notice there are no timetables or end dates. We have said only what we will do in a broad-brush approach. We cannot as yet pin down the organization about specific deliverables. That's the job of goals and projects, as we shall see.

Goal:
A singular or multiple sets of specific measurable statements of intent that define when the desires (or portions thereof) of the objective(s) will be accomplished and bounded by time. In an industrial setting, the time frame is seldom longer than 1 year.

Example: Develop a database for heat-transfer theory, formula, and examples of applications by December 1, 2002.

Now we have specifics. The vision said we were going to create a database. The mission statement told us we would set up an organization to do so. And the objective directed the company to set up databases that were useful for the customer's work application. But none of the broader-based statements specified exactly what and when the databases would be done. The goal does that in spades. Everything about it is pin downable, to create a phrase. We can easily tell if we've done what we said we would do. We know the topic of the database, what has to be in it, and when it has to be delivered to the customer. We can measure our performance against completing the goal.

Projects:
A specific plan with measurable steps, bounded by time, that leads to the accomplishment of a goal. One or a set of projects may be required for the accomplishment of a single goal.

Example: A set of steps for releasing the heat-transfer knowledge database is shown. Note: Each step contains a measurable quantity to judge for completeness and a due date.

1. By 7/15/02, investigate heat-transfer literature and list the theories that have use in current mechanical engineering practice.
2. By 8/15/02, write a short descriptive or explanatory summary of items selected in Step 1.
3. By 10/1/02, add addendums to the summaries of Step 2 to contain working formulas with examples.
4. By 10/15/02, select a minimum of 25 common heat-transfer problems and develop examples applying the working formulas, singularly or in groups.
5. By 11/15/02, debug the database and beta test to ensure ease of entry and direction to the proper subheading.
6. By 12/1/02, have available for release and sale.

Goals tell us what has to happen. Projects state how to make it happen. Notice how very task-oriented a project is vs. how lofty a vision is. But we need both. Visions and mission statements are necessary to get the process to the right "tree in the forest." However, picking the fruit and everything else about the tree is the domain of objectives, projects, and goals.

Anyone who has an idea for a business needs to have that visionary glow. But to be successful, the entrepreneur has to know how to translate from the glow to the pragmatic reality of the real world. Figure 4-1 shows how we can do just that by ensuring that our desire can be accomplished by defining the hard facts and tasks to do so.

FUNCTIONS OF A BUSINESS

To accomplish what we have established as hierarchy of purpose, the vision through project scenario described above, we need to have organization structure. Businesses need organization, with specific departmentalized tasks, to carry out the various activities necessary to achieve their purposes. Even more so than business teams of larger companies, the small business must recognize what are the necessary functions within the organization and make sure they're attended to. Many would-be entrepreneurs are tripped up because they do not realize what functions are needed to manage their business. So they do some management tasks spottily at best or without any forethought of reason, and frequently fail.

If we look for reasons why new businesses fail, most often it's because of lack of financing and lack of a cohesive plan of how to conduct their affairs. We can consider the former as a symptom, not really a cause. If the entrepreneur had planned properly then, barring any unforeseen events, the financing would have been sufficient for the needs. So actually, the reason a new business fails is only because it didn't have a cohesive plan to succeed. With a good plan and a team organized into an effective structure to carry it out, a startup venture can succeed.

To carry out a cohesive plan, we need to have an organization structure. Without an organization structure, the plan can get done only in an ad hoc manner, meaning whoever is available at the time does the task that needs to be done. This could work, but only if everyone involved in the company has equal skills and can effectively communicate status to everyone else, sort of like a collective mind that knows everything anyone is doing at the time it is being done. This is the picture of "merry but busy startup," one from folklore. What folklore fails to relate is that virtually every one of these chaotic startups fail. Those that survive are the ones that had the foresight to create an organization at the time of their inception. They may have had 1 person assigned to more than 1 task. But that person knew what job he or she was doing at any particular instant of time. The person may have been wearing more than 1 hat, but never more than 1 at a time, and certainly knew when he or she was the design engineer and when he or she was the purchasing agent.

The point of this discourse is to lead into the discovery of the basic organization structure every company, large or small, needs to have. Entrepreneurs need sound ideas and the wherewithal to implement them with minimal risk. The proper organization, put into motion with the initiation of the desire to form a new business, will allow the venture to be successful. It doesn't guarantee success, but to have no organization virtually guarantees failure. Let's look at the recommended organization for the entrepreneurial startup company.

THE BASIC ORGANIZATION OF A BUSINESS

Businesses have to start with the premise called the *basic tenets of manufacturing*. We use the noun "manufacturing" to indicate a maker of products. This can be physical products such as software on a CD. Or it can be a purveyor of services, such as teaching people how to use the software. The basic tenets of manufacturing include the following:

Know how to make the product.
Know how long it takes to make the product.

With these 2 "knows" it is possible to structure an entire business organization. We need exterior communications to the outside world that lets the potential customers know that we are there to offer our product to them. This is the Marketing function. We will also need to do the work of tailoring our product to the customer's needs and actually making and delivering the product to the customer. This is the Operations function. Finally, we will need a scorekeeper and supplier of capital to do the work of the other two functions. This is the Finance function, sometimes called the Comptroller function. So we have the whole organization of a company boiled down to its core essentials:

Marketing
Operations
Finance

We hear of titles such as Engineering, Sales, Quality Assurance, Human Resources, Production Control, Manufacturing, etc. All of these are subsets or subsubsets of the three core essential functions. Every company must do the core essentials effectively to be successful. An entrepreneur has an additional burden to bear. She has to hit the ground running at the very beginning of the venture and do all three functions well. Entrepreneurs have little time to gain experience and learn how to do it on the job. If procrastination is allowed to flourish, the opportunity that spawned the idea for the venture may be extinguished and superseded before the entrepreneur got his act together. Let me give an example.

Example: An entrepreneur has an idea for software that allows Mac-based computers to interact seamlessly with PC-based systems. The entrepreneur goes ahead and develops it, putting money and effort into developing the software. The entrepreneur starts up a company to bring his or her dream into reality, using the philosophy "build a better mousetrap and they will beat a path to your door." All the entrepreneur's efforts are focused on developing the product. Finally, he or she has a product to offer to the market and hires an agent to peddle it. Soon the entrepreneur finds out he or she doesn't have the funds to continue the effort. Manufacturing costs were not considered and besides, the marketplace isn't interested.

What went wrong? Two basic things. First the entrepreneur failed to test the market for acceptance and salability of the product. If he or she had, the entrepreneur probably would have found that the need had been filled already, even though his or her method may have been a paradigm change in the capability for transparency. It would take extensive messaging to get the market to want this product over existing products.

Second, the cost of doing the positioning of the product was ignored. The entrepreneur spent money only to do the engineering phase of creating the software. No money was spent for selling or producing the product in commercial quantities.

How did these blunders occur? Probably through ignorance of what kind of business structure was needed to launch a product. The failed entrepreneur didn't recognize the need to have a Marketing function do the positioning of the product for future sales to customers and do it in a way that the customers were salivating to get their hands on it. The entrepreneur also failed to establish a Finance function to keep tabs on the costs and measure what they were getting for their money. Neither did he or she have Finance constantly judging the magnitude of funding needed and making provisions to get it. Finally, the entrepreneur failed to do the whole Operations function. The only thing the entrepreneur did was R&D. Planning how to make the product in commercial quantities and figuring out how to distribute it were ignored. Also, we could say turning the product over to an agent to sell was like giving it to foster care for rearing while the parents went about other pursuits.

So we see that understanding the nature of necessary organization—the core essentials—is important to entrepreneurial success. Let's explore how the core essentials work.

USING THE 7 STEPS OF THE MANUFACTURING SYSTEM

The entrepreneur, having launched a new business and setting in place the core essentials organization, needs to orchestrate the work in the most efficient manner possible. I have found that the best way to do that is to visualize how work should flow through an organization. Obviously, work can flow through an organization in many ways, and we would think that some flows would be better than others. Intuitively we would expect that, and happily that is so. Just like a process has a start, middle, and end, so too does work because work is also a process. (When I speak of work I am using the industrial definition, not that of physics.) We know that for work to begin, a set of instructions is needed that tells people and/or machines what to do. We know that work is complete after it is deemed to have met the requirements for the finished product and/or service. We know a lot about work if we stop to think about it because virtually all of us have been involved in performing work of some kind all of our lives. The entrepreneur can expect that the work of performing the new service or making the new product will follow a pattern that is not entirely unique to the new product or service. In fact, there will be no uniqueness at all to the pattern of doing the work. The most efficient manner of doing work always follows the same pattern, like it or not.

We already know that we have external, internal, and measuring tasks that have to be done in any effective business. The most efficient way of doing the work is in that order: external, internal, and measure. This is very abstract, so let's put some pragmatic realism into it; what we come up with is called the 7 steps of the manufacturing system, which was introduced in Chapter 2. I know, you say work is more than manufacturing and some companies don't even manufacture, they provide services. So how can the pragmatic expansion of external, internal, and measure have manufacturing in its title? Probably because people doing manufacturing thought of it first and because most of industrial engineering theory has evolved around manufacturing—only as of the last 3 or 4 decades has it expanded to include service companies. This is a long way of saying the title doesn't matter; its' universally applicable.

Even though it wasn't explicitly stated in the discussion about education, Figure 2-3 is arranged in the order in which work is done. To be optimal it is necessary to follow the sequential order. If you do not, you are literally putting the cart before the horse. It would be very difficult to deliver a service before it is planned and scheduled. How would you know where to go and what to do? So the sequence makes a lot of sense. The following is an example of a service company violating this simple rule of doing things in sequence, which cost them dearly.

Example: An air conditioning service and repair company is called to patch a broken underground refrigerant line. Apparently someone dug a hole next to the compressor and much to his chagrin, he severed the copper tubing containing the refrigerant (and probably scared 3 lifetimes out of himself when it went "whoosh!"). The repairman shows up. He has no shovel to excavate down to the broken line. He borrows one from the distraught homeowner. Next he finds the broken line, decides he needs to patch it and then purge the line of water and dirt, then complete the job by recharging the system with refrigerant. He goes to his truck and finds that his oxyacetylene torch kit is depleted of gas. So he can't do the job. He gives the homeowner a bill for showing up and borrowing his shovel to find the problem, and he says the dispatcher will have to send out another truck tomorrow or the next day.

Is it any wonder that the customer didn't pay the bill and hired a competitor to do the job? The company had done no planning beforehand to make sure the truck was properly stocked, and the schedule step was destroyed because we had a service truck that couldn't complete a service call. "Obtaining the product specification" may have been known by virtue of the customer calling and saying he had cut a line and gas escaped. But the second step, "design a method of producing," that is, fixing the problem, was not planned. There was no forethought about the refrigerant line being buried beneath the surface and a digging tool being required. So the truck went on the service call ill-prepared. Then to compound the error, the "purchase raw material" step was also compromised—there was no acetylene on the truck. This is a true example of the consequences of trying to perform steps out of sequence.

Take my word for it, doing work in the proper sequence counts and organizing to do so makes it easy to happen. (To learn more about the 7 steps of the manufacturing system as it relates to optimization, I refer you to my book *Fundamentals of Shop Operations, Work Station Dynamics,* published by the American Society of Mechanical Engineers and the Society of Manufacturing Engineers, and to my article "Making it in the Competitive World," in the May 1998 issue of *Mechanical Engineering,* published by the American Society of Mechanical Engineers.) Trying to freelance outside of the system often leads to results similar to the air conditioning repair service's fate.

Let's match the seven steps to the core essential organization. Step 1 is Marketing (external). Step 6 is Finance (measurement). The remaining 5 steps are Operations (internal). The exact names of the subfunctions doing the work is not important. What is important is to know what core essential activities have to occur and make sure they do happen.

Obtaining product specification, Step 1, is a Marketing task as defined by the core essential definitions. We need to know what the customer wants, match our capabilities with customer needs, then design the product or service to meet those needs. The common denominator is interrelationships with the customer. In a fully flushed-out organization, we would have a Marketing Department analyze customer needs vis-à-vis our products, and then do its utmost to get the word out to the defined customer base as to what we can do for them with our product line. We would also have a Sales Department focus on selling the products or service to customer segments identified by the Marketing Department. These would be defined as the customer base having a potential interest because of a good match-up from a quality functional deployment analysis. The marketing task would also include an R&D and Design Engineering group who actually develop and design the company's products. R&D and Design Engineering would also be closely allied with Manufacturing Engineering of the Operations Task. This dual alliance shows how the 3 core essentials are linked together. The 7-step sequence demonstrates this linkage in a pragmatic sense.

Monitor Results, Technical Compliance, and Cost Control is the heart of the Finance Task. Like the other 2 tasks, it is entirely feasible to have shared subfunctions and sub-subfunctions. Technical compliance implies checking and ensuring that specifications have been met. This can be a Quality Control subfunction, or perhaps the doers of value-added work, the operators and service providers, who are required to check their work quite extensively. Normally this work is done under the Operations Task umbrella. We also have the Management Information System reporting into the Finance Task because of the predominance of financial information maintained through and within the databases. This leads to direct finance subfunctions such as Accounts Receivable, Accounts Payable, and Payroll. But keep in mind, the emphasis on all the work done under the Finance umbrella is a check of how well Finance, Operations, and Marketing did their jobs.

The truth is, in the dynamics of day-to-day running of a business, there is no true separation of work between the 7 steps of the manufacturing system. It's sometimes very difficult to know when 1 step ends and another begins. If we are designing a method to produce a product for a customer, where does that become Step 2 vs. Step 1: obtain product specification? They are distinct steps but when we're discussing alternatives with the customer, is that Step 1 or 2? Perhaps we can think of this in an allegorical sense, borrowing from quantum physics. There we know it is possible to have a particle in 2 places at the same time, based on how it is observed.

Perhaps the same is true for the manufacturing system step segmentation. A nice way to show this is go back to the 3 core essentials and portray them as shown in Figure 4-2.

The 3 circles representing the 3 core essentials have overlapping areas that signify that there is no distinct and sharp boundary between steps. A sharp boundary would rarely exist in nature. There is no reason to expect organizational dynamics to exhibit it. After all, organizations are simply contrivances for allowing people to work effectively with each other. In working within a company, an employee needs to know what the steps are and have a reasonable understanding of where the boundary areas or zones are. But don't expect any series of steps to be so completely defined that there are always tight boundaries. Organizations go astray when they become too

**Organization Core Essentials
Spheres of Influence**

Outside Triangle: *Singular*
Inside Triangle: *Blended*

Figure 4-2. The blending of the core essentials.

bureaucratic. They try to rely on firm, definable boundaries that really don't exist. This is what leads to catch-22 situations.

Example: The following is an example of when trying to abide by fine demarcations of rules going from 1 step to another creates chaos rather than order. Recently I saw an advertisement in a newspaper soliciting applications for a high-level government position to set utility rates. We can say that the work of applying for the job as defined by the 7 steps would be as follows:

Step 1. Obtain product specification:
Identify requirements and if a match is apparent, then get an application.

Step 2. Design a method of producing:
Determine how to fill out and send in the application.

Step 3. Schedule to produce:
Get the information required to complete the application.

Step 4. Purchase raw materials in accordance with the schedule:
Get the information to fill out the schedule in a timely manner.

Step 5. Produce in the factory:
The actual filling out of the information.

Step 6. Monitor results:
Check the form for errors.

Step 7. Ship the completed product to the customer:
The application is submitted to the agency soliciting it.

The advertisement did not specify qualifications other than stating some experience was needed in some of the following areas: engineering, legal, finance, and customer relations; this is okay because it gives a lot of leeway to hire people who would complement existing employee strengths. So far, it is not bureaucratic and is easy to work with. Now comes the crusher. The advertisement says forms may be obtained from the Internet, filled out, and e-mailed back. So far, okay. But in fine print it says the returned form must be an original, filled out in ink, and signed by the applicant. Now what's wrong here? We're putting very fine demarcation rules on Step 1, the need to get the application. It says that in reality, only a requested application mailed from the agency and filled out by the applicant with a pen will suffice. Applications sent via e-mail are not valid unless the applicant has some signature authorization software that would be accepted by the clerk.

The point of the example is that if any of the demarcations boundaries are set too sharply, too black or white, it puts severe constraints on the organization's abilities to function efficiently. Anyone who has had to deal with any activity requiring filling out of legal forms with signatures, notarizations, embossed seals, and the like knows it takes longer to do so and can always require another cycle to get it right. We certainly do not want this for the entrepreneurial organization. We want the organization to be an aid in gaining efficient performances vis-à-vis providing products and services to customers. Therefore, it is necessary to make sure that the boundaries between the 7 steps are seamless. We want the flow to be unidirectional and we want iterations to be minimized. But by no means do we want the prevention of iteration to be championed by roadblocks and tollgates from one step to another. The government position advertisement is very efficient in preventing iteration, but it creates a roadblock to Step 7, sending the finished product to the customer, for any applicant who tries to send the application via e-mail.

THE INTRICACIES OF ORGANIZATION IN APPLYING THE 7 STEPS OF THE MANUFACTURING SYSTEM, USING THE BASIC MANUFACTURING RESOURCES PLANNING (MRP II) SYSTEM

The core essentials functions already expanded to more specific steps via the 7 steps of the manufacturing system need to be expanded to an even more specific task-oriented sequence of events for the entrepreneur to gain control. So we go from abstract philosophy to a pragmatic "how to do it" approach. Figure 4-3 demonstrates this for a company that is making a job shop–based product, *job shop* meaning a product that is not mass-produced and one that tends to be tailored to specific customer needs. An example would be making store fixtures for the Ralph Lauren line of clothing sold in major department stores. There may be lots of fixtures tailored to the desires of that customer only. But no other customer would purchase the production overruns. Figure 4-3 is an excellent model for the entrepreneur who must expand from R&D to real live commercial production. It shows how data would flow for a computer-integrated manufacturing module called MRP II (Manufacturing Resources Planning System) laid out to emphasize the basic scheduling algorithm path and organized along the 7-step approach.

Notice that we show all 7 steps of the manufacturing system in the sequence where and how they are going to be done. A close look at the figure may cause you to wonder whether I misled you. We see some arrows going backward and not indicating a 1-direction-only process. The answer is no, I didn't. Arrows going back to previous steps represent information being extracted from a completion of a phase of a step that needs to be recycled back to another step to do planning for future work. For example, job complete in the Monitor Results (Step 6) column shows arrows back to Purchase Materials (Step 4) and Schedule to Produce (Step 3) activities. This is necessary to tell these steps that the work is done so more work can be scheduled and materials ordered for it. The feedbacks are information, not action activities.

You may still be a bit baffled as to the complexity of Figure 4-3. Why is it portrayed in such detail if the purpose is to demonstrate the 7-steps sequence,

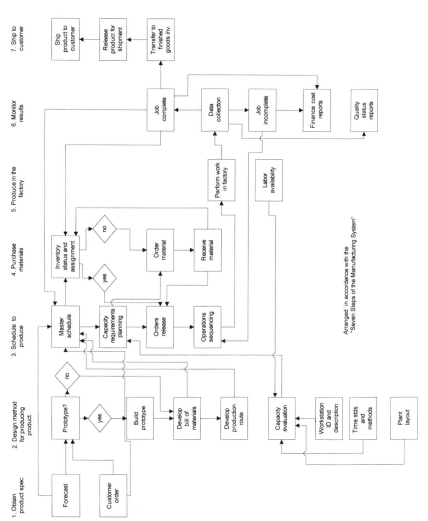

Figure 4-3. MRP II and master scheduling system.

not necessarily any other control mechanism? The answer is yes it is complex, but it's relevant to the training of the novice engineer entrepreneur. The 7 steps is a way of seeing and conceiving of process flow, but a system is also needed to make it happen. The Manufacturing Resources Planning System is one of the most common ways of using computer databases for controlling production or process flow. It is the dominant member of what we commonly term enterprise resources planning, sometimes known as computer-integrated enterprises. I don't want to go off on a side trail exploring these fascinating management philosophies, but I do think it is necessary for the entrepreneur to understand the basics of how MRP II works. This is necessary because as sure as we know a company needs to be profitable, we know a controlling system is needed for accomplishing the 7-steps process. So let's take a little side trip and explore the rudimentary facets of MRP II. If you would like to become more conversant with the entire subject of computer-integrated enterprises, I unabashedly recommend you read my book *Computer Integrated Manufacturing, Theory and Practice*, published by Taylor & Francis.

MRP II is a computer-based system that coordinates materials, technical resources, and labor resources to produce products (this coordination can also be done for services with slight modifications) in accordance with a schedule. Under Step 3, Schedule to Produce, we see the heart of the MRP II systems modules. It starts with a master schedule, goes through an analysis of the capacity to do the jobs, creates a sequence of when the jobs will be done, and at the same time queries availability of materials and labor to do so; if the materials are lacking, it releases requests to Purchasing to replenish, then releases the orders to the shop floor for production, and finally gets feedback from Operations telling it how well it is doing vs. schedule so it can modify future schedules. There, in one very long run-on sentence, is a description of MRP II. You can see from Figure 4-3 that many inputs come into the system that are allied but are not "officially" part of the MRP II system. These are the "hooks and handles" to the outside world. They are similar to the sensors that cause the MRP II system to react. Let's delve just a bit deeper to explain what each module of my run-on sentence does, then get back to the main trail.

Master schedule:
This is the repository of all orders received with their due dates and contains links to the design database. This module sifts the information to form a sequence of scheduled due dates based on inputs of factory capacity cycle time information. It works on the basis of the "two knows"—know how to make the product and the time to do so, to set the sequences and the time durations at each workstation. It also checks availability of materials to ensure that the scheduled arrival of materials will coincide with a need date. This macro schedule is sent to the next module.

Capacity requirements planning:
Here the schedule is modified by the exact capacity of the factory from a database maintained by factory engineers or other personnel who

continuously monitor the company's ability to perform processes and the rates of performance, similar to monitoring traffic on major arteries. If capacity is cut in half because of a breakdown of a machine, this module will alter the schedule to reflect that. The module also sends orders to Purchasing, which orders materials, and again can alter the master schedule module's output to reflect changes in material arrival dates. The schedule altered to reflect current abilities and materials availability is sent to the next module.

Orders release:
This module serves as a check valve to ensure that jobs get released for manufacturing only if there is a match of labor and available materials. This all-important filter makes sure that jobs that get released can actually be done, that material and labor is actually reserved for them. It knows what the previous module has arranged for and releases jobs as those planned for kits of materials and labor come together.

Operations sequencing:
This is the detailed daily scheduling of each workstation based on the jobs released by the previous module. These are the schedules that actually get dispersed throughout the factory. They are typically updated daily to reflect actual performance vs. schedule. Data collection is sometimes linked to this module; but in a theoretically pure system, the data collection output is used as input to the master schedule, where it then gets dispersed down the hierarchy to this module. The information from data collection is used to update the next issue of the daily schedule.

Many more intricacies are involved in making MRP II work efficiently, but basically they are only performance enhancements to the modules described above. Now let's return to the main trail once more.

DEVELOPING AN ORGANIZATION COMPATIBLE TO THE 7 STEPS OF THE MANUFACTURING SYSTEM

With a detailed list of what needs to be done under each step, as shown in Figure 4-3, the entrepreneur can decide what subfunctions need to be created to carry out all of the necessary activities. Looking at the figure once more, we can imagine how the entrepreneur would mark it up to create a sensible organization, guaranteeing that all the core essentials are being done. Figure 4-4 shows how the list is marked up.

The entrepreneur created 9 jobs to do the necessary work. This work being done represents the core essentials as specified for the company. Notice how the Shipping and Receiving subfunction does work under more than 1 step. That's okay because we can see how all the work is accomplished and there

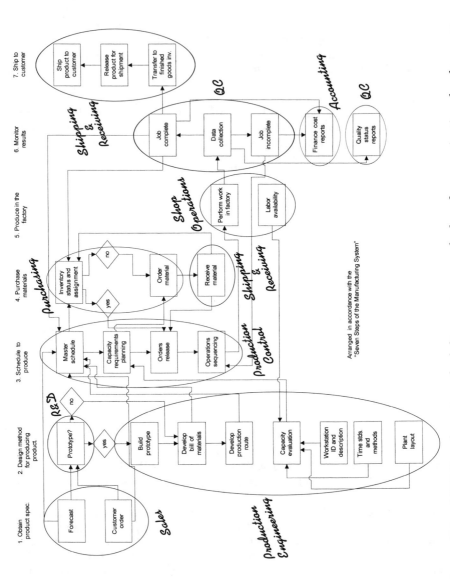

Figure 4-4. MRP II and master scheduling system; marked up for organization development.

are no loose ends. The markup doesn't take the place of an organization chart. If the entrepreneur had only 9 people, each assigned to perform the work of 1 job title, then an organization chart wouldn't even be necessary. And if the entrepreneur had only 7 people, and some were doubling up in tasks, then an organization chart still probably would not be required. Say, for example, that all engineering, R&D, and Production were being done by 1 person, also, that Production Control and Purchasing were being combined. As long as everyone knew about the "2 hatters" and the incumbents could distinguish when they were doing one task or the other, it could work without an organization chart.

However, once we have a situation where people do not report directly to the boss, we need an organization chart to see what the chain of command is. Without it, we create confusion and find that subordinates are given conflicting instructions.

Example: A company makes sailboat sails. It has an operations manager who runs all activities except sales and finance. The entrepreneur owner comes into the shop one morning and tells the seamstresses to work on order 21, although yesterday the operations manager had to delay order 21 because Finance told him that the customer hadn't made the required cash deposit. So the operations manager had instructed the seamstress to work on order 22, which was late and would earn the company a nice amount of money when completed, enough to pay a lot of bills. The entrepreneur is not informed of the change or the reason, so in a bit of a power trip he orders the seamstress to change and get back to working on order 21. Result: confusion, hurt feelings, and loss of production. The seamstress reported to the operations manager, not to the entrepreneur. The entrepreneur shouldn't have countermanded the production sequence directly. He should have asked the operations manager why the change was made and if he disagreed, have the operations manager change it with the seamstress. If an organization chart existed, there would have been a decent chance that the entrepreneur would not have gone directly to the seamstress, but instead to the operations manager. Also, if the interpersonal chemistry is okay, the seamstress, who has a copy of the organization chart, could have told the entrepreneur that except for an emergency, she's suppose to take her direction from her immediate boss, the operations manager.

This scenario is repeated over and over again in companies that have no organization charts but are big enough to require one. The definition of *big enough* is when the firm has intermediate reporting levels, as in the previous example. Getting back to our entrepreneur who has marked up the system flow diagram: if the entrepreneur has intermediate reporting, let's say she or he has 20 people total, the entrepreneur should take the time (probably no

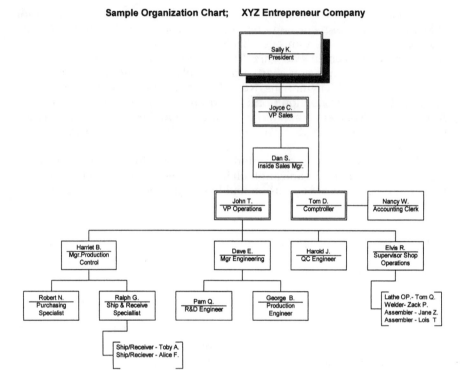

Figure 4-5. Organization chart for a small startup manufacturer.

more than 2 hours) to draft an organization chart. It would be similar to that shown in Figure 4-5 and would be in total concurrence with the core essentials. Marketing, Operations, and Finance tasks would all be accomplished by the organization and everyone would know what their job is and how it relates to the others.

DIFFERENCES BETWEEN A SMALL BUSINESS AND A BUSINESS TEAM

I've alluded to the paradigm shift in thinking when involved in a small business vs. being part of a business team. While the premise of this book holds that significant similarities exist and affect how engineers work in 21st century industry, it is not exactly the same as going off on your own as a risk-taking entrepreneur. Let's examine those differences because before you make the leap, you need to be aware of where you may land.

There are real differences, material and mental differences, and the latter are mindset differences. Material differences come down to the fact that the business team will always have access to a larger resource base. The mindset ones are the most critical, and if you think about it, quite evident. When working on a business team for a company you do not own, you may have a feeling that what you do affects the ability of others to earn a living. But you also know that if you do not do your job adequately, the first line of defense that the company has is to relieve you of your responsibilities, thereby protecting the enterprise and your fellow team members. When you're an entrepreneur, I'm sure you would still have the same empathy for your team. However, in this situation, the remedy for poor performance is not so readily available. You can fire yourself, but that means the company goes under and fails. So the only solution is to make sure you do a better job. This means putting more pressure on yourself to succeed. It means the desire to succeed needs to be very strong, much more so than if you were a team member or even the team leader. The entrepreneur has no safety net for himself, none whatsoever. This creates a mindset that is much different than when participating on a business team. Entrepreneurs have to be more focused, more conservative, more strategic in their thinking, and at the same time perform tactically in a superior manner. The mindset is much sharper. It is like a batter batting with 2 strikes against him instead of none. Each and every action is very black and white: either win or the game's over and you lose. The team leader on the enterprise's most risky team does not share these finality decisions. Or if he does, he subconsciously rationalizes that they're really not so. He cannot conceive that he, the "wunderkind" of the company, would be severely punished if the team fails. The entrepreneur, on the other hand, cannot harbor such subconscious fantasies. The entrepreneur knows there is no safety net under his or her tightrope; therefore, the entrepreneur needs to be ever-vigilant and focused.

With this harsh realism, what would ever make anyone want to take the risk and be an entrepreneur? I don't have any sure-fire answers for this. But in my case, it's the shear drive to create something that's all mine and to realize that I've had the ability to create my own destiny, to march to the beat of my drummer, not yours. It's also a belief in one's own ability to succeed. I can remember as a lowly 4th class cadet, a "swab," at the U.S. Coast Guard Academy, watching many fellow cadets that first summer break under the pressure and resign, I had to develop a new personal philosophy. And it was simple: "I can do anything I have to do to succeed; all I have to do is apply myself to the task." I did graduate, so I guess my philosophy worked. But realistically, entrepreneurs have to think like that. They have to be confident in their abilities and go pursue their quest. They have to realize the steps to success are the same whether there is a safety net under them or not. And they have to want to do it in a manner that is so strong that if they fail, they still think they've succeeded and all they need is to fine-tune their approach and go at it again.

The mindset difference between an entrepreneur and a team leader is probably the most important factor in the decision to start your own business. But one also needs to be cognizant of the resource differences the entrepreneur faces when deciding to launch out alone. Let's say you're convinced marching to the beat of your own drummer is for you. Don't go ahead and launch that venture until you have a good idea about resource differences that you have experienced in your working career up to now compared with what your resources will be for quite a while in the future. Remember, a successful entrepreneur is not headstrong enough to do something foolish. She or he should be very calculating and know just when to launch the new adventure. Understanding these differences will give the entrepreneur just one more tool in the arsenal to create a better chance for success. Let's look at these resource differences.

COMPANY SUPPORT FOR BUSINESS TEAMS COMPARED WITH THE ENTREPRENEURIAL APPROACH

Companies try to emulate the philosophy and drive of entrepreneurship by forming business teams. In fact, a new word has been coined to identify these so-called internal entrepreneurs. They're called *intrapreneurs*. Intrapreneurs are supposed to have all the characteristics we deem desirable in entrepreneurs except they're working for and within a larger company, and one where their equity stake, if any, is miniscule. Their task is to cut through large company red tape and create new products and/or services for their corporations as speedily and with as much zeal as an entrepreneur working independently. If we could get intrapreneurs to have the same philosophy and drive as an entrepreneur, then we would get the best of both worlds. We would have the drive and creativity of the self-motivated entrepreneur match with the financial and other resources of the larger company. This would be an ideal situation for the large company if they can make it work.

Let's look at the positive aspects of intrapreneurship. Financing for the startup should be readily available. So too should resources to assist in evaluating risks and creating contingency plans. The intrapreneur business team leader would have a vast array of mentorship talent available to help him or her. Also, special equipment can either be subsidized by the parent company or delays in getting access to the needed resources minimized. Personnel needed to do specific tasks, such as legal and accounting, can be "loaned" from the parent company at the first indications of need. People can also be recruited from within the company to join the team without having to really leave their "permanent" job. This way, it is possible to get superior skilled people to join the venture without them having to put their careers at risk.

With all these positives going for it, intrapreneurship should be an outstanding success, much more so than entrepreneurship, but it's not. In fact, the best business teams have smaller success rates than entrepreneurial startups in developing new products for market. Xerox and the invention of the computer screen icons is a great example of this. Let's look at the reasons.

Big companies have procedures and rules to follow to keep them organized and relatively efficient. Without these rules, you would end up in gridlock. In fact, the procedures and rules probably make the larger company very efficient for its size. Do you ever wonder why companies like FedEx or UPS get letters and packages to their respective recipients within the time limits quoted? They do so because they make sure every member of their staff understands the procedures to follow and are measured for their ability to comply. Following rules and regulations is a good thing for most tasks. It creates the most effective pursuit of accomplishing an "average type" task. However, although the majority of the entrepreneur's tasks may be "average type," the entrepreneur is different because she or he has a special few tasks that need to be done differently and on an ad hoc basis when needed. And the entrepreneur goes ahead and does so with virtually no receipt of permissions. The intrapreneur, on the other hand, may want to do that same thing, but inevitably is delayed by the company's procedures and rules, which would need to be waived.

You see, no matter how free intrapreneurs' bosses tells them they are to do what they want, intrapreneur are still accountable to the company's fiduciary management requirements put in place by their superiors. Entrepreneurs are also accountable to themselves and to their financial supporters. But financial supporters are not superior to their entrepreneurs; they are their partners. They usually are not anxious to put bureaucratic ties on their partners because they know entrepreneurial ventures are risky by their very nature. Companies trying to instill entrepreneurship values within their teams often stop short of authorizing unbudgeted funds to be spent unless a thorough review is held first and approved by a higher authority. This takes time, so while the intrapreneur is fussing with this administrative requirement, the entrepreneur counterpart is simply informing his or her backers and proceeding. The difference is in authority. The entrepreneur has as much as he or she wants. The intrapreneur is limited by company policies.

This limitation experienced by the intrapreneur also has a telling effect on aggressiveness and decisiveness. No matter how often and with how much praise the intrapreneur is told she or he has the full confidence and backing of the bosses, there is always the unsaid message that the intrapreneur will be accountable for failure if it occurs and will be second-guessed. The entrepreneur is also accountable for failure, but the only second-guessing is going to be done by the entrepreneur for what he or she did and could have done differently to get a better outcome. The entrepreneur also has a better chance at getting a second chance, which will probably mean just finding additional financial resources, although this may be difficult. The intrapreneur has virtually no chance at getting a second chance because if the intrapreneur fails, he or she will be relieved. The idea may get tried again but under different leadership. I've heard, read, and experienced statements by senior management types that they give managers a chance to do things, to take risks and not be punished for failures. But this not intrapreneurship. These so-called risk-takings, which are small compared with the entrepreneurial mode being emulated by the intrapreneur, have been vetted by the authority-granting management level and deemed to be okay beforehand. So everyone is in bed together, and blame for failure is widespread. In that manner, this phony, bold, management risk-taking permission strategy is allowed to go forward. These types of risks are minimal compared with the risks the intrapreneur needs to take. The truth of the matter is that intrapreneurs who take risks and fail lose their jobs. This definitely stifles the desire to take a chance. Therefore, as a general rule, intrapreneurs are more risk-aversive than entrepreneurs. So, the game is most often won by the entrepreneur, and the desire to emulate their performance within a larger company is not significantly successful. Entrepreneurs bring new products and services to market faster and more frequently than their intrapreneur counterparts.

THE CONTINUING ROAD TOWARD ENTREPRENEURSHIP

Entrepreneurship is spirit, a mindset, coupled with skills that make risk-taking a sensible choice of behavior. We've just finished a section that spells out the differences between a business team member intrapreneur and a true entrepreneur. Two things come to mind: First, skills necessary to be a successful entrepreneur are shared with, and in many cases are identical to, those required of a business team member. Second, what sets the entrepreneur apart from the intrapreneur is the strong desire to own the fruits of one's labor, rather than getting paid well to be successful for the company's benefit. Since entrepreneurs want to own their businesses, they tend to be adverse to the corporate culture of growing to maturity within its confines (as much as that's possible in the early years of the 21st century). Entrepreneurs are loners, which means they have to learn their lessons very well. There is no safety net. Whereas business team members may get by quite well not having master-level competence in project-management skills, this would never suffice for the entrepreneur. The business team member may be able to lean on the shoulders of fellow members when necessary. This luxury hardly exists, if at all, for the entrepreneur. Learning skills well is a "must" for the entrepreneur; it is only a "high want" for the intrapreneur.

The rest of this book is about skills that entrepreneurs need to master: introducing a new product or service; learning what a business plan is and why it is critical to success; developing a mastery of financial funding and the managing quagmire; learning what's important to measure and what's not, and how to do it; selling and marketing your grand idea; raising capital in a manner that keeps the venture capitalists (some say vultures) at arm's length; becoming familiar with commerce and supply-chain management; learning the soft skills of effective communications and leadership. You could say we will cover everything you need to know to run a business but perhaps you didn't even know enough about it to ask. The only major topic we will skirt is the hard technology of your product or idea and whether it is sufficiently developed to have a market for its commercial disbursement. We'll cover the marketing aspects, but I leave it to you to determine whether your idea is a good bet for a successful entrepreneurial venture. (Don't fret. Throughout the remainder of the book will be plenty of examples to help you measure your idea against proven successes.)

I will identify the skill sets you'll need to be a successful entrepreneur, or at least a good team member intrapreneur. It's your choice as to how you use them. I have no doubt that if you use the skills explained in the succeeding chapters, you can become a successful entrepreneur. If you choose not to, but want to remain an active team member, the technical member, of a company business team, these skill will be vital for you too. They will help you maintain your contract of skills offered to your employer in exchange for employment.

Chapter 5

All Communications Are Good, Some Better Than Others

It doesn't matter whether you're an entrepreneur managing your own company, an intrapreneur of a business team, or a member of a business team. Being able to communicate effectively is a keystone for success. I'll even expand my claim to state that effective communications is the hallmark of success no matter what your occupation or avocation. This chapter is devoted to learning how to communicate, both sending messages and receiving messages: written messages, electronic database messages, oral messages, and even some thoughts on body language messages. There are ways to communicate to be to be most efficient and we will explore those methodologies.

Unless you're a hermit living an isolated existence on a desert island and want to remain that way, you need to be able to communicate effectively to be successful in any activity you pursue. In the business world, this is absolutely fundamental. Whatever our avocation is, we all have 1 thing in common: We need to communicate. We are social beings and we believe isolation can cause mental harm. This human condition drives us, so it's not surprising that it is extremely important for business success.

Business leaders, as well as individual contributors, need to receive inputs pertaining to the status of their activities. Based on those inputs, they need to transmit instructions and requests to others. These instructions and requests become inputs to others, up and down the real or perceived organization. Entrepreneurs have the same needs for receiving and sending information as anyone else. The only difference is the speed at which communications takes place. Entrepreneurs, by the nature of what they do, require faster and more accurate communications than most business organizations. When you're in charge of everything, communications will seek a path to your desktop for all and every issue related to your venture. Learning how to deal with those communications in an effective way is a skill worth mastering.

You as an entrepreneur may be individually brilliant. But if you can't get your thoughts across to others so that they can respond, then you can't manage your company. It's like a computer not being able to input data or send data, and everything is contained within its hard disk incapable of being shared. The phrase "No man is an island" is an appropriate philosophy when it comes to understanding the need to be an effective communicator. This means we have to strive to practice "communications excellence." It's the only standard that is sufficient to clear the path for success.

THE COMMUNICATIONS PROCESS

Communications can be defined as the process of transmitting a thought from one mind to another. It is a 2-way process that requires a sender and a receiver (there can be multiple parallel receivers). Both the sender and the receiver share responsibility for the success of the communications.

Figure 5-1 illustrates the closed feedback loop that successful communications require. This is true regardless of the type of communications being used: verbal, written, or electronic. We see that the communications cycle is a continuous and closed cycle. Think of it as the looped wires of a very long coiled spring with a hazy beginning and end. I say this because communications is never, in the real world, a onetime cycle. As we send a message, we're probably thinking of the next bit to be added; and when we get a reply, we modify our reply to the received reply and send it out again. We're always modifying our database as more "real" information is received. In business, we make plans, communicate plans, receive data as a result of our plans, modify plans, communicate the modified plan, receive data as a result of our modified plans, and so forth over multitudes of cycles. Figure 5-1 shows only 1 cycle, but it illustrates what's occurring. Let's take a look.

We enter the cycle with the communicator, seen in Figure 5-1 under the Sender column. The communicator is responsible for putting her thoughts into understandable symbols. Symbols are words spoken or written, or data such as digital data used by a fiber-optic cable. If the communications is face to face, that is, within sight of each of the participants, then body language and gestures also convey meanings as modifiers of the main message.

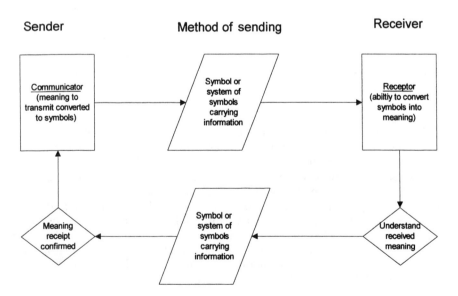

Figure 5-1. The communications loop.

The communicator has the responsibility of transmitting these symbols through an appropriate medium to the sender. This is represented by the middle column of Figure 5-1: the Method of sending. I call it a carrier, a scheme for carrying interpretable symbols to the receiver.

The receptor is responsible for receiving the carrier, then interpreting the information. The receptor is located under the Receiver column. The receptor has an additional responsibility of understanding what the communicator has sent and verifying it back to the communicator. This may be something you haven't thought of before. As a receptor, it's your responsibility to understand what the communicator sent. If you don't, then the action step is to ask the communicator to be more precise. The communicator must explain what she or he means. This responsibility abdication is so very apparent all around us and is something the entrepreneur must work with his or her team to eliminate. The biggest cause of mistakes in business is due to people saying something like "I thought you meant something else. I didn't think that's what you meant when you told me." When I give seminars on achieving communications excellence, I typically start with a warm-up or icebreaking exercise to illustrate this abdication problem. It goes something like this:

Exercise: I play the following parlor game: Pass a message around the room from one person to another in serial fashion to see if we can communicate as effectively verbally as with a written form. First, I hand a written message to the start of the chain (I'm always the end of it). Each person simply reads the message then passes it on to the next person until it gets back to me. When I get the message back, I randomly ask a few attendees to come to the board and write down what they had read. I then read what I wrote and invariably it's not what they wrote on the board and sent. Casually close but not the same. Here's an example of the type of message:

> " A typhoon struck the Philippines Island of Mindanao on August 23, 1939, resulting in $2.65 million in property damages and 143 deaths, including 3 U.S. citizens. The President of the United States is dispatching emergency aid via the U. S. Navy's Pacific Fleet."

Typically the words written on the board would get the fact that a typhoon had struck, but no location. The date would inevitably show the current year, not the correct one. And casualties and damage would be all over the ballpark of being correct.

Then I do the same exercise again, but this time we do it verbally. I whisper a message, similar in content to the one used for the written example, into the ear of the first person, who in turn does the same to the next person, and so on around the room. Finally, the last person hears the message and I ask him to repeat it aloud. I then read to the group the message I whispered into the ear of the first person of the chain. As you would probably guess, the message returned to me is much different than the original and often totally wrong.

Why does this happen? Primarily because the receptor doesn't bother to confirm what he or she has been sent, therefore abdicating his or her responsibility. During my parlor game, I do not tell anyone that he or she is prevented from questioning the communicator about the content of the message. Occasionally some do, but it's the exception, not the rule. The written message usually come through significantly better than the verbal message, I think because most of us do know how to interpret what we read even though the retention time may not be very long. However, we do not know how to listen. I'll have more about that later. But the prime message here is that when we're in a receptor role, we're constantly abdicating our responsibility to confirm that what we've been sent is the actual communication. The most frequent excuse is, "Well I assumed..." This is not a valid excuse. It brings to mind a saying, "Those that assume without trying to check it out, make an ass out of you and me (assume = ass – u – me)".

> Don't be afraid to make assumptions when facts are not verifiable. Making an assumption is not the same as assuming. Assumptions are meant to be realistic approximations of facts when facts cannot be obtained or verified. They are based on the best probable set of facts that exist and can be interpolated to fit the current situation. Assuming is jumping to conclusions without checking any corollary facts at all.

Communications is more than writing and speaking. We know there are many ways of communicating meaning. If I desired a canker-infested orange tree to be cut down, I could write my directive in 2 ways:

1. Please cut down the diseased tree, or
2. Cut down the diseased tree.

Which sentence implies a stronger demand for action? Number 2, of course. There's no asking as implied with "please"—Only a directive to do so; discussion over.

The same is true with spoken words, but more subtle. Tone of voice can indicate questioning or authority: "Cut down the diseased tree?" is less authoritative than "Cut down the diseased tree," and this is less authoritative than "Cut down the diseased tree!"

Coupling this with passive or aggressive body language such as posture, use of hands, where eyes are focused, etc. can imply meaning that the receptor has to understand. Here are some tips on body language vis-à-vis achieving "communications excellence":

- Eyes cast down or not looking directly at the other person implies the speaker is not sure of his argument or may be fudging it, or perhaps is embarrassed by it.
- Tightened jaw muscles signify defiance and disagreement.
- Pointing with hands or fingers implies aggressiveness in defending a position.

- Speaking low and or slow while stairing directly at the other person indicates a festering anger and that the person is trying to control his emotions during the response.
- Pacing back and forth while speaking indicates an emotional tie toward the subject that would be taken as a personal affront if the listener didn't agree.
- Laughter at inappropriate times indicates nervousness and not being sure of one's position on the subject matter.
- Heavy breathing, especially exhaling, indicates frustrations with the contents of the communications being received.
- Smiling and nodding while listening indicates agreement with the speaker or getting a reply that was expected.

Obviously good acting can cover true feelings about a communication by being aware of and masking body language. One of the best ways of doing that is to force yourself to relax and smile. So its important how you verbally communicate, as well as what you say. This is especially so since people tend to abdicate their receptor responsibilities, and observing their body language will give a good indication as to whether or not they understand the content of the message.

BARRIERS TO EFFECTIVE COMMUNICATIONS

The goal of communications is to provide exact transmittal of a message between the sender and the receiver. If we do this we're practicing communications excellence, and this depends on minimizing the barriers of effective transmittal. There are many barriers to effective transmittal that we have to be aware of. We are not machines. We deal with many subtleties of meaning. Our written and spoken language is full of nuances of meanings, and sometimes these are contradictory. Before we can minimize these problems we need to know what they are. Let's look at some common barriers to effective communications.

There are 3 common barriers to effective communications you should be aware of:

- Lack of common core experiences
- Confusion between the symbol and the item being symbolized
- Overuse of abstractions

I'll expand on each of these.

Lack of Common Core Experiences

Dissimilar physical, mental, or emotional experiences lead to misinterpretation of symbols. These symbols can be words or drawings or any other abstract way of expressing thoughts. Words do not always have precise meanings. In fact, we know words can have more than one meaning, depending on how they are used.

> **Example:** The word engineer can mean the driver of a railroad engine. It can also mean one who practices the art and science of engineering in applying scientific principles to solve problems.

This is an exact definition from a dictionary. What happens when words convey meanings that are based on our personal experiences? This can cause even more confusion because we will put the emphasis of meaning on the word or phrase based on our previous experiences, which are completely unknown to those we're trying to communicate with. This is especially true with people conversing from different cultural backgrounds.

> **Example:** Two people here the phrase "I'm hungry." One is an American from the great middle class socioeconomic status. The other is an African refugee who has fled from a severe drought area. Here's how the 2 of them could react to that phrase:
> American middle class: "I'm hungry. When do we eat?"
> African refugee: "I'm hungry. Will I die?"

We see lack of common core experiences can have detrimental effects on being able to communicate what you really mean. How many times have you heard about international treaties being written in ways that satisfy both parties, even though they really don't agree, but the words are such that each side can interpret them relative to their own core experiences. And then we wonder why treaties are violated causing each side to display righteous indignation against the other. The reason is so obvious: They never agreed in the first place. They chose to obfuscate just to cool the situation at the time so they could come back and argue at a later time. The ability to convey what you mean so the other person understands it is critical for business success. Business situations rarely, if ever, need to have both sides of a negotiation convey different meanings to an agreement. Indeed this is so antithetical to good business practices that we go out of our way to ensure there is only 1 meaning understood by both parties for any 1 phrase. This means finding some common core experiences to use as a vehicle to gain understanding is critical for achieving communications excellence.

Confusion Between the Symbol and the Item Being Symbolized

Words are representations. They are symbols corresponding to anything that exists, experienced, or talked about. Spoken or written words do not necessarily convey accuracy, and inaccuracy in how we use words can cause communications breakdowns. We sometimes use words to try to trick people to do what we want them to do. A perfect example is the "sweet talk" technique.

> **Example:** I want you to do a task and you're not entirely sure you want to do it. So I say, "You're a good engineer." Am I speaking the

truth? If we're speaking the literal truth, I would know you're a good engineer because I checked your record and references. If I'm sweet talking you to convince you to do what I want, then I'm not speaking the truth. I may have no way to tell if you're a good engineer or not. But I know by saying so I may be flattering you and you'd be more inclined to do what I want. So the symbol I'm conveying may or may not be true. I'm creating confusion between the word symbols and the message I'm conveying.

We often purposely create confusion between the word symbols and the message we want to convey because we think it will be advantageous in gaining the upper hand in our negotiations. Every interaction between people is a negotiation, however innocent it may appear, and most of them are indeed that. The danger occurs when we really want "X" to happen but "Y" does because the receptor gets confused. In business situations, we need straight talk within the team and this requires there be no confusion between the symbols we use to communicate and the meaning we intend to convey.

Overuse of Abstractions

Abstractions are words or group of words that convey a general sense of meaning. They are excellent shortcuts to effective communications as long as the communicator and the receiver share the same interpretation.

Example: Here's a statement made by the leader of a project team to her team: "As the project manager, I assure you we will employ 'proper measures' to understand the component failure." The abstraction is "proper measures." Does everyone on the team share the same interpretation of its meaning? Maybe, maybe not. Here are some typical reactions she should expect:

Front line worker: "She's looking for someone to blame."
Front line engineer: "What kind of tests can I use to find out why the component failed?"

If the project manager believed the failure was mechanical, not human, error, then she should have said, "As the project manager, I assure you we will employ 'proper measures' to understand the component failure. I'm not trying to place blame that's irrelevant because I know for a fact that everyone followed the process explicitly."

Abstractions are great for making messages shorter and quicker to transmit, but there is always the danger that interpretation, call it decoding, will be in error. Engineers in particular, have to be careful. We are notorious for talking in abstracts:

> "The psi value of valves being manufactured under the MRP II system is hard to change because the JIT system doesn't allow for easy modifications."

I know what that means because I wrote it. Do you? It depends on your background. If you're a manufacturing engineer with an understanding of systems, you probably can interpret it correctly. What I wrote means:

> "The pressure value, measured in pounds per square inch, of valves manufactured in the factory controlled by a manufacturing resources computer system, and where the staff uses industrial engineering techniques to minimize waste by strictly evaluating design changes, requires significant data for a change before it is allowed."

You get the point. Abstractions shorten the communication cycle but could be dangerous if the interpretation standards used between the communicator and receivers are not the same.

SOME GUIDELINES FOR EFFECTIVE COMMUNICATIONS

At this point, you probably appreciate that communicating effectively is not as simple as it appears. Creating and maintaining communications excellence requires lots of attention to detail. We all know that we have to be careful of what we say and how we say it. In our daily lives, how we interact with other people bears this out. How often do we hear of, or are part of, a rift between groups or individuals based on what someone thought he or she said or wrote or implied. This is the gist of the gossip mill. The innuendo, the false facts contrived from poor communications, certainly make life interesting. If this weren't the case, then the entire entertainment industry would perish. Think about virtually every drama or comedy presented in the theater, as a movie, or on television. There is always some sort of conflict. And most of the time the conflict is caused, or at least significantly abetted by, a misunderstanding. The misunderstanding manifests itself as a communications failure of some type. In business we're not interested in entertaining as a main outcome of our efforts. We want to be profitable. So we need to do our best to minimize communications misunderstandings.

The entrepreneur is particularly vulnerable to communications failures. The entrepreneur's business is a dynamic activity with virtually every action or inaction being a result of a communiqué between allies, competitors, and customers. To do well in this environment takes skills to be at least slightly ahead of the competitors. And it should come as no surprise that communicating effectively is a great tool to have in your arsenal of weapons. To prevent misunderstandings we must improve communications skills. What follows are some basic building blocks we need to use to communicate effectively.

Prepare What You Want to Communicate Before You Communicate

Remember the Boy Scout and Girl Scout motto, "Be prepared"? Understanding what you're going to communicate before you engage your vocal cords is a manifestation of the motto. Most often, when communications fail it is because we haven't thought out the consequences of what we're going to communicate. Another perhaps crude way to put it is "Engage brain before mouth." Plan before you issue the memorandum. Here's a checklist to guide you through the communication process.

- Understand the meaning of the information you intend to communicate to the recipients. You have to understand what you're saying to them so you can verify that they received it correctly. In addition, knowing the meaning lets you anticipate questions and have reasonable explanations available. This understanding gives you the ability to have strong arguments available to sway the receivers into taking measures favorable to your position.
- Know who the recipients are and how they will probably react to the communications. You have to put the message into language and terminology they will understand without misinterpretation. If you expect them to take an action based on the communication, you need to do everything possible to communicate in a way they will absolutely understand.
- Know what you want the recipients to do with the contents of the communications. All communications need to have an action of some type associated with them. But all actions do not have to be direct. You can tell someone a status, a situation report of which there is no directly related action to consider. But this information may be such that if the situation develops requiring action, the recipient has the information to guide him or her to the proper decision later on.

Always Favor Written Communications Over Verbal

Unless you believe all recipients of your communiqués have total memory recall, the preference is for written messages over verbal. This is just common sense. Since most of us use mnemonics to help us remember things, it is hardly surprising that we depend on written symbols over verbal ones. Think of the parlor game I described earlier. It is very evident that written messages are deemed important while verbal messages are not, when it comes to defining accuracy.

If Verbal Communications Are Necessary, Follow up with a Written Memo

People will forget the details of verbal messages shortly after receipt. And not too long after that, they will forget they ever received a message on the subject matter of your communiqué. By following up with a memo, you are also leaving an audit trail that will verify that you did indeed communicate the specific message content. This will be very helpful, especially if the communiqué required an action on the part of the receiver.

Always Request that Recipients of Communications Restate the Content Before the Communications Session is Ended

Referring back to Figure 5-1, recall that it is the receiver's responsibility to inform the sender that he received the message. The easiest way to do that is to repeat back the message to the sender. If it comes back as originally sent, then there is no doubt that the communications has been received as it was meant to be. It is interesting to note that IBM, in the design of their computers, always has circuitry to perform this function. When information is sent to the memory bus to be executed, it in turn sends the information back to ensure it is acting on the right data.

When the Communications Requires the Receiver to Take an Action, Make Sure the Receiver Understands That he is Obligated to Report Back on Status and Completion

While this may appear to be simple common sense, it is one of the most frequently violated principles of good communications methods. A communiqué requiring an action virtually always comes from an issuing source that is depending on that action to be carried out. Furthermore, it is usually important that the originator of the communiqué know that the action has been completed. For these reasons it is vital and just plain common courtesy for receivers to feed back status to senders, not as a nice thing to do but as a requirement for successful business communications.

I learned this axiom as a fourth classman at the United States Coast Guard Academy. As "swabs," we were drilled to do just that, report back, when given an order to do something by an upperclassman or an officer. Some of the tasks were extremely trivial, such as reporting back the weather status when all the upperclassman had to do was look out the window, as would be expected in a military academy indoctrination setting. But that triviality didn't matter. The training point was that no matter what, when you're directed to do something, no matter what you think of it, you have to report back status to the originator of the directive. Any good communications system requires a feedback loop. Reporting back is the feedback loop.

Never Assume That the Other Person(s) Understands the Nuances of the Communiqué. Be Thorough and Complete

Remember: assume = ass+u+me. And that's exactly what happens when you think but haven't checked whether the recipient knows what you're trying to communicate. I learned, again in my Coast Guard Academy training, that the originator of the message must make an effort to question the receiver to make sure he or she understands exactly what you mean. Otherwise, if you receive a pizza instead of a tuna fish sandwich, you have little reason for being angry except at yourself. My advice: Simply ask if the receiver understands what you mean and have him or her state it back to you so you understand that the receiver got the message correctly. Again, the feedback-loop principle.

Spend as Much Time Listening as Speaking to Ensure That Your Message Has Been Properly Received

The danger is that you're so intent in framing the message correctly and so relieved that you transmitted in a timely manner to the receiver that you're daydreaming when the receipt of transmission comes back and you don't check it for accuracy. Listening is one half of communicating. Unfortunately, many of us do not give it enough conscious thought to make sure we understand what we heard. Lots of times we have preconceived viewpoints of what we're going to hear so we hear what we want to hear, not what was really said. This is another reason for preferring written over verbal communications. We must be on guard against this tendency to not listen effectively. Listening effectively is just as important as speaking effectively for communications excellence.

LEARNING HOW TO LISTEN

Remember in those foregone days when you daydreamed a lot and your pleasant escape from reality was interrupted by an angry question, "Are you listening to me?" How many times did your mother or other lesser figures in your life say this to you? Hundreds? Thousands? And you answered in some derivative of "yes." But were you really listening or simply hearing? There is a difference as we shall soon see. You probably weren't listening. The words were going through one proverbial ear and out the other without any comprehension taking place. What you were doing was hearing. You were sensitive to acoustic energy but you derived no meaning from it. Listening is deriving meaning, comprehension, from verbal communications.

Most of us take listening for granted. It is a necessary part of being a good communicator and is often overlooked. As we saw in Figure 5-1, the need to close the feedback loop is synonymous with successful communications. It is the responsibility of the receiver to close the loop, and when communicating verbally the listener is the receiver. The listener must be prepared to receive and interpret the message.

Being a skilled listener is an acquired skill that takes some practice. What follows is based on United States Air Force Manual *Guide for Air Force Speaking*, Air University, AU-2, printed May 1955, revised February 1960. I was fortunate to have attended the Air University's Academic Instructor's Course in the summer of 1965 where I was exposed to the world of listening from people who took listening very seriously for the success of their business. They stated that most businesses fail not because the staff didn't know what to do but rather because they couldn't execute due to an inability to verbally communicate. They amply demonstrated that to me through many exercises in speaking and listening. Most often, communications broke down because the listener couldn't comprehend what he heard. When we learned how to listen, we overcame the real hurdle toward effective communications and were successful. While I can't replicate the intensity of that training experience, I can relate the main points of mastering the skills of effective listening.

There are 7 points to master to become a successful listener. I will present them in the order of the process.

1. Get Ready to Listen

This requires physical and mental preparation. To comprehend what you hear you need to be physically comfortable and mentally alert. You can't be daydreaming, nor can you be distracted by discomfort. Sitting on a chair that causes backache or some other malady will definitely affect your listening acuity. You have to be ready to tune into the speaker's mental wavelength. To do so means you can't be fidgeting in your seat or thinking of far-off places.

Try to imagine what the speaker will say. This part of your preparation lets you, the listener, match core experiences with the speaker. In business situations, you already have reduced the potential range of speaker subject matter significantly. You know the speaker will be engaging you in conversation about what you and she have in common about the business. So even before the speaker speaks, you can set up an imaginary 2-way communications loop.

You might say, "Wow, this is a lot to do before the speaker even utters one word." Perhaps. But keep in mind that you can think at a speed at least 4 times as rapidly as you can talk. So that gives plenty of time to get ready to listen. If you are prepared to listen, you will have 2 reactions when you hear the speaker's words. It will be either "That's what I expected to hear and I'm ready to comprehend the message." Or, "That's not what I expected to hear, but my sensors are at peak efficiency to comprehend what I'm hearing because I'm ready."

2. Take Responsibility for Comprehending

A successful listening situation requires the listener to relate the speaker's ideas and thoughts to his own. To do this, you may have to restructure the speaker's sentences and phrases into ones you feel more comfortable with. For example, the speaker may say, "I'd like to express my gratitude for the new watch all the employees have given me for my retirement." If you're 20-something hearing this, perhaps you would rephrase the speaker's words to: "It was real cool of you guys to give me this great watch for my retirement." But the meaning is the same. The key is comprehending. It's like a translation process, only we're all speaking the same language (perhaps different idioms) and there's no need for a third-person translation.

3. Listen to Understand Rather Than to Refute

Not taking statements made by the speaker to be de facto correct is a proper course of action. But don't start a mental argument with the speaker before the speaker is through talking, for example at a natural breakpoint in the conversation or after the speech has occurred. Remember you can think more than 4 times faster than the speaker can talk. This gives you plenty of time to compose a reply.

Silent argument with the speaker while he talks stunts your ability to comprehend the totality of his argument. A response based on listening to the entire argument will be stronger than one focused on incomplete data. It is better to listen for the whole message and to see if it is logical before tearing it apart. By listening and not refuting as you listen to it, you can comprehend the nuances of the speaker's points more effectively because you will have been listening, not simply hearing. You will also ensure that your eventual counterargument is not being skewered by incomplete data.

4. Control Your Emotions

No one who is listening to a good speaker or receiving a verbal communication (a telephone call, for example) can fail to react positively or negatively and remain positively calm. But strive to do so to prevent emotional blocks from being raised between speaker and listener.

Try to listen to what the speaker has to say, not your emotions. Try to block out thoughts about the speaker and force yourself to concentrate on the words and their meaning. Many times when we're in visual range of the speaker, we get distracted by his or her appearance and think more about that than comprehending what we're hearing. When we can't see the speaker, his or her tone of voice or accent may distract us, if we let it. When we're doing these things, we're engaging in a form of daydreaming, which obviously is not what we want to be doing.

5. Listen for Main Ideas

Don't concentrate on facts. Concentrate on main ideas the speaker is endorsing. Try to build a mental structure of main and supporting ideas the speaker is presenting. This is sort of like dissecting paragraphs of written communications for what they mean and how they support each other. Only now you're getting a verbal input and it not reading it, and that's more difficult to do. This takes practice. You have to be continuously asking yourself, what are the main ideas and what are the supporting ideas? Here's an example, unfortunately written because there's no audio with this book:

Example: The sales manager of a Chevrolet dealership during his weekly sales meeting tells his sales people: "We have 7 red Corvettes to sell out of 25 other assorted colors. Chevrolet is offering an inducement of $500 to salespeople for each red Corvette we sell because they've overproduced. I know you find it gauche to push colors but you need to do so. And in order to incentivise you guys, the dealership will give you 10% additional quota points for each of the red vettes you move out."

What's the main idea? Sell the red Corvettes and earn a bigger commission.

What are the supporting ideas? GM has overproduced and wants to move red inventory. Sales people don't like pushing colors.

What are the facts? $500 salesperson incentive for red Corvettes. Ten percent additional quota point for selling the red Corvettes.

If you had concentrated only on facts, you may confuse the issue and expect to get the extra $500 for selling any Corvette. Period. It's very easy to hear only $500 and not the rest of it. We call this *selective listening* and most often occurs when you're deciphering for facts, not ideas. If this had happened to you, then you would be very disappointed when the sales manager says, "Thanks for selling a Corvette, but yellow ones don't qualify for the incentive."

Before we go on to Point 6, I'd like to reiterate that ideas and facts are only relative to what the verbal message communicates. They need not be true at all.

6. Be Mentally Agile

Concentrating throughout a speech or other monologue type of communication is a challenge because we think more than 4 times faster than we talk (some experts say it's up to 10 times faster in certain instances, especially when the listener is bored). This is the reason listeners drift during speeches, especially if it's a topic we'd rather not listen to. But if you're going to get the most out of any verbal communications as the receiver, you cannot allow this to happen.

To keep your comprehension at a high level, try this. Continuously discipline yourself to review what the speaker has said. Then try to predict what the speaker will say next. This exercise enhances retention of what ideas the speaker is conveying. By forcing yourself to concentrate on the speaker, you are creating ample mental time to repeat, summarize, and paraphrase the speaker's remarks, all of which aid in retention.

7. Take Notes

I know taking notes is a discipline that's hard to maintain. But mental concentration alone is not sufficient for good retention of a speaker's comments.

So for any verbal communication lasting longer than a minute or so, get in the habit of taking notes.

The priority for note taking is to first look for main ideas with some supporting details, then the subideas with some supporting details, and finally, the facts to support the ideas. Again these are subjective ideas and facts that are endorsed by the speaker and may or may not be true. When you're taking notes, leave plenty of space between your words and phrases for later additions. Also, there is no need to be grammatically correct with your notes. As long as they're clear enough for you to understand what you wrote, say, a day or 2 later, it is sufficient. Notes are great memory joggers, and supporting data from the conversation may be spontaneously recalled later. When that happens, insert them into your notes.

Don't try to quote the speaker on what she said. Instead, paraphrase the essence of the idea because it takes too long to get it all down. Remember you're not trying to be a court stenographer.

Watch for road signs that the speaker is moving from one point to another. Such phrases as, "Let 's look at another issue," or "Also, we should consider," are dead giveaways that he's changing subjects. If you are concentrating, you'll easily pick up these verbal paradigm shifts. By doing so, your notes will become an outline of the verbal communication. In fact, if you're listening to a speech, a good set of listener notes should be virtually coincidental with the outline of the speech. Sometimes, to keep my listening skills sharp, I will take notes and compare them with the written version of the speech. I once did this for a Presidential State of the Union Speech and then compared it with the New York Times reprint of the speech the next day. I did it because I was brushing up to teach effective listening, and it was a chore. However, I certainly understood what Mr. Nixon said, whether I agreed with it or not.

A final point on notes: Keep them as short and concise as you can. The test is whether you understand what you wrote and the context of the issues 1 or 2 days later. If the answer is yes, you've got it just about right. If no, then you're being too skimpy.

And now I'm going to give you an imaginary listening signal that we're transferring to another communications skill. We need to consider how good communications can be applied to the advantage of the entrepreneur. We're going to look at 2 special forms of communications, both interacting with the customer. They're Total Quality Management and Quality Functional Deployment.

COMMUNICATING WITH THE CUSTOMER

There are 3 phases of communications with customers, each of them distinctly different in the manner you wish to communicate. Most often you want to be excellent at both giving and getting information. However, sometimes it's necessary to be very good at deciphering incoming information and going after more information but not giving out information. But never do you want to be good at giving information and not receiving information. The 3 phases of customer communications are:

- Communications with potential customers
- Communications with potential customers during negotiations
- Communications with customers after receipt of orders

Most of the time, we want to communicate with customers with the same effectiveness as we do internally. This means we want to practice an excellent method of giving and getting information. We will do everything in our power to eliminate all barriers to effective communications. However, there are times when this is not to our advantage. When we're nurturing a relationship with a potential customer and when we're negotiating a contract with that same customer, we need to communicate effectively but be biased in 1 direction. We want to be excellent in receiving information and at the same time be very controlled as to what information we transmit to the customer. This amounts to selective applications of the principles for removing barriers to information. These are special cases and are not the normal course of communications practices. However, they are necessary in the competitive world and the entrepreneur needs to be well versed in their techniques. What I'm about to discuss could be construed by some to be devious. It is not. It is simply using information for competitive advantage, both in receiving and releasing of information. In no way is this a license to cheat, lie, or steal. Let's look at these special cases of dealing with potential customers.

Communicating with Potential Customers and While in Negotiations

Communicating with customers while they're still potential sales and then while negotiating a sale are 2 distinct activities. But it's very difficult to tell when 1 ends and the other begins. So let's look at them as a continuum, first learning about the customer and then trying to sell him our product or service.

Communicating with potential customers is akin to the ritualistic mating dance of the saber-toothed tiger. Since none of us have ever seen this fabled beast, we have no idea what that ritual comprises. All we know is based on reputation; we don't want to get too close, we could get hurt. Communicating with customers, especially before they become customers, is similar and can cause an equal amount of pain. We don't know the ritual in its entirety; we only think we know what our role should be toward this customer, so we also instinctively don't want to get too close. But unfortunately we have to. We have to communicate as efficiently as we can to discern what the customer really needs and what he thinks of our ability to comply with his needs.

In my mind, discerning customer needs to gain an advantage in selling is the most difficult communications task there is. We are circling each other, seeking information, but for many reasons are unable to ask for or give totally complete information. We know that the communications loop (Figure 5-1) needs to be complied with, but we're afraid to let our guard down and confirm that the message has been understood or the message has been delivered as sent. We use diplomatic phrases more than we like, which means we're allowing the other party to interpret content as he likes, not as it really is. We do this just to keep from reaching closure until we're sure we've really won or lost. Why is this happening? Why can't we be straightforward and totally transparent? Because it is an adversarial situation and we are parrying to gain an advantage. In parrying, information about how the other party will react is strength. The more strength we have, the better the outcome will be for us. In business negotiations, we want to get the best deal possible for us. We want to take as much from the other guy as we can without mortally wounding him. So that means we need to learn as much about his needs and at the same time giving out as little as possible about our reciprocal needs. This may seem cutthroat, but for an entrepreneur without a large bankroll it is the only strategy that's available.

Example: You are the developer of the quark 35 razorblade machining center. It has taken your entire life's savings to fund your startup company. Your machine allows your customer to produce razorblades 100 times faster and reduces metal strip waste factors by 5 to 1. It is a breakthrough technology. You finally have secured an appointment to meet with the purchasing agents of one of the major razorblade manufacturers in the world. You want to sell machines to them. You give your sales speech and show a video of the machine-tool prototype. Here's what you want to learn as a result of the presentation:

- Is the client impressed with your technology and willing to buy now?
- Is the customer willing to buy the machine without tying you up with exclusives and not allow you to sell to his competitors?
- Is the customer willing meet your price? But you have to learn how much he is really willing to pay and whether that is below or above your price.
- Is the price you're setting optimal for you?
- Is the customer aware of your precarious financial situation?

Here are some things you do not want the client to know:
- How the technology works, to the point he can duplicate your machine himself
- The status of your patent
- Your financial staying power to resist low ball-prices
- Your ability to simultaneously negotiate with the customer's competitors
- The technical depth of your organization and how loyal they are to your company

What you want to achieve is a sale as soon as possible for the largest number of machines you can get at the highest price. And then you want to turn around and sell the same type of machine to the client's competitors. Barring that, you are willing to sell machines to this customer at a premium and refuse to sell to his competitors for a fixed period of time. You are David going up against Goliath, and you know the negotiations on their part will be an attempt to wrestle the technology away from you at the lowest price they can get it for.

The example shows ample reasons why communications is going to be difficult. Unlike the internal communications within your team, here we're using communications to ferret out information to gain a negotiating advantage. A good tool to use in this situation is Quality Functional Deployment (QFD). In Chapter 2, I explained how QFD could be used to determine whether a business team had the capability to meet a customer's needs on a technical basis. In negotiating, the same tool can be used to determine whether the customer's needs are sufficient enough for him to acquiesce to enough of your needs to create a deal, about which both sides feel they came out the winner. Let's look at how that's done.

Quality Functional Deployment as a Communications Information Gathering Tool

For discerning customer needs to match with our ability to deliver the required product or service, we need to have some cooperation from the customer. In a negotiating sense, this will not be straightforward. Let me demonstrate by first going through the theoretical and practical aspects of QFD, show where they come up short for getting information we will need for improving our negotiating position, and then how we can use the negotiation version of QFD to get closer to our needs. I'm talking about nuances here but they are very important. Remember what we learned about barriers to effective communications. We want to overcome them so the other side will give us the answers we need without even realizing it. We want them to be a partner in removing barriers to effective communications for giving us information without even knowing they're doing it. At the same time we want to minimize the information they request from us about our capabilities and intent.

The general approach to QFD:

- Look at the needs of the client company.
- Assess the client's needs with your firm's ability to solve them.
- If yes, prepare a strategy to do so:
  - Marketing and sales
  - Operations
  - Contract/pricing
- Sell the product or service to the client.
- Perform the manufacture or service.

We have to do this 3 times, first, to determine if there is a good enough match to warrant going after the client's business. Are our capabilities sufficient to successfully perform the service or make the product the customer wants? Here we're looking to see if we can construct a sellable package for the customer. If our capabilities match up well with the client's needs, we will be able to construct an enticing offer for the customer, one we hope he or she will not be able to refuse.

The second time we do QFD is to determine whether the client can and will pay the fee we want or to determine what fee we can charge that the client will pay. We only do this QFD if the client has been sufficiently enticed to want us to quote for our services or product offering.

And the last time the QFD process is done is to specifically match our current abilities with the needs of the client. This way we will know where we need to gain strengths to perform the work we've contracted to do. This third part is what I described in Chapter 2 as part of the practical training of student engineers (Figures 2-4 and 2-5).

These 3 segments of QFD have names. Applying QFD to determine whether the company can match up with the needs of the potential customer is called the *Theoretical QFD Process*. A variant of that process used in gaining competitive information during the negotiation phase is known as the *Negotiation QFD Process*. And the QFD process used to define which skills or processes may be missing in gaining optimal performance while doing the work for the customer is defined as the *Practical QFD Process*. Figure 5-2 illustrates the similarities and differences between these 3 QFD processes.

Of the 3 QFD approaches, the Practical QFD Process is the most communications-friendly. The Theoretical and Negotiation versions exclude getting involved in normal communications with the customer. You virtually have to adopt a clandestine approach: lots of market research in libraries, interviewing the client's customers, and just plain sniffing around. You're trying to discern information that the client may think of as company confidential. And the reason you're doing this is because you're trying to get a competitive edge. You can hardly ask the client what he's paying the XZY company for parts you want to sell him. Any answer you get directly may not be accurate because the customer wants to buy your product as cheaply as possible, while you want to sell it at the highest price you can get away with.

The Practical version is easier to do. It's like buying an automobile. After you purchase it, it's to both you and the dealer's advantage to want the new car to perform perfectly. In this approach we're actively working to minimize the barriers to communications. Notice in Figure 5-2 that we have the customer actively involved, while in the other 2 version, we do not chose to do so. As you would expect, this version moves along at a much more rapid pace than the other two. We don't have to surreptitiously dig for needed information, we simply ask.

You might ask why there is a Negotiation and a Theoretical version. There need not be, but I include it to bring out the differences is communications policy. Also if you look at the process steps used in the Negotiations

Theoretical QFD Process

Use:
Identify customer needs to construct a sellable package.

Process
- Identify customer desires.
- Rank these desires.
- Based on the ranking, assign a weight for each of these desires.
- Develop measurable parameters for each of these desires in direct relationship with the business team's capabilities.
- Include the highest measurement parameter into the product being offered to the customer.

Communications Policy
One Way - get information
Objective - eliminate all barriers to effective communications for incoming information.

Practical QFD Process

Use:
Identify customer needs to construct a workable package.

Process
- Identify customers' desires by customer visitations and discussions.
- Have customers rank these desires.
- Rank the business team's capabilities using "ask the experts" techniques.
- Match the project team's highest capabilities with the highest customer desires.
- By subjective evaluation, determine whether the match is high enough to warrant a concurrent-engineering approach study. If yes, gain customer concurrence and proceed.

Communications Policy
Open - same as internal
Objective - eliminate all barriers to effective communcations.

Negotiation QFD Process

Use:
Identify customer needs to extract the optimum price for providing a sellable and workable package.

Process
- Identify customer desires.
- Rank these desires.
- Rank the business team's capabilities using "ask the experts" techniques.
- Match the project team's highest capabilities with the highest customer desires.
- Include the highest measurement parameter into the product being offered to the customer.

Communications Policy
One Way - get information, actively block giving information.
Objective - eliminate all barriers to effective communcations for incoming information. Set up controls for outgoing information.

Figure 5-2. Quality functional deployment communications strategies.

version, you see that the third and fourth steps are the same as the Practical version. This is so because as we are negotiating; time is a much more significant constraint than it is when we're simply studying to determine whether there is a good enough match between our company and the targeted potential customer.

In the Theoretical approach, we're taking our time to get good data on ranking our capabilities vs. the intended need. And we're using a set of parameters that are as objective as possible to measure needs vs. capabilities. Having already done that, the Negotiating QFD exercise can cut corners by simply verifying the previous results through an "ask the expert" routine.

Communications policies for all 3 QFD versions are different. The Practical QFD communications policy is the same as we would expect for internal communications. We want to remove all the barriers to excellent communications so we can do the job for the customer as quickly as we can. Time really is money and anything we can do to minimize communications misunderstandings is money not spent. On the other hand, the purpose of the Theoretical and Negotiation versions are quite different. Here we want information for our internal use, but that information is not going to be used directly or in the near term to benefit the customer. So in both of these cases, the objective is to only eliminate barriers of communications for incoming communications. We are gathering information for intelligence purposes to prepare a competitive position vis-à-vis this potential customer. In the Negotiation QFD, our communications policy even goes 1 step further. We actively strive to filter all outgoing communications to the potential customer so he does not gain a competitive advantage against us.

Communicating with a customer can cause a company to have a split personality, especially if you have an ongoing relationship with that customer. When that happens over time, the Theoretical and Negotiation versions of QFD gradually fade away and a more open relationship develops whereby each company allows the other company to achieve some acceptable level of profit and cost controls based on discussions of their mutual needs. In fact, you see situations where they help each other lower their internal costs so they can both improve their bottom profit line. This is especially true for companies engaged in a supply-chain relationship, such as selling weapons systems to the U.S. Department of Defense. To put it another way, for companies with supply-chain relationships, only the Practical QFD is possible. The other 2 versions are totally counterproductive and should not even be attempted.

Communications with Customers After Receipt of Orders

We've seen how communications can be different at different times in the customer-vendor relationship, and we've taken it to the point of actually making the product. Now we'll move on to a powerful communications-related technique used during the performance phase. It's Total Quality Management. I hope you'll be pleased to learn that this popular "buzz word" is based on good communications practices, such as eliminating barriers to communications excellence. Its purpose is to ensure that the job being done for the customer is enhanced, not hindered, by good communications skills.

Total Quality Management as a Communications Tool

We use this tool as a communications strategy while we're in the process of performing work for a customer. Before we go forward, I want to discuss the concept of customer because it's important for the overall Total Quality Management (TQM) philosophy. There are many levels of customers, both internal and external to the company. Everyone has a cadre of customers, regardless of their job function. This is the heart of the TQM concept. Everyone has a customer and you, the vendor, has the responsibility to do the best job possible for your customer. That is TQM, pure and simple. The key is to identify your customer and make sure the customer knows you recognize him or her as your customer.

Of course we know who the ultimate customer is. That is the authority figure who pays the bill for the work your company does. But there are many other intermediate customers. They are defined as the recipient of the team's work at the next stage. They are internal and external. For example, the vendor's customer is the purchasing agent. The purchasing agent's customer is the assembler of the prototype in the laboratory. The assembler's customer is the test technician who will put the product through its paces, and so forth until the last person in the chain is the customer who exchanges his firm's cash for your company's product or service. It is vital that each intermediate customer be satisfied. To do that we have to practice communications excellence, and as I've said before, that means striving to eliminate all barriers to effective communications.

Back to the main path again. TQM requires identifying one's customer and doing right by him or her. The TQM process is a structured communications regime. It is symbolized by the TQM Triangle, as shown in Figure 5-3.

The TQM Triangle helps us focus on servicing the customer's needs. During the process of doing work for the customer (this can be ongoing or project-style work), the performer of the work needs to continuously monitor performance against the customer's needs, which means we have to know precisely what the customer needs. Even in a factory production line, needs are never static; they never remain constant. Change is always going on due to the inevitable variability of materials and labor performance. Since variability is always with us, we have to remain in constant contact with our customer to ensure that we know how much variability is acceptable and where the cutoff point is for go-no-go acceptance. We have to know what the true points of need really are. We want to be able to compare our results with the needs expressed by the customer. This way we can begin to understand variance causes and make corrections, then compare again. We repeat this cycle until there is a convergence between need and performance.

The communication part of this process is what is shown on the TQM Triangle. Notice the labels at the 3 corners: CUSTOMER, DATA, PROCESS. The

Relationship of customer, data, and process in the pursuit of continuous improvement.

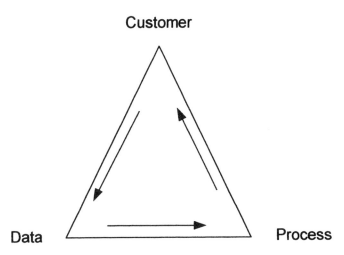

Figure 5-3. The TQM triangle *(from: Daniel T. Koenig,* Manufacturing Engineering: Principles for Optimization, *2nd ed., Taylor & Francis, Washington, 1994).*

next paragraph explains how the communications process works to direct work toward continuous improvement. It is a simple variant of the venerable scientific method: propose a hypothesis; test the hypothesis; gather data to see if the test worked; if not, modify the hypothesis by what has been learned; test again; and iterate as many times as possible until the hypothesis is satisfactory for all conditions within your spectrum of interest. By understanding this flow, you will have mastered the technique of TQM.

The communications process starts with the customer, who tells us what is required; this becomes the data. You and your team do the work to achieve the need; this is the process. The results are presented to the customer, who in turn tells our team how well the process did in meeting the requirements; this becomes the revised data. The work team revises and improves its model; this becomes the revised process. The results of the revised process is once again presented to the customer. The customer evaluates this latest effort and again tells the team how well they met the need. And this cycle continues until the convergence is satisfactory to the customer.

The key to successful TQM is practicing communications excellence. You have to make sure you are communicating effectively with your immediate customer. There can be no assumed nuances of understanding. Every bit of communications has to have singular meaning and not be open to many correct but invalid interpretations. Let me illustrate. When I say "How are you?" and you answer "I'm fine," we both have to be sincere and honest. I have to care about whether or not you are fine, like a father to a child. And you have to answer honestly and not just be making a knee jerk response. Most of us when asked that question will say they are all right, while in truth they may have a headache, be worried about a potential expensive automobile repair, and so forth. TQM communications depends on truthful answers, not on polite white lies. If the transceiver doesn't exhibit the static resistance it's suppose to, even though it may be close, the vendor needs to be told that. You cannot cover up discrepancies because the vendor tried hard, hoping that they will go away in the next round of production. They won't because there is no information, no incentive to make it better if a need isn't expressed. Cold hard truth needs to be the hallmark of TQM communications.

SUMMARY

In this chapter, I have highlighted the need for effective communications. The entrepreneur especially has to practice communications excellence, simply because inept communications cost money and add time to the development of new products, and entrepreneurs are usually short on both. I have purposely gone over the theory of good communications, including listening, so you would be enlightened to the importance of understanding what your team and customers are saying concerning your ultimate success. Sometimes this is especially hard for entrepreneurs because they are so focused on the right of their cause and their belief in the overwhelming superiority of it that they become blinded and deaf to the reactions of the rest of the world. Good communications skills are a remedy for this malady. Good communications skills contribute to making the entrepreneur a good manager, which he or she needs to be to nurture their dream into reality. Good communications always leads to more efficient operations with less recycling of work. This in turn leads to superior morale within your team, and this is the intangible you need to have to be successful.

Here are some last action thoughts about communications excellence. Practice them well.

- Preciseness in expressing ideas and giving directions is mandatory at all times.
  - Avoid using phrases that can be interpreted many different ways.
- Listen to understand.
  - Hearing is only acknowledging energy entering your auditory canal.
  - Listening is comprehending meaning from that energy.

- Compare body language to verbal language—they should agree for the communicator to be telling the truth.
- Follow up on what has been communicated.
  - Report back to the communicator about results of the communications.
- Demand feedback to make sure the receiver understands.
  - I tell, you tell me what I told you.

Chapter 6
Going from Raw Emotions to a Polished Commercial Offering; The New Product Introduction Process*

The entrepreneur is different from mere mortals in one significant manner: Persons blessed with the entrepreneurial spirit have a burning desire to see the idea of their dreams achieve the reality of commercialization, probably to make a significant profit, but that's not the key ingredient. They're more desirous of just seeing their creation become real, just like an actor dreaming to see his name on the theater marquee. It doesn't matter if the show's a hit. That's not the priority. But to have your name in lights as the headliner is a priority. The same is true of the entrepreneur. The thrill of being recognized as the inventor of the "thingamajig" or the driving force of the management process that's revolutionizing industry is the key driver for what makes entrepreneurs go. Of course in back of it all is the dream of being rich and famous, with the emphasis on the latter because we've all been brought up to expect fame to generate riches, such that we can have everything we've ever wanted from the material world. The entrepreneur dreams of fame and works with a driven intensity to make it happen. But having an idea and making it real is a perilous journey. In this chapter, we will see what it takes to make that journey and come to a happy ending.

USING THE SCIENTIFIC METHOD TO INTRODUCE A NEW PRODUCT OR SERVICE

Entrepreneurs are said to have "a fire in the belly." They have a drive to succeed with their idea. The fact that so many fail never seems to deter these dreamers. I believe approaching the idea in a rational manner is the way to minimize failure, and there is a way to do it: the engineering approach of using the tried-and-true scientific method. As you will see, the scientific method, which has so many uses in management as well as scientific pursuits, is the basis once more of nurturing an idea to reality.

---

* Note: this chapter is based in part on the author's paper "Introducing New Products" printed in *Mechanical Engineering* magazine, August 1997, published by ASME International.

1. Make Observations

2. Develop a Hypothesis

3. Test the hypothesis

4. Make Revisions to the Hypothesis Based on the Test

5. Test the Revised Hypothesis

6. Iterate Steps 1 Through 5 as Necessary to Reach a Workable Conclusion

Figure 6-1. The scientific method.

The scientific method applied to a new product goes something like this. It starts by requiring the development of a hypothesis, then a to test it to see if the hypothesis is correct. If by chance (like 1 in a million) it is, the game's over and the idea is commercialized immediately. Most likely the idea isn't sufficient and data coming back indicates areas of weakness, or even of downright failure. This data is the basis for a revised hypothesis. We build modification 1, which again tested. If these results are better than the first try, then we're making progress. It depends only on how good it has to be before we say it's good enough. As you can surmise, developing a new product or idea, no matter how good it seemed originally, takes considerable effort and discipline. Like most other factors in a successful business, it just doesn't drift into being. It takes objectivity, coupled with diligent attention to details to reach a successful conclusion. Let's start that journey.

REASON FOR A BUSINESS—A GREAT IDEA THAT CAN BE COMMERCIALIZED

A great idea may have tremendous benefits for profit, but only if it has the capability to be commercialized. A product can become a commercial success when it can be produced in large enough quantities and can be sold to those desiring it at a price they can afford. Let me illustrate with an example:

> **Example:** Here is a goal for a product: Create an automobile that can get 200 miles per gallon, would be safe in a 100-mile-per-hour crash, is plush and comfortable, is capable of accelerating from 0 to 60 mph in 3 seconds, and can cruise at highway speeds for a range of 1000 miles before needing to be refueled. Add to that, that it be totally environmentally friendly. All these are great virtues and you would think the automobile would be instantly successful, but only if it's affordable. Let's say that such a vehicle could be built but would cost $1,000,000 per automobile.

Is it commercially viable? No, of course not. That's the difference—it's a great idea but not marketable. A business needs to have a great idea combined with the ability to commercialize the product.

Let's see how we get to the commercialization level. The following list is a general chronological order that steps must happen for an idea to be commercialized.

1. Idea is conceived that is technically feasible.
2. Idea is accepted by the intended customer base.
3. We have ability to produce it.
4. We have funds to produce it.
5. Product is priced at a level that will generate sales.
6. Revenues are received that are greater than costs to produce.
7. Resources are generated for further development of the idea.
8. Product can Become Commonly Known to Customers

If the idea fails for any reason at any point on the chronological list, it is not a marketable idea. A business is established to manage commercial ideas for profit. We discussed previously what constitutes profit and all of those comments apply. So the idea the entrepreneur possesses needs to pass this test to be a successful candidate for a business. If it fails at any step, the entrepreneur has to overcome that failure point and go on. This is where the work of the entrepreneur comes about. If the entrepreneur has the "fire in the belly" to proceed, he or she must clear this hurdle to be successful.

How do we prove commercial viability? If there were a simple formula to apply, life would be beautiful and all of our fondest dreams would be realized. The fact is, there is no way to prove commercial viability beforehand but there is a process that gives us a much better than happenstance chance of being successful. The chronological list is an indication of what that process is. We know what we have to do to get from an idea to a marketable product. We have to attack that list through the 3 aspects of any business: marketing, operations, and finance, by having the functions pay attention to meeting all 7 steps of the manufacturing system successfully as applied to the proposed product. When we get done, we will have completed as thorough a job as possible to discern whether the product can be successfully commercialized. This methodology is known as the New Product Introduction process, commonly called NPI.

But even the NPI process doesn't start in a vacuum. The entrepreneur must have some initial indication that her idea is commercially viable. Usually this comes about because the person with the idea has been vexed by not being able to do something. It can be as mundane as a not being able to haul in dried laundry off a clothesline because a pulley froze. Some mechanical genius could have been driven to develop a sealed bearing pulley that would never freeze. Now is that commercially viable? We know it is and we know history has shown that there are applications way beyond that of clotheslines.

The message is that commercial viability probability needs to be almost intuitively known at the very beginning, but with a very big "but." We still have to comply with the chronological list. Even though we know the idea is sellable, it has to be marketed in a manner that makes other people agree with us so they'll spend money for it.

There are lots of ideas that are great and when we see them we say, "Gee that's neat!" But would we part with our hard-earned cash to have it? Have you ever gone into stores like The Sharper Image? In my mind, they are retail product venture capitalists. They fund and sell development of entrepreneurial gadgets, the better mouse traps. They obviously put forth all sorts of nice ideas that they hope will attract buyers because they're so intriguing. And like venture capitalists, they would be supremely satisfied if 1 in 10 of the items they sell are satisfactorily profitable to the point that they recover the initial investment and other failed products plus something left over to put in their pockets. They know that 1 hit can cover the cost of hundreds of failures. Here we have great ideas that definitely provide value to the buyer but is it enough to successfully complete the chronological list? Maybe, maybe not. The point is that the start of an NPI has to be based on something that will at least get us this far. Keep in mind that most products The Sharper Image's of the world offer are flashes in the pan and are not very successful products. They have no staying power and do not become available in multiple venues. The NPI process helps us get beyond this point. It helps us evaluate whether or not the product can reach the multiple venue stage, ensuring staying power and going beyond the gadget level. I don't believe gadget-makers are really profitable businesses for the long haul and the dream of fame and fortune is quickly dashed. Let's now look at the NPI process.

THE NEW PRODUCT INTRODUCTION PROCESS PHILOSOPHY

The most challenging task the entrepreneur will take on is to maintain the discipline of objectivity during the NPI process. This is equally true for more established companies trying to grow their businesses through the introduction of additional products. Perhaps I should qualify a point here. Entrepreneurship is often associated with startup companies—those that are introducing a new product or service for the first time from a zero baseline; it is equally applicable to a company that already has a product and is expanding. The entrepreneur time line is:

    Entrepreneur—new startup company
    Entrepreneur—new company
    Entrepreneur—aggressive small company
    Entrepreneur—aggressive company
    Intrepreneur—larger company

Somewhere along the line we go from entrepreneur to intrepreneur in the larger company. Where that is, is hard to say. The point is that the NPI process is the same where ever we are on this developmental time line.

NPI demands an accurate look at the marketplace the company serves or would like to serve. It demands an objective evaluation of where the company is in that market with its customers and the true strength of its engineering and manufacturing abilities. In addition, NPI demands a strong commitment to continuous improvement. Finally, those firms fortunate enough to have current products must attend to the well-being of these products while they develop their new products. There must be a strong resolve to dedicate a multifunctional team to the development of the new product, even though the payback period may not be within the current fiscal year.

Because NPI can take considerable time, it is necessary to use project-management techniques to ensure all the activities are carried out as quickly as possible. Using project management, we can monitor all steps for status and take corrective actions as required. We also know that smaller companies, out of necessity, require their employees to carry out more than 1 task; they wear many hats. The engineer may also be the shop supervisor, etc. This is an even more powerful reason for using project management to control the activity.

Concurrent vs. a serial approach must be evaluated. Since time is of the essence for NPI, especially for startup companies, a concurrent schedule for the tasks is preferred over a series schedule. As a general rule, a series schedule should be used only if there are predecessor and successor steps necessary. An example would be if a design must be complete before a machine is ordered, then a predecessor serial schedule is necessary. However, if the design will not affect the machine selection, then a predecessor serial schedule isn't necessary.

NPI is synonymous with change, and successful change requires people to accept risk and be confident in their abilities. This is normally the case with startup companies but begins to wane as companies mature. Here an individual's security may be as important as potential rewards for a successful NPI. When staffing an NPI project, it is important that the cross-functional team be willing to take risks. Therefore, it behooves the entrepreneur manager or owner to make sure his or her team is cognizant of the risks of failure and that the reward justifies the level of risk. Coupled with the ability to accept risk is the competency of the members of the team. They must be at the top of their game to ensure the best efforts are put forth to make the NPI a success.

Like other project-management-controlled activities, each step will be led by the team member best qualified to do it, as we shall soon see when we describe the steps. However, unlike other projects, NPI uses virtually all team members for each step. This means that although the design review activity may be led by the design engineer, we would expect the human resources representative also to be involved, offering perspectives on how to obtain the necessary skills needed to do the job. So while active leadership may vary from step to step, the entire team is involved in all steps. It is remarkable how many times a different perspective can aid in making a better decision. One that comes to mind in my experiences was the color of cosmetics sales cases for upscale department stores. The color of the final fixtures was a marketing-led step. They favored gold leaf—although it was more expensive than gold paint, it was acceptable to the customer. However, what wouldn't be acceptable to the customer was wear from contact with retail customers as they leaned on the counter to look at products. I demonstrated that the gold leaf would wear off at a rate 10 to 12 times faster than gold paint. Of course my main concern was the manufacturing problem of applying gold leaf and having it stay secured on the product during fabrication. I knew it would be a costly problem that probably would cause significant production delays. By pointing this out, coupled with the poor product performance at the customer's stores, I was able to get the decision changed. This is an example of a manufacturing representative on the team driving a marketing decision.

NPI teams do have officially assigned leaders. In fact, the entrepreneur or owner is usually the leader in a startup company NPI. Leadership responsibility may vary from NPI to NPI, depending on the project. The important factor, however, is that the leader act as a facilitator rather than as the ultimate decision-maker to the greatest degree possible. This makes it more plausible for consensus to occur, which is necessary if buy-in for the ultimate project results will be strong enough for a successful project launch. Keep in mind that these same people doing the NPI will ultimately be responsible for initiating its commercial success. To be successful, they and all the team members must be fully committed to the product.

THE NEW PRODUCT INTRODUCTION PROCESS TECHNIQUE

Now let's look at the mechanics of the NPI process. Remember, I said it's an application of the 7 steps of the manufacturing system being done by the functions of the company. So as would be expected, it is derived from the 7 steps. As a refresher, Figure 6-2 lists the 7 steps.

Again, please note that the 7 steps make up the entirety of the business process. Companies that recognize this truism usually are efficient, hence profitable. They understand that to do well, their specific business system

1. Obtain product specification.

2. Design a method for producing the product, including the design and purchase of equipment and processes to produce, if required.

3. Schedule to produce.

4. Purchase raw materials in accordance with the schedule.

5. Produce in the factory.

6. Monitor results for technical compliance and cost control.

7. Ship the completed product to the customer.

Figure 6-2. The 7 steps of the manufacturing system.

needs to account for each of the steps. Therefore, they recognize that it is in their best interest that they do each step as best as they can. Conversely, companies that are oblivious to the existence of this natural law of business flounder from pillar to post and are usually in a state of crisis one way or another.

Figure 6-3 shows the nominal 11 steps of the NPI process and shows which of the 7 steps they were derived from. We will go into considerable detail of what happens at each of these 11 junctures of NPI.

Another way to look at the NPI process is shown in Figure 6-4.

The 11 steps need to be performed in a logical manner, as described in Figure 6-4. The purpose is to make sure all avenues are explored to ensure that the new product is commercially viable before larger sums of money are expended to produce it in quantity. Note also in Figure 6-4 that all 3 aspects of a company's organization are employed: Marketing, Operations, and Financing. Let's go through each of the 11 steps in chronological order, as shown in Figures 6-3 and 6-4. Keep in mind that the idea for a product must meet all 11 steps if it is to be a successful product. Failure at any step of the analysis literally requires going back to the drawing board, that is, starting over again. This is a very pragmatic approach for entrepreneurs because they normally have limited funding and need to minimize risks before expending their precious resources. Let's look in detail at the context and content performed during each of the NPI steps.

1. Define Customer Requirements

Defining what the customer needs is a 2-part activity: opportunities identification and a corresponding technology capability analysis. The former is most often led by the marketing member of the team, whereas the latter is more often than not led by engineering.

Opportunities identification involves gaining an intensive understanding of the market the entrepreneur wishes to compete in and ways to uncover selling opportunities. Often the entrepreneur, and the marketing team member if the role is not being filled by the entrepreneur, will have a good feeling for the market situation. This is usually the case because the idea for the entrepreneur's "better mousetrap" is due to his or her familiarity with shortcomings of the current product offerings servicing the market. It is up to the entrepreneur and the team to make sure that they are really looking at the market opportunities for their new product objectively and not through overly optimistic hues generated through rose-colored glasses.

Technology capability to produce the product needs to be done as soon as an idea for a product is generated. Granted it takes creativity to come up with a product idea. But I think it takes equal creativity, but perhaps of a different kind, to develop a way to bring the idea from concept to a plausible plan for how the product can be made, particularly in the facilities the company may own or have access to.

In the very beginning of the NPI process, we purposely go through a process to determine how a proposed product can be made. If we can't figure

1. Define customer requirements. (1)*
    a. Opportunity identification.
    b. Technology capability analysis.
2. Determine how these requirements can be met. (1, 2)
    a. Product concept and definition.
    b. Product specification.
3. Produce engineering drawings and instructions. (2)
    Design specifications.
4. Define a method for manufacturing the product. (2, 4)
    Prototype build and test.
5. Develop a bill of materials (BOM) and routing, and determine total costs (materials, labor, and overhead). (3)
6. Evaluate capacity to make this product with relations to all other products requiring the same capabilities (time phased capacity analysis). (2, 3, 6)
7. Compare capacity (delivery date to the customer) with the needs of the customer and see if it is suitable. (6, 7)
8. Determine the required margin, calculate profit (or loss). (1, 6)
9. Take corrective actions to achieve the required margins and meet customer needs. (2, 3, 4, 5, 6, 7)
10. Release for manufacturing after all internal and external requirements are determined to be achievable. (5, 7)
    Volume manufacturing.
11. Monitor production and sales progress, take corrective action as required, to the intent of item 9. (6)

* Governing step(s) of the 7 steps of the manufacturing system.

Figure 6-3. The new product introduction system.

out how to make the product, there is no reason to continue. This seems so obvious, but countless sums of money are wasted every day on schemes that have no chance of reaching commercial viability because virtually no thought is given to how to translate a lab bench device into a real product.

The methodology for New Product Introduction (NPI) must be
* Thorough but non-complex.
* Simple to follow.
* Specific enough so we know when a step has been completed.

The basic steps are as follows and are done as far as possible in chronological order. The steps are all completed in a concurrent team approach. For clarity purposes, the lead function is listed.

Step	Lead Function
1. Define customer specification.	Marketing
2. Determine how those requirements can be met.	Design Engineering
3. Produce drawings/instructions.	Design Engineering
4. Define a method for manufacturing the product.	Manufacturing Engineering
5. Develop a BOM and route, and determine total costs (materials, labor, and overhead).	Manufacturing Engineering
6. Evaluate the capacity to make this product with relation to all other products requiring the same capabilities (time phased capacity analysis).	Manufacturing Engineering
7. Compare capacity (delivery date to customer) with the needs of the customer and see if suitable.	Marketing
8. Determine required margin, calculate profit (or loss).	Finance
9. Take corrective actions to achieve the required margin and meet customer requirements. Iterate steps 1 - 8 as required.	All
10. Release for manufacturing after all internal and external requirements are determined to be achievable.	Manufacturing Engineering
11. Monitor production and sales progress; take corrective action as required to meet the intent of item 9.	All

In essence, the classic steps of the manufacturing system have to be complied with. If they are, the company will not be in danger of releasing product for manufacture that cannot be made or sold for a satisfactory profit.

Figure 6-4. Generic plan for new product introduction.

The idea may have been great in a laboratory demonstration, but the yield may have been so low for good parts vs. rejects that the cost is simply prohibitive. Yields in semiconductors have been chronically prone to this problem. Designing a chip with many orders of magnitude more circuit density is usually many product generations ahead of reality. Therefore, the need for technological capability evaluations is critical for NPI in the semiconductor field. This need is also very much in need for virtually any product where manufacturing capability is more of an art form than a science.

Unfortunately, most advanced technology products fall into this category. Many required processes that have been newly invented require significant trial-and-error development time to get the process yield up to levels that are commercially viable. This is one reason why new products, when first released to the market, usually cost more than when they have been available for a while. Obviously part of this is due to product competition. When competitors come into the market, the laws of supply and demand will lower the prices. However, in addition to this, a large part of the price drop is due to a maturing of the manufacturing process. It becomes more robust, thus generates higher yields. This is necessary for the true competitive nature of business. We lower our prices because our cost are going down due to higher productivity so we can lower the price to put more pressure on the competitors.

Understanding our ability to make a new product, even though we are certain it would be a marketing success, is critical to ensuring that it will also be a commercial success. This is absolutely critical for the success of the entrepreneur's business venture. For that reason, it is the first step of the NPI process. This assessment cannot just be a superficial one; it must be thorough. So all types of product-offering opportunities must be evaluated. This ranges from skills of the workforce to size ranges of equipment the company has access to. The more intense the review is at this point, the more successful the eventual implementation will be. This is definitely not the time to be a full-fledged champion of the new idea at the expense of reality vis-à-vis the ability to produce it.

Along with the capability to make the product there must also be a realistic vision as to its desirability from the potential customer's viewpoint. Often an entrepreneur's idea, which he or she has fallen in love with, is met with indifference by the marketplace. So the dual aspects of this step need to be truly achieved if we are to have a successful product launch.

2. Determine How Customer Requirements Can Be Met

Design engineering and marketing are the natural choices for leadership roles in this phase. They will use tools such as brainstorming and quality functional deployment techniques to match company capabilities with the perceived market needs. The goal of this step is to flesh out the plan of how the product will be made within the factory the company owns or has access to.

It is not enough to just design the product. It is necessary to also design the process for making the product. This is where small companies may have an advantage over larger companies—where tasks are more compartmentalized. If a design engineer does not have to make his own creation, he or she is more apt to gloss over the methodologies needed to manufacture it. On the other hand, if the designer is also the fabricator, he or she is going to be more attuned to the needs of manufacturing to get the yield necessary to be commercially viable. Machining tolerances, how things are assembled, and packaging requirements are all going to be more compatible with the true capabilities of the factory.

This means there must be an objective assessment of product concept and definition that is compatible with current company manufacturing capabilities. This includes the ranges of what its facilities are capable of and the rate at which it operates. The latter is extremely important. For example, making 1 computer on a bench top is far different than making 1 computer every 15 minutes off a production line. The rate of production sets the volume available for sale and its sale price. The rate has to be sufficient to satisfy the market need and the price must be at a level to attract enough buyers to cover all expenses and generate a profit.

So the concept of the product and the definition of how it will be produced is critical. Once that is done, we can provide a set of specifications covering materials used, geometry, chronological assembly requirements, and performance characteristics. All but the last item really defines the factory that it will be built in. The last item defines the market need it will satisfy.

Note how closely the factory of origin needs to match the market it will satisfy. Any idea that will require physical manipulation must be totally compatible with the factory it is intended for. This is sometimes ignored, much to the dismay of the entrepreneur. I am aware of computer-chip manufacturers who depended on others to make devices to cool the chip in operation. They designed a wonderful chip capable of working at an extremely fast rate but they failed to understand the heat-transfer needs of such a chip. The design of the heat sink was inadequate, primarily because it was designed under the assumption it could be made in a vendor's factory while in truth it couldn't. The designers committed an unpardonable sin. Since they weren't going to make the heat sinks, they failed to adequately design them. They didn't understand the mechanical engineering technology and thought it was a trivial pursuit that could be handled by the vendor. They left that element of design up to the vendor and didn't check on it until it was too late. The point is, the total design must be totally compatible with all of the factories that will be working on the product. This oversight cost the chip maker over 1 year of delay in introducing the chip, and I believe a good share of the market.

Creating the proper specification is exceedingly important. And the sooner it's done, the better. With a specification in hand, the marketing representative can test to see if it meets the perceived needs of the market. Some would argue that specifications shouldn't be finalized before enough prospecting has been done to ensure maximum potential business is lined up. This is a valid argument. So I would suggest that specification finalization and prospecting be done in parallel as much as possible. But keep in mind there could be some potential deal killers if a product specification doesn't exist.

Example: A manufacturer has the potential to sell some motors. The company has the required technology, and the team is excited about the prospect before a specification is defined. Then the specification work reveals that ISO 9001 certification is required, which the company is 6 months to 1 year away from obtaining. The window of opportunity to get the sale is only 3 months. Without a product specification in existence, a lot of effort could have been wasted.

Creating a product specification also forces a definition of what the company will offer to the marketplace. This minimizes the chance of making an offer that will be difficult to fulfill. It also tends to preclude misunderstandings between the company and its customers as to what constitutes successful delivery of the product. This last point is especially critical for a startup company trying to build a reputation. Too often entrepreneurs over-promise because they are supremely confident in their "wonder product" and tend to downplay reasons for less than stellar performance. By having a proven specification in place, the entrepreneur is checked from portraying unbridled enthusiasm for his or her product. This keeps the entrepreneur out of trouble and truthfully portrays what the product will do. Customers buying the product on the basis of its specification will be happy with it if it meets the specification criteria. And that's what the startup company needs to do—make its customers happy.

3. Produce Engineering Drawings and Instructions

The abstract phases are complete and now the team needs to produce a first iteration of a design that is relatively producible in the factory. I use the adjective "relatively" because it will still need to be tried. Simulation may avoid some trial and error, but real manufacturing with real production equipment is still needed to test the process. The production of engineering drawings and instructions will be done concurrently with the next step—designing a method of manufacturing.

As with the other steps, all team members need to get involved. Don't expect the design engineer to have all the answers on making the design as good as it can be. Remember, a design engineer understands physics and chemistry and can do the engineering calculations to ensure that the product will do as it is intended to. But he doesn't have an exclusive on ideas. Ideas come from everyone. The manufacturing people certainly can contribute ideas that reduces the time of manufacture and makes it simpler to do. The marketing people certainly will be able to relate what features are desirable from the user's viewpoint. Finance will certainly be championing anything that reduces costs. Human resources will probably add ideas that make the product easier to make so their burden of finding skilled operators is eased. The point I'm making is that good ideas come from many sources if we let them, and we must. But since the engineering types will have to put all of the design data together in a useful format, they typically lead this and the methods development efforts.

Preplanning here can minimize the iterations required to get a mature design. One difficulty may be that there may not yet be a firm customer. Unless the product is universal, that is, used exactly the same way by everyone and with the same exact geometry, the final nuance of design will be influenced by the intended user. So it is incumbent on the team to try to get a buyer for the prototype run.

4. Define a Method for Manufacturing the Product

The first task will be to create a road map for making the product and then build a prototype using that method. Here the team will be heavily influenced by the capabilities of the company—what type of equipment and strengths it possesses. For the entrepreneurial startup, the questions become what type of equipment can we afford to buy and is its delivery time within the time constraint for the product offering? The entrepreneur also has more leeway in selecting processes because he or she is not encumbered with pre-existing equipment.

Whether it is a startup or an existing company expanding its product offering, the need to follow the core strengths the company possesses is paramount. A company that is strong in metalworking would be-ill-served in trying to use plastic as a base material. This would be true even if plastic were cheaper than metal. Even if plastic also had a sales advantage that metal had to be overcome it would still be a bad choice because of the learning curve the company would have to pursue in order to develop competency. It is always better to produce a product of an alternative material of which you have superior ability and knowledge than to go with another choice. The unfamiliarity would more than likely lead to significant quality problems that could doom the new product before it could get a sufficient customer following. The only time a new technology or material usage should be considered is if the product introduction lead time is sufficiently long to gain competency with it.

The development of methods for producing the product is a concurrent activity with the design process (see my books, *Manufacturing Engineering, Principles for Optimization,* 2nd edition, Taylor & Francis, Wash., DC, 1994, and *Fundamental of Shop Operations Management, Work Station Dynamics,* ASME Press, NY, NY, 2000, for details of the concurrent engineering processes and methods development). Continuous checking and iteration with the design development is needed to maximize optimum producibility. When building the prototype, it is vital that whereever possible, production equipment be used. For production runs it is very comforting to know that the methods required were actually proven beforehand. Doing otherwise could lead to some disappointing surprises. A prototype design and manufactured product must represent the features of the production run product; otherwise it isn't worth too much for the well-being of the business.

5. Develop a Bill of Materials (BOM) and Routing, and Determine Total Costs (Materials, Labor, and Overhead)

This is the stage of testing economic viability for full production, led by manufacturing engineering, materials, and shop operations representatives. The prototype is structured for manufacture, and the necessary vendor supply chain relationships are set up. (Details of how BOMs and routings are developed are discussed in my 2 books mentioned earlier.)

To compress cycle time to market, the entire logistics system must be developed and be capable of producing the product even if the prototype may not be the same as full production models. There are lessons to be learned that are still valid. The exercising of the supply chain will strengthen the final design through feedback about what works best. This experience will also validate the cost estimates. While volume production will probably be less expensive, the actual costs for the prototypes, factored for reductions due to volume material purchases and operator learning curves, yield a very good cost model.

Some practical advice on BOMs for the startup entrepreneur: Keep them simple. A BOM should be of the indented variety and be as flat a structure as possible. The highest level should be the completed product. The next level down should be the major assemblies. The third level is the subassemblies that make up the major assemblies. Finally, the lowest level is the individual parts that make up the subassemblies. The BOM should represent how the product is put together. All items should be shown, even though there are no new parts being added at some assembly activities. Also, the BOM should show units of measure for how parts are purchased. And since not all parts are purchased, the BOM should show if a part or assembly is a make or buy item. Many parts are made from previously arrived raw materials so they are listed as make items.

Some advice on routes: They too should follow the "goes into" pattern. A route needs to show the trip that the raw material goes through to become a finished part or assembly. It shows the operation number, what happens at that operation, the materials used, and the amount of time it should take to do the operation. Sometimes it will also list the quality requirements to be met at specific operations. Again, the key is to list information that is necessary to do the job, no more.

6. Evaluate the Company's Capacity to Make the Product

Capacity is an abbreviated way of saying, how many of the *item(s)* we want to produce can be done over a specified period of time? Notice I've underlined the word "item" and also indicate it could be plural. For a new company with its first-ever product, "item" is singular. However, for an existing company that is adding to the list of product offerings, "item" is plural. The reason we must know this is to make sure that when we calculate capacity, we're not inadvertently "borrowing" facilities that are used to produce existing products and at the same time assigning the new product to them. Unless you know how to modify the laws of physics, it is impossible to have a facility make 2 of anything at the same place at the same time. So we are forced to look at the rate of production of the new product along with the existing products.

In this step, we use the method for making the products along with the route structure to define the times for each step. It is a task compilation activity to determine the rate of output at each workstation. Due to the technical nature of this step, it is most often done by the manufacturing engineer member of the team (again, keep in mind this may be a multi- individual, especially for a startup company). The outcome of capacity is critical because it sets the number of product units the company can sell over a period of time. That number, multiplied by the selling price, sets the upper limit of sales to be expected. If this number is not sufficient to justify going forward with the project, then the project ought to be stopped. No matter how laudable a new product is, if the company cannot see its way clear to making a profit, it must be dropped.

The process needs to take into account this new product and the existing products requiring all or some of the same resources. Let's look at an example to demonstrate the typical thought processes and decisions that must be made at this point in the NPI process.

Example: Suppose we are a small manufacturer of plumbing supplies such as pipe fittings and valves, some made of plastic extrusions, some from cast iron, and some from stainless steel. Let's say we have a new revolutionary design for a dripless valve. The design promises to be a tremendous sales leader for our product line. We find that we have extra machine capacity to apply 3 full shifts of manufacturing, 1 machine each shift (we have a 3-shift factory) except for making the ball valve seat. That takes precision machining on a CNC milling machine to get the contour required. That machine is used to the extent that we have only one half of a shift (4 hours) available to dedicate to this new product. However, it is a very productive machine. In 4 hours, including setup and teardown, we can make 100 ball valve seats. This means we can produce 500 during a normal workweek. All other facilities can make significantly more components during the same workweek period. In fact, the least productive machine can make 5000 component units for the new dripless valve during that period of time. So the bottleneck is the CNC milling machine

Here is the first question: How many ball valve seats do we need to have to satisfy the NPI plan? If the answer is less than 500 per week, we're okay. Proceed to set up the factory to accept the new product. If the answer is that we need less than 1000 per week, then perhaps we need to look at running the bottleneck machine on Saturdays, exclusively dedicated to the ball valve seats. If we can make 100 in 4 hours, simple arithmetic tells us we can make 200 in 8 hours and 600 in 3 shifts. It would be costly but we could get upward of 1100 in a week of production. If, however, the answer is that we need more than 1100 ball valve seats and probably closer to 5000, then we have no immediate solution other than to find more capacity somewhere else, or bring in more capacity.

This leads to the second question: How much time do we have to meet the target output rate? Suppose we feel that our toughest competitor may be only slightly behind us in developing a similar product, and if we don't reach the market first, our product will not be readily accepted by the industry, if at all. We feel this is true because once a standard is set, its very difficult to "unseat" it, like Microsoft's Windows vs. IBM's OS2 and Apple's Mac OS X. It really doesn't matter which one is technically superior; what matters, primarily, is who got there first to set the standard. Now understanding this predicament, marketing has to tell manufacturing engineering what the window of opportunity is. If it's 6 months, the actions taken will be different than if its 18 months. Let's say it's 6 months and we need 2000 units per week to create a lock on the standardization issue. This eliminates the possibility of purchasing more CNC milling machines. It would take 9 months minimum to buy, have delivered, and put into production. That means the manufacturing engineers would join forces with the purchasing team member to find suitable machining farmout sources. Of course this poses other logistics problems and cost problems, but it is a solution.

This leads to a third question: Are there any other ways of making the parts that would incorporate the company's capabilities to reach the desired capacities? This is the thinking-out-of-the-box scenario. Remember, the company has plastic extrusion capabilities. Usually the extrusion rates are far greater than machining rates. So if the design could accept a plastic ball valve seat, we would be home free. It's utilization would not compete with equipment being used to machine cast iron and stainless steel. The drawback would be the length of time to make the extrusion dies, and are there enough extrusion presses available to not compete with other company product? Similar questions apply to what questions were asked with the metalworking facilities. We would also have to know if the design of a plastic ball valve seat is compatible with a stainless-steel body. So here the manufacturing engineer and the design engineer need to do producibility reviews.

The example shows we have an iterative process that could involve the entire team. We need to know how much of the new product we need and by when. We then try to find solutions and see if the company can live with the solutions available. We would go with the one we have the most confidence in doing successfully. If none of the solutions instill sufficient confidence, the project ought to be abandoned. Abandoning a project is a hard thing to do because so many egos are involved. But for the ultimate benefit of the company, the decision must be based on the ability of the company to be successful. Let me give another admonition: It's very easy to become emotionally involved with a project. In fact if you didn't, then the compassion to be successful would be lacking. What we call "fire in the belly" would be missing. Having this attachment is a good thing, to a point. But when it clouds the ability to make objective decisions about any given situation, then it becomes detrimental. You must believe in your project to have any chance at all to succeed. But your beliefs must be based on real-world facts, not through the world as viewed through rose-colored glasses.

7. Compare Capacity With The Need Dates Of The Customer

This is usually done when we do the internal capacity evaluation (see previous step). The new product needs to be introduced in a timely manner to take advantage of the market opportunity. All too often, timing of the product introduction is not considered as carefully as it should be. The entrepreneur and his or her staff are so caught up with making the product a reality that they dismiss the fact that the market has to be ready for it before it will be commercially acceptable. Have you ever heard the phrase "He was before his time"? What that implies is that the innovation or new product was a work of genius but it couldn't be accepted by the public because the public didn't have the knowledge base to accept it. This is why products go through the typical "S" of inception to demise. A product must have achieved a critical mass level of acceptance by the public before it can be commercially acceptable.

At different levels of the "S" curve, different capacities need to be available for providing product to customers. In the earliest stage, only prototypes need be available for demonstration purposes. As the product gains acceptance by potential customers, the ramp up of product availability begins. The rate of change of the ramp up depends on 2 things:

1. The degree to which the customers want the product
2. The capacity of the company to supply the product

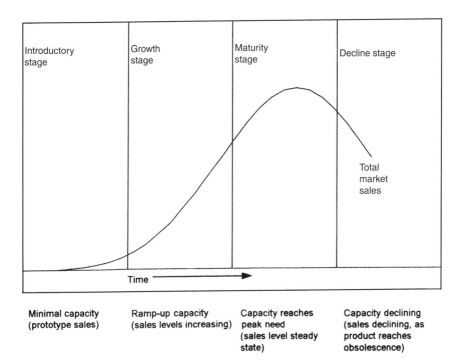

Figure 6-5. Product life cycle "S" curve.

Neither of these issues are permanent. The degree to which customers want your product depends on the type of need it fulfills. If it is absolutely critical, for example, it cures cancer, then the amount of time that the degree of want is high remains for a long period of time. If there is a real clamor for the product, then customers are willing to be patient to wait for capacity to reach par with demand. If the product is a fad rather than a relatively stable need, then the demand will drop off quickly, even though there was a spike in need for higher capacity. Such things as hulahoops would fall into this category, as would last year's "must have" Christmas toy.

We see that capacity is more than the available physical equipment. It is also people and their emotional needs for your product. At the very beginning of the "S" curve, the company must be engaged in research to determine the desirability of its product. Just because the entrepreneurial team believes totally in the importance and worth of their product doesn't mean the general public will. Knowing the level of initial desirability sets the initial capacity needs. Having a marketing and sales plan to grow that desirability establishes the rate at which capacity needs increase as the product reaches maturity. Having an idea about when the desirability of the product will wane will determine the capacity deceleration needs, as will happen with any "S" curve-type product. All except basic commodities products follow the "S" curve path.

Keying customer needs to capacity planning is a critical factor in the success of any NPI. Any issue that prevents shipment of the required quantity of product at any time during the product's "S" curve life cycle will reduce the amount of profit downward from the theoretical optimum. And unfortunately it is rare for that lost ground to be recovered. Therefore, capacity is a critical team issue that needs to be addressed from a marketing and sales viewpoint just as strongly as the technical manufacturing and engineering issues.

8. Determine the Margin and Calculate Profits

Notice I don't say "profits" or "losses" because it goes without further elaboration that if the product results in losses, it is not worthy of further consideration. Of course it is permissible for a product to be in a loss position for a period of time. But a product must show a profit, and it must be sustainable, after a reasonable period of time for it to be viable.

What is viable and what isn't depends on the company's financial position. Entrepreneurs most often work on a shorter turnaround to profitability than an established company. The established company has deep pockets. It has cash reserves to allow a new product to be in the red longer than the typical entrepreneur can allow. This is why in folklore and drama we see the erstwhile entrepreneur scrambling for cash from venture capitalists and angel investors to keep the entrepreneur's dream alive. This makes for good entertainment. But believe me, it is really a drain on the creative juices of the entrepreneur and his or her colleagues. The way to avoid this is have a well-thought-out NPI, covering all the steps we've discussed. Then have the necessary cash reserves available to deploy, just as a general deploys his reserve troops as needed.

How do we calculate the required margin and calculate the profit? It's an iterative activity of inputting financial data based on the NPI planning process and seeing where the bottom-line net profit is likely to be. If it is unsatisfactory, as it often is in the first passes of this exercise, adjustments are made and then tested to see if the financial adjustments can be made in reality. This iterative process continues until the NPI-induced planning can create a financial picture that gives satisfactory results.

The basic tool is the classic pro-forma profit and loss (P/L) statement. The basic line items of a P/L statement are shown in Figure 6-6. The difference between pro-forma P/L and a normal P/L statement is that pro-forma is an estimate, whereas the P/L is normally done after the results are in; this means it's done after an accounting period has been completed and results of sales and expenses are known. To complete this portion of the NPI process, we have each member of the team predict what they think the financial inputs will be for their specific area. The data is then inputted into the pro-forma P/L by the financial member of the team, who then calculates net profit. Based on the results, other members of the team are asked to look at their areas of responsibility to see if changes can be made that will lead to further improvements. In this iterative manner, the optimization process takes place. Let's look at an example.

Example: Suppose we have an idea to introduce a new electronics storage device that's based on the entrepreneur owner's new patented application that makes using CD RW disks as easy to use as floppy disks. We have gone through all the NPI steps and are now ready to evaluate our P/L status. To do so, we'll need to input data to a pro-forma P/L form set up to show what the sales and expenses will be and the resulting profits or losses for several reporting periods. Since this is a new product, we know that the engineering R&D, manufacturing production prototypes, and marketing efforts will be extensive for a considerable period of time before sales can catch up and cover these initial costs. So we know there will be periods of losses before we experience periods of profits. Therefore, the pro-forma P/L will cover several accounting periods. Let's say our seat of the pants estimate is that the startup will cover a year before we see some profits. So we will structure a pro-forma P/L for a 2-year period as a first iteration. Now we will input numbers to our form. The basic pro-forma P/L form we'll use is shown in Figure 6-7.

In this version of a P/L, we see space for 5 different forecast periods. This differs from a normal P/L, which would usually only show 1 period actual and a comparison column for the budget or prediction. The pro-forma is entirely a prediction. Note also that the accounts are the same for both the

Account description	For period: Date:		
	Period budget	Period actual	Period variance
Income			
Sales			
Products			
Other			
Total sales			
Cost of goods sold			
Direct labor			
Direct material			
Total COGS			
Gross profit			
Expense			
Advertising			
Automobile expenses			
Bank service charges			
Contributions			
Depreciation			
Dues & subscriptions			
Equipment purchases			
Insurance			
Liability			
Property			
Workman's comp.			
Total insurance			
Interest expenses			
Finance charges			
Other			
Total interest expenses			
Lease fees			
Licensing fees			
Payroll			
Bonus			
Indirect labor, hourly			
Exempt salary			
Non-exempt, salary			
Payroll expense, other			
Total payroll expenses			
Postage and delivery			
Printing and reproduction			
Professional development			
Professional fees			
Accounting			
Consulting			
Legal			
Other			
Total professional fees			
Recruiting expenses			
Rent			
Repairs			
Building			
Equipment			
Other			
Total repairs			
Shipping expenses			
Supplies			
Factory			
Office			
Other			
R&D			
Total supplies			
Taxes			
Federal			
Local			
State			
Total taxes			
Telephone			
Travel & entertainment			
Uncategorized expenses			
Utilities			
Electric			
Gas			
Oil			
Water			
Total utilities			
Total expenses			
Net profit			

Figure 6-6. Profit and loss statement format.

Account description	Period: month( ), quarter( ), year( ); dates				
	Period 1	Period 2	Period 3	Period 4	Period 5
Income					
Sales					
Products					
Other					
Total sales					
Cost of goods sold					
Direct labor					
Direct material					
Total COGS					
Gross profit					
Expense					
Advertising					
Automobile expenses					
Bank service charges					
Contributions					
Depreciation					
Dues & subscriptions					
Equipment purchases					
Insurance					
Liability					
Property					
Workman's comp.					
Total insurance					
Interest expenses					
Finance charges					
Other					
Total interest expenses					
Lease fees					
Licensing fees					
Payroll					
Bonus					
Indirect labor, hourly					
Exempt salary					
Non-exempt, salary					
Payroll expense, other					
Total payroll expenses					
Postage and delivery					
Printing and reproduction					
Professional development					
Professional fees					
Accounting					
Consulting					
Legal					
Other					
Total professional fees					
Recruiting expenses					
Rent					
Repairs					
Building					
Equipment					
Other					
Total repairs					
Shipping expenses					
Supplies					
Factory					
Office					
Other					
R&D					
Total supplies					
Taxes					
Federal					
Local					
State					
Total taxes					
Telephone					
Travel & entertainment					
Uncategorized expenses					
Utilities					
Electric					
Gas					
Oil					
Water					
Total utilities					
Total expenses					
Net profit					

Figure 6-7. Pro-forma profit and loss statement format.

pro-forma and the normal P/L. You may ask why, since the mundane items such as postage and rent are not necessarily specific to the new product. That is so but each product a company produces has to carry its share of the entire cost of the company. Similar to sharing expenses for a taxi between several passengers. In its aggregate we call this the *burden* or *overhead expenses*. Every product needs to be charged for their specific percentage of the overhead costs to get a fair picture of profit or loss for the product. It is unreasonable to expect an existing product to carry costs for new products. It dilutes the record of true profit of the existing products and gives a false sense of security to the new products' team.

Let's look at some other aspects of this measurement tool before we go on with the example. All P/Ls distinguish between income and expenses to show a net profit (or loss) as the final number. This the famous (or infamous, depending on your viewpoint) bottom line. This is just plain common sense. We know how much we sold; subtract the expenses to make those sales, and the results are the hoped-for profit. So far, so good. But we do a little more to define direct expenses from so-called indirect expenses. Let me explain how and why.

Note the top section of the P/L titled "Income" consists of several categories of sales, identified as "Product" and "Other." "Product" is the sales related to the specific NPI item for the pro-forma. "Other" is incidental sales that should be accounted for but rarely amount to a significant percentage of the total. They are usually 1-time things. For example, let's say we had to buy a test device to use in development of a portion of the new product, and the use is complete so the company sold it. This cash received would end up in the "Other" row. "Product" and "Other" are totaled and we have the sales for the period. Now comes the unusual or perhaps unexpected twist. In addition to sales shown under income we have another subcategory called "Cost of goods sold" (COGS); this is a subtraction from sales in the Income section of the P/L. The COGS comprises all of the direct costs associated only with the making of the product. We know that product costs in their entirety consist of labor, material, and a fair share of the Overhead. COGS is the labor and materials associated only with the product. The overhead which is more difficult to discern is the entirety of the Expense portion of the P/L. COGS is a very directly calculated expense that belongs only to the sales for the period. Let's say we made 1000 CD RW assist devises for the period of this P/L, then the COGS, which consists of direct labor and direct materials, would be for the labor expended and actual materials used to make the devices. In this case, it would be the labor hours multiplied by the hourly rate plus the costed bill of materials for that number of devices. The COGS is then subtracted from total sales to give us the gross profit. Note we have income expressed as gross profit. In essence, what we have is sales minus labor and materials. Some people call this the above-the-line number. All sorts of financial ratios are derived from it with equally numerous pontifications as to what the spread between sales and COGS must be to have a successful business. Most of that means nothing. Obviously the less the spread, the less overhead a company can maintain. You may be interested to know that selling price sometimes is linked to this spread. We know that companies that run very lean can afford to set a selling price lower because their overheads are lower than larger companies with lots of staff support, hence more overhead expenses.

To summarize then, the Income portion of the P/L gives us a gross profit. How much of it becomes a net profit depends on the fair share of overhead expenses that will be tallied in the Expense portion of the P/L, and subtracted from it.

Now let's look at the Expense portion of the P/L form. As you see, there are many more categories than in the Income portion. Also, there is no exact set of rules saying what the descriptions of these items are. There is no specific table or dictionary saying what needs to be included. What I've shown are the traditional items of expense that most manufacturing and service companies would have as expenses, and I've listed them in alphabetical order simply for clarity. Every item shown shares 1 thing in common. The cost of them needs to be carried to some degree by the product represented in the sales. For example, banking expenses need to be shared with all products of the company that require some sort of financial support. If loans are taken specifically to support the CD RW device and nothing else, then the banking expenses would be entirely allocated to that product alone. On the other hand, if loans were general in nature to support the entire business, then the CD RW device would be given a pro-rated proportion of the cost, in this case, probably on a ratio of what percentage of total company sales are derived from this product.

We allocate expenses in a similar manner for every expense item category the company chooses to maintain. In that manner, the fair share of burden is placed on all products. Some items are easier to determine fair share than others. An easy expense item category to calculate is Exempt Salary. This is the team's base compensation. While working on the project, each individual's salary is charged to the new product. If they are assigned for 100% of their time, their entire salary is charged. If only part-time then only what percentage they worked on the project is charged. Quite often, time cards are used to allocate charges between different jobs for those who do not participate full-time on any 1 project. A more difficult area for allocating costs is in the area of items like utilities. How do you know how much of the cost of electricity should be charges to the new product? This is abstract and accountants try to come up with ratios based on real estate space occupied by the new project as a percentage of the whole or some other subjective optimization.

However we do it, we will tally the items in the Expense portion of the P/L then subtract it from the Gross Profit to identify Net Profit. We are adding in overhead to complete the cost triumvirate of labor, materials, and overhead. This gives us a very good picture of whether the project is profitable or not. Now let's flush out this explanation by continuing on with the example.

As I said, the pro-forma P/L is the primary documentation tool used for analyzing data related to decisions on new product introductions. Figure 6-8 show the first pass at gathering financial data for the CD RW Assist Devise. You can imagine that all 3 segments of the entrepreneurial team,

Business plan pro-forma: CD RW assist devices
Period: month(), quarter(x), year(); dates QTR 1-02 through QTR 1-03
$ =(000)

Account description		Period 1	Period 2	Period 3	Period 4	Period 5	Notes
number of units sold		1000	2000	4000	8000	16000	a. unit sale =$150.00
Income							b. DL/unit = $45.00
Sales							c. Mat/unit = $52.50
Products	a.	$150.0	$300.0	$600.0	$1,200.0	$2,400.0	d. Bank loan = $2.0mm/60mos.@6%
Other							$1,000.00 initiation fee
Total sales		$150.0	$300.0	$600.0	$1,200.0	$2,400.0	e. four auto leases
Cost of goods sold							@$400.00/ea.
Direct labor	b.	$45.0	$90.0	$180.0	$360.0	$720.0	f. Use proprietary process
Direct material	c.	$52.5	$105.0	$210.0	$420.0	$840.0	in manufacturing
Total COGS		$97.5	$195.0	$390.0	$780.0	$1,560.0	g. Staff: 5 Exempt, 1 Non-
Gross profit		$52.5	$105.0	$210.0	$420.0	$840.0	exempt 1 IDL
Expense							h. benefits and payroll tax
Advertising		$10.0	$25.0	$25.0	$25.0	$25.0	@25%
Automobile expenses		$0.3	$0.3	$0.3	$0.3	$0.3	i. Shipping @$3.00/unit
Bank service changes		$0.1	$0.1	$0.1	$0.1	$0.1	j. non-depreciable PC's &
Contributions		$1.0	$1.0	$1.0	$1.0	$1.0	SW
Depreciation		$0.8	$1.5	$1.5	$1.5	$1.5	k. State sales tax @6%
Dues & subscriptions		$0.1	$0.1	$0.1	$0.1	$0.1	
Equipment purchases		$30.0	$30.0	$0.0	$0.0	$0.0	
Insurance							
Liability		$2.3	$2.3	$2.3	$2.3	$2.3	
Property		$0.9	$0.9	$0.9	$0.9	$0.9	
Workman's comp.		$0.8	$0.8	$0.8	$0.8	$0.8	
Total insurance		$4.0	$4.0	$4.0	$4.0	$4.0	
Interest expenses							
Finance charges	d.	$30.0	$30.0	$30.0	$30.0	$30.0	
Other		$1.0					
Total interest expenses		$31.0	$30.0	$30.0	$30.0	$30.0	
Lease fees	e.	$4.8	$4.8	$4.8	$4.8	$4.8	
Licensing fees	f.	$5.0	$5.0	$5.0	$5.0	$5.0	
Payroll							
Bonus		$25.0	$25.0	$25.0	$25.0	$25.0	
Indirect labor, hourly	g.	$10.0	$10.0	$10.0	$10.0	$10.0	
Exempt salary	g.	$62.5	$62.5	$62.5	$62.5	$62.5	
Non-exempt, salary	g.	$10.0	$10.0	$10.0	$10.0	$10.0	
Payroll expense, other	h.	$20.6	$20.6	$20.6	$20.6	$20.6	
Total payroll expenses		$128.1	$128.1	$128.1	$128.1	$128.1	
Postage and delivery		$0.6	$0.6	$0.6	$0.6	$0.6	
Printing and reproduction		$5.0	$5.0	$2.5	$2.5	$2.5	
Professional development		$2.0	$2.0	$2.0	$2.0	$2.0	
Professional fees							
Accounting		$3.0	$3.0	$3.0	$3.0	$3.0	
Consulting		$35.0	$20.0	$15.0	$5.0	$0.0	
Legal		$10.0	$5.0	$2.0	$2.0	$1.0	
Other							
Total professional fees		$48.0	$28.0	$20.0	$10.0	$4.0	
Recruiting expenses		$20.0	$5.0	$5.0	$0.0	$0.0	
Rent		$3.0	$3.0	$3.0	$3.0	$3.0	
Repairs							
Building		$0.5	$0.5	$0.5	$0.5	$0.5	
Equipment		$0.5	$0.5	$0.5	$0.5	$0.5	
Other							
Total repairs		$1.0	$1.0	$1.0	$1.0	$1.0	
Shipping expenses	i.	$3.0	$6.0	$12.0	$24.0	$48.0	
Supplies							
Factory		$3.0	$3.0	$3.0	$3.0	$3.0	
Office		$1.5	$1.5	$1.5	$1.5	$1.5	
Other	j.	$14.0	$0.0	$0.0	$0.0	$6.0	
R&D		$30.0	$30.0	$15.0	$10.0	$5.0	
Total supplies		$48.5	$34.5	$19.5	$14.5	$15.5	
Taxes							
Federal							
Local							
State	k.	$9.0	$18.0	$36.0	$72.0	$144.0	
Total taxes		$9.0	$18.0	$36.0	$72.0	$144.0	
Telephone		$8.0	$8.0	$8.0	$8.0	$8.0	
Travel & entertainment		$25.0	$25.0	$30.0	$35.0	$50.0	
Uncategorized expenses		$2.0	$2.0	$2.0	$2.0	$2.0	
Utilities							
Electric		$3.0	$3.0	$3.0	$3.0	$3.0	
Gas		$2.0	$2.0	$2.0	$2.0	$2.0	
Oil							
Water		$0.5	$0.5	$0.5	$0.5	$0.5	
Total utilities		$5.5	$5.5	$5.5	$5.5	$5.5	
Total expenses		$395.8	$373.5	$347.0	$380.4	$486.0	
Net profit		–$343.3	–$268.5	–$137.0	$40.1	$354.1	

Figure 6-8. Pro-forma profit and loss statement example 1.

sales/marketing, operations, and finance, have submitted data to the team member responsible for financial analysis to put together the pro-forma P/L.

As you can see, we have estimates of sales and costs for 5 consecutive quarters of the year we expect the product to be introduced. We see that we expect to sell 1000 units during the first quarter and double each quarter up to 16,000 units 15 months later. We know what we want the selling price to be so we can calculate the sales revenues and see that it will grow from $150,000 to $2,400,000 in the first 15 months. This is all the cash we expect to generate from this product during its introductory phase. This should be a very sobering fact because without a readily available source of funding, everything we do would need to be supported by the sums of revenues over that period, in this case $4,650,000. And not only that, expenses would have to be planned to coincide with sufficient cash receipts to pay the bills as they become due. Fortunately, the notes on the P/L show the company has a bank loan of $2,000,000 to cover the startup period. So cash flow should not be a problem. If you go all the way to the bottom of the P/L, we see that the initial losses are $343,300 the first period through $137,000 during Period 3 for a total of $748,800. Thereafter the cash flow is positive. But what this analysis shows is that we will be operating at a loss for the first 9 months of this endeavor. A very nervous time indeed! But according to our plan and with the bank loan we should be all right.

The sales information is supplied by the sales/marketing team members. This includes what they believe to be the selling price for the product. They would have been working with market surveys to make sure the selling price is being set correctly. If not, the cash-received portion of the P/L would be in error. This puts a very large burden on that segment of the team. If they can't sell the number they forecasted for the price they forecasted, then the whole plan is seriously flawed.

The next part of the P/L, the COGS, is the responsibility of the operations team. Remember the 2 basic tenets of manufacturing:

- Know how to make the product.
- Know how much time it will take to do so.

We need to know how many hours it will take to make each device we intend to sell, at the volume proposed. The operations team will set the volume for a mature sell-through period. After all, they have to be in that position as soon as the product is launched. So the volume is set at the highest volume the company anticipates for the life of the product. In this case, we would have set it for 16,000 units per quarter plus a reserve margin of approximately 25%, a nominal 20,000 units per quarter in total to be produced. Using this number, the same number that the operations team has previously designed its facility around, the team calculates the associated throughput rates and labor hours. This is converted to labor costs and is reported to finance as the direct labor dollars. So we match up units of product with the labor hours necessary to produce them and derive direct labor dollars.

We also calculate the materials cost for each unit. Previously we established a bill of materials based on the design. The materials subteam of operations had developed costs for each line item on the BOM. So we have a materials cost for each unit. The values for 1000 units for the first period and values for the respective number of units for the succeeding quarters is input to finance for inclusion in the P/L.

We can calculate the gross profit by subtracting the period values for COGS from total sales. Theoretically this is the total of all the funds we have available for all indirect expenses. This says that in order to break even (financially akin to the satisfaction of kissing your sister instead of your wife), this is the total amount of money we can spend. Another way of expressing this is via the gross profit ratio. We see from this example that for Period 1, sales are planned to be $150,000. We know that for that amount of sales we need to spend $97,500, leaving a gross profit of $52,500. Dividing the gross profit by the sales give us a ratio expressed as 35%. This is the so-called margin many salespeople talk about. Out of the margin needs to come all expenses and what's left over is net profit—the real stuff you can put in your pocket and share with the Internal Revenue Service. Is 35% good or bad? That depends on your fixed expense levels. I can say that most companies try to have at least a 25% margin. Anything below that makes it difficult, if even possible, to generate a net profit. It would be like scoring 10 runs in the bottom of the ninth inning to overcome a 9-run deficit.

So now that the pro-forma P/L has shown us how much money we have to play with to generate a net profit, we go to the expense section of the P/L. Here we'll be interested in determining what are fixed expenses, that is, expenses that are extremely difficult to reduce because we have ironclad commitments to spend the money for those services; without them we couldn't exist. We will also want to understand the nature of the variable expenses. Think of fixed expenses as "musts" and variable expenses as "wants." Remember the axioms: Musts have to be. Wants are nice to have and can be rated by how desirable they are.

As a practical matter, fixed expenses are those that cannot be changed. These would be contractual relations of any kind. Leases, rent, contracts to buy fuel, etc. Even salaries can be considered fixed if there is a contract involved. Some other items might as well be fixed. Repairs to buildings and equipment, depreciation, and insurance are examples. Sales taxes also falls into this category.

Then there are semi-variable items to consider. While they are strictly speaking variable accounts, the fact is we don't have much room to maneuver to affect costs—items like telephone and electricity fall into this category. We can restrict telephone usage and turn up the thermostat setting on the air-conditioner. But at what price? Probably lower morale and consequently lower productivity. This in turn leads to higher COGS, which is just the opposite of what we want to achieve.

So, the variable account pool becomes very limited. And for all practical purposes it is limited to items directly affecting the ability to design, produce, and sell the new product. In fact, it comes down to a throttling process on the velocity of implementation of the new product. The question becomes how fast to we need to get to maturity of the product to ensure a safe market

Business Plan Pro-forma: CD RW Assist Devices
Period: month(), quarter(x), year(); dates QTR 1-02 through QTR 1-03
$=(000)

Account description		Period 1	Period 2	Period 3	Period 4	Period 5	Notes
number of units sold		2000	4000	8000	16000	32000	
Income							a. unit sale =$150.00
Sales							b. DL/unit = $45.00
Products	a.	$300.0	$600.0	$1,200.0	$2,400.0	$4,800.0	c. Mat'l/unit = $52.50
Other							d. Bank loan
Total sales		$300.0	$600.0	$1,200.0	$2,400.0	$4,800.0	=$2.0mm/60mos.@6%
Cost of goods sold							$1,000.00 initiation fee
Direct labor	b.	$90.0	$180.0	$360.0	$720.0	$1,440.0	e. four auto leases
Direct material	c.	$105.0	$210.0	$420.0	$840.0	$1,680.0	@$400.00/ea.
Total COGS		$195.0	$390.0	$780.0	$1,560.0	$3,120.0	f. Use proprietory process
Gross profit		$105.0	$210.0	$420.0	$840.0	$1,680.0	in manufacturing
Expense							g. Staff: 5 Exempt, 1 Non-
Advertising		$10.0	$25.0	$25.0	$25.0	$25.0	exempt 1 IDL
Automobile expenses		$0.3	$0.3	$0.3	$0.3	$0.3	h. benefits and payroll tax
Bank service charges		$0.1	$0.1	$0.1	$0.1	$0.1	@25%
Contributions		$1.0	$1.0	$1.0	$1.0	$1.0	i. Shipping @$3.00/unit
Depreciation		$0.8	$1.5	$1.5	$1.5	$1.5	j. non-depreciable PC's &
Dues & subscriptions		$0.1	$0.1	$0.1	$0.1	$0.1	SW
Equipment purchases		$30.0	$30.0	$0.0	$0.0	$0.0	k. State sales tax @6%
Insurance							
Liability		$2.3	$2.3	$2.3	$2.3	$2.3	
Property		$0.9	$0.9	$0.9	$0.9	$0.9	
Workman's comp.		$0.8	$0.8	$0.8	$0.8	$0.8	
Total insurance		$4.0	$4.0	$4.0	$4.0	$4.0	
Interest expenses							
Finance charges	d.	$30.0	$30.0	$30.0	$30.0	$30.0	
Other		$1.0					
Total interest expenses		$31.0	$30.0	$30.0	$30.0	$30.0	
Lease fees	e.	$4.8	$4.8	$4.8	$4.8	$4.8	
Licensing fees	f.	$5.0	$5.0	$5.0	$5.0	$5.0	
Payroll							
Bonus		$25.0	$25.0	$25.0	$25.0	$25.0	
Indirect labor, hourly	g.	$10.0	$10.0	$10.0	$10.0	$10.0	
Exempt salary	g.	$62.5	$62.5	$62.5	$62.5	$62.5	
Non-exempt, salary	g.	$10.0	$10.0	$10.0	$10.0	$10.0	
Payroll expense, other	h.	$20.6	$20.6	$20.6	$20.6	$20.6	
Total payroll expenses		$128.1	$128.1	$128.1	$128.1	$128.1	
Postage and delivery		$0.6	$0.6	$0.6	$0.6	$0.6	
Printing and reproduction		$5.0	$5.0	$2.5	$2.5	$2.5	
Professional development		$2.0	$2.0	$2.0	$2.0	$2.0	
Professional fees							
Accounting		$3.0	$3.0	$3.0	$3.0	$3.0	
Consulting		$35.0	$20.0	$15.0	$5.0	$0.0	
Legal		$10.0	$5.0	$2.0	$2.0	$1.0	
Other							
Total professional fees		$48.0	$28.0	$20.0	$10.0	$4.0	
Recruiting expenses		$20.0	$5.0	$5.0	$0.0	$0.0	
Rent		$3.0	$3.0	$3.0	$3.0	$3.0	
Repairs							
Building		$0.5	$0.5	$0.5	$0.5	$0.5	
Equipment		$0.5	$0.5	$0.5	$0.5	$0.5	
Other							
Total repairs		$1.0	$1.0	$1.0	$1.0	$1.0	
Shipping expenses	i.	$6.0	$12.0	$24.0	$48.0	$96.0	
Supplies							
Factory		$3.0	$3.0	$3.0	$3.0	$3.0	
Office		$1.5	$1.5	$1.5	$1.5	$1.5	
Other		$14.0	$0.0	$0.0	$0.0	$6.0	
R&D		$30.0	$30.0	$15.0	$10.0	$5.0	
Total supplies		$48.5	$34.5	$19.5	$14.5	$15.5	
Taxes							
Federal							
Local							
State	k.	$18.0	$36.0	$72.0	$144.0	$288.0	
Total taxes		$18.0	$36.0	$72.0	$144.0	$288.0	
Telephone		$8.0	$8.0	$8.0	$8.0	$8.0	
Travel & entertainment		$25.0	$25.0	$30.0	$35.0	$50.0	
Uncategorized expenses		$2.0	$2.0	$2.0	$2.0	$2.0	
Utilities							
Electric		$3.0	$3.0	$3.0	$3.0	$3.0	
Gas		$2.0	$2.0	$2.0	$2.0	$2.0	
Oil							
Water		$0.5	$0.5	$0.5	$0.5	$0.5	
Total utilities		$5.5	$5.5	$5.5	$5.5	$5.5	
Total expenses		$407.8	$397.5	$395.0	$476.0	$678.0	
Net Profit		–$302.8	–$187.5	$25.1	$364.1	$1,002.1	

Figure 6-9. Pro-forma profit and loss statement example 2.

niche. This becomes a judgment call. What the pro-forma P/L does is show us how the costs lay out for various judgments of how fast to proceed. Example 1 shows a plan that has losses of $748,800 before it becomes positive. Do we believe it? Has the market been scoped properly? Can we afford to have only 31,000 units in service by the end of the fifth quarter to have gained a sufficient market share? These are just a few of the questions we may ask that we now have the figures to ask about due to doing a pro-forma P/L.

Pro-forma P/Ls are forecasting tools. We had a series of questions posed above. So now we can use the tool to help get some answers. Figure 6-9 shows a second iteration of the pro-forma P/L for the CD RW Assist Device.

Here we've done a "what if." The question was asked if it was possible to sell 2000 instead of 1000 units in the first period. And if we could still double sales for every period through the first 5 periods. With a few keystrokes into the Excel program that constitutes the pro-forma P/L, we get some answers to what that would mean. The calculation shows that the losses are still with us but only through Period 2, and the aggregate total is significantly less, only $490,300 vs. $748,800. But more significant, the net profit in the fifth quarter is at the million dollar level. This is a much more appealing forecast. In Figure 6-10 I've illustrated the cash flow profiles of Examples 1 and 2 for comparison purposes. There is definitely a lot more net profit to play with, which would make this project much more attractive to potential investors, if that was our plan.

But all is not peaches and cream. We need to know if the change in sales is doable. We have to ask the question of whether we could have started with 2000 instead of 1000 unit sales in the first period. Why didn't we say so in the first place? Is it due to reluctance to promise anything but a conservative sure-fire bet? Or is it the cold hard truth that we need to build a market for the product, and selling 1000 units during the first quarter is a push target. These and other items would have to be resolved before either version of the pro-forma would become the plan. The "what if" games can be virtually endless, but the tool is available to handle various scenarios including going about it backward, starting with desired net profits and see what the sales needs to be to achieve them.

The important point of this example is that there is a very usable and viable tool that can be used to derive information about whether or not a new product introduction can be profitable and how long it will take to get there. It points out what many of the risks are and also the opportunities. It is a must-use tool for any NPI project.

9. Take Corrective Action to Achieve the Required Margin and Meet Customer Requirements, if Necessary

If unsatisfactory results are indicated by various iterations, the pro-forma calculations of profit, the team must reconsider all of its options to see if the project can be reconfigured. If changes cannot be made that will result in a satisfactory profit then it will have to be abandoned.

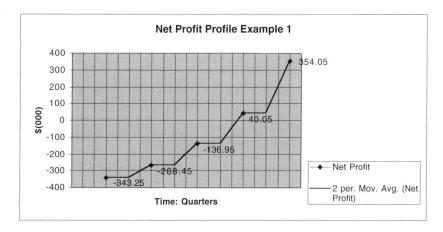

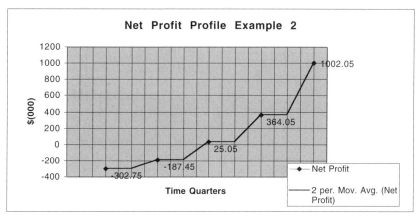

Figure 6-10. Net profit comparisons.

This is serious business when a project reaches the stage that profit forecasts indicate it is a poor risk to continue. Therefore the lead for the evaluation has to be taken by the entrepreneur or his designated team leader and requires top priority for every team member's schedule and attention. The evaluation is a review of the project to date and needs to be done in its entirety, regardless of any shortfalls that have occurred at any step of the review. If the project can be saved, it will get its redemption during this phase. I'd also like to point out that even if a project doesn't require a review because the profit forecast is unsatisfactory, a full-fledged review still should be done.

Keep in mind that after the pro-forma profit exercise is completed and the results are satisfactory, the next step is launching the product or service in the marketplace. Therefore it is prudent to hold and do a once-through thorough review of all the previous steps before final authority is granted to release the product. So this step of review will take place regardless of the results of the pro-forma, satisfactory or unsatisfactory. If the results of the pro-forma are unsatisfactory, the review's aim is to find weak spots and evaluate for the practicality of corrective actions. If the results were satisfactory, then the purpose of the review is to ensure us that the data is competent and believable.

10. Release for Manufacturing/Distribution After All Internal and External Requirements Have Been Met

This means we have demonstrated that a viable product or service is within our capability. At this point, for larger organizations the team leader will be turning the project over to the company's line organization for responsibility for production and sales. In the entrepreneurial organization it usually means the team leader must switch hats from a development role to a steady-state production role.

Obviously the larger corporation with resources for steady-state operations and project development will have the resources to make this transaction. It has people to manage both very successfully, but does the lean startup firm? This is always an emotional question. Instead of answering the question directly, let's look at the traits that are necessary for success in each venue.

The startup NPI activity requires an ability to manage divergent talents and let each of them have the lead when they are the primary focus of a task. It requires the ability to manage a budget, apply resources, and be quick to take corrective action when necessary. Well, the same is true for the line manager engaged in large production runs or setting forth resources to deliver services on large scales. So what's the difference? Very little. Perhaps a willingness to pay attention to the same type of details day in and day out in large-scale manufacturing. The NPI project manager must be adept at handling lots of different types of problems day in and day out. This means that the key to success is being willing to handle the type of problems presented well and with enthusiasm. In other words, have the fire in the belly to be successful.

The steady-state production manager needs to get his satisfaction from solving all those daily problems that will allow output to be at the volume required. The NPI project leader similarly needs to get satisfaction from solving all those problems that impede making progress in achieving the projects goals. Both leaders require excellent management skills. They should both be equally successful in the other's position. But that rarely happens. Why? I'm convinced it's the "fire in the belly" syndrome. It all boils down to preference. If the startup company's NPI leader does not like going down the same road day after day, even though the challenges vary (production management), he will not be able to make the transition and production will suffer. Lots of people feel this way; they want constant change, or perceived constant change. Perhaps they do not realize that production management offers the same challenges of dealing with constant change: broken machinery, people not showing up for their shift, material delivered late, operator error, etc. But if the person who led the NPI does not feel comfortable with this type of challenge, then he won't be successful; he won't have the "fire in the belly."

To summarize, the ability to lead the NPI or the steady-state production is based on the same management skills. The only difference is the type of problems that need to be solved. I believe a good NPI manager has the skills to be a good production manager. The specific technologies needed to lead the product's introduction aren't going to change when the shift is made from prototype to steady-state production, just the emphasis. Therefore the answer to the question is that the lean startup firm has the ability to make the transition; whether it has the deep-down desirability to do so is the real question. Is there the fire in the belly to expand from a brash startup to lean tough producer?

11. Monitor Production (or Delivery of Service) and Sales Progress, and Take Corrective Action as Required

This is the post-NPI step. Each member of the team needs to follow the early progress of commercialization to ensure that all processes and procedures are running in accordance with expectation. They need to be keen observers to make sure things are running smoothly, even if they no longer have an official capacity with the product. The NPI team is the keeper of the industrial culture for their product because they gave it life. They have more than a passing interest in its success. Psychologically speaking, they have an emotional attachment and the company is counting on that to steer the new product to prosperity.

Invariably some deviation from the plan will develop that the former NPI team members will be best equipped to handle. The strategy would be to reconvene the NPI team to evaluate and plan corrective actions if required, then have the company's resources implement the fix. These deviations should be minor if the team had done a thorough job during the course of the project. In fact, most of these minor deviations should not be much of a surprise to the team. A deviation would be more on the order of arriving at the wrong street, rather than in the wrong country. If it is subjectively more than this, then the team had not done a sufficient job.

So we see that this last step in the NPI technique is really a transition step. It is the evolution of control from the team to the permanent operating component of the company, who now has the role to rapidly commercialize the product and make the targeted profits. You may ask, what of the small entrepreneur? Who can she or he hand off the NPI results to? Probably no one at first, and the team has to make the transition from prototype development to commercialization. The aspects of doing so were discussed previously, where the most important factor is mindset change and the need to be comfortable with repeated tasks, all needing to be managed as if they were being done for the first time. Sort of like an actor giving a creditable performance for the umpteenth time. And just like this allegory, the production team needs to realize that while they're doing something they've done over and over again, the customer is most likely receiving the product for the first time. The team wants to make sure that the enthusiasm for doing the job is just as high for the thousandth time as it was for the first. By doing so, the customer is assured that his or her specific product is at the highest quality level. This makes customers for life and keeps the profits coming in.

I've strayed a bit in answering my hypothetical question. To summarize: An entrepreneurial startup company's staff goes from NPI team to commercial production team in a seamless manner. The team must understand that they will need to do this to make all their efforts worthwhile. Most entrepreneurial startup failures are due to lack of ability to make this transition.

This concludes our look at raw prototype to successful commercialization. The NPI process was indeed initially designed for entrepreneurial startups but has been used even more successfully by the small to mid-size companies, usually launching their second product. This second product often was an upgrade of the first, and the first suffered because the company had no NPI guidelines. NPI is an intensive investigation. It will time and time again uncover deficiencies in how the company does business. The team members, being relatively senior in the organization, are the right people to make fixes and help their organization iterate to a higher level of performance. NPI acts as a cleansing process that allows companies to learn how to perform better. It is done not through the direction of an outside expert, or perhaps not even an internal one, but through a self-taught help-one-another learning process. This bootstrapping process teaches team management and is a catalyst for expanding that knowledge throughout the entire company.

To be successful in transiting from a raw idea to a polished commercial product takes extensive attention to details. Above all it takes commitment, commitment, commitment to the idea.

Chapter 7
Financial Potpourri

It is rare to have a combination of traits residing in 1 person for both idea creativity and the logical approach to finance. But either skill can be learned. Starting up a new business requires both traits to be successful. Entrepreneurs, more often than not, exhibit creativity traits for envisioning new product or services, rather than being well-versed in finance. The most common scenario for the entrepreneur is to initiate an idea for a business and fill out the details of what it's all about, sell the idea to friends and backers and start working on it, and then find out in short order that he or she is a "babe in the wilderness" when it comes to knowing whether the company is progressing satisfactorily with respect to money issues or is having financial troubles. At that point, the entrepreneur either learns the basics of business accounting or hires someone who does understand them. In the process, the entrepreneur starts to worry about money issues instead of creativity issues, which if not overcome spells doom to the fledgling business. This need not happen and can be overcome.

Entrepreneurs do not need to be expert financial analysts to succeed. But they do need to know how to safely pass through the twisted and often perilous path of business finance to succeed. In this chapter, I will be your guide along this sometimes confusing and often intriguing journey. Businesses keep score via the finance reports. So to learn whether we have a good or bad score we need to understand and interpret the rules. The goal for this chapter is to provide an understanding of the operational characteristics of financial management of a business. We've already discovered some of them in our explorations of project management and new product introduction. Now we'll put that together with budgeting and financial measurements basics to show how the tools of finance work to evaluate the performance of a company.

THE CLASSIC WAYS OF MEASURING FINANCIAL STATUS

Three common measurements are used to define the financial status of a company:

- The balance sheet
- The profit or loss statement
- The cash-flow statement

Why are these measurements important to understand? The flip answer is because the business world uses them to determine how well a company is doing. If you never have any dialogues or discussion with others pertaining to the financial status of your company then you can elect to not get an understanding of these measurements. However, this is highly unlikely. If you ever decide to get a bank loan or line of credit, or need to evaluate whether you want to do business with a potential vendor or customer, or file income tax, you'll find the language they speak entails these common measurements. The real answer is that they are rational tools that really do spell out the financial viability of a company. They describe how much the company is worth, whether it is going to be profitable in the near-term future, and whether it will be able to pay its bills when the invoices arrive. We use these 3 ways of measuring the financial health of a business because no 1 measurement by itself is sufficient to tell the whole story.

The balance sheet presents a snapshot of the company's financial strength at any particular time (usually at the end of a significant measurement period such as the end of a calendar quarter or the year) by playing off its assets vs. its liabilities. This shows what the company is worth if it were to be liquidated at that point in time.

The profit or loss statement shows whether the company made money or lost money over the reporting period. It is usually represents the same period as the balance sheet. The P/L was explained in detail in Chapter 6 with its use as a forecast tool. The methodology is the same as for a pro-forma when used as a measurement except we use real "historical" data rather than projections of what is likely to happen. Since the details making up a P/L have already been covered in Chapter 6, I will not repeat them here.

The cash-flow statement measures the up-to-the-minute cash receipts and dispersements. It is akin to a general ledger or the balance in a checking account. Just like a personal checking account, the cash-flow statement portrays our status up to the last entry of the period and tells us how money came in and how it was spent. Its main strength is that it tells us if we have enough money to pay the company's current bills.

All 3 measurements are needed to display the financial state of a company. The balance sheet tells us how much a company is worth. The P/L statement tells whether it is currently profitable. And the cash-flow statement tells us the rate at which money comes into and out of the bank accounts. A healthy company would have a significant net worth, be profitable, and be able to pay its bills.

While we like to think of finance as being very objective and unemotional, it really isn't. Balance sheets and P/Ls are very subjective and are at the mercy of the observer as to whether they are satisfactory. Likewise, being able to pay bills via sufficient cash flow may be a laudable accomplishment, but at what price? Does it mean there's no cash reserve for emergencies? Or are there more than sufficient funds to pay all bills and the company is being shortsighted in not investing more for labor, material, and associated overhead? So we see each of these measurements while portraying that an objective state of being needs to be evaluated on the basis of standards that the owners and those with interests in the company set.

Are there guidelines? Sure. Are they important? They probably are but only in a transitory nature when they're used to further a specific goal such as obtaining a line of credit. Rather than listing various goals for each of the 3 measurements from different accounting and management organization sources, I think it's more important that the values portrayed by these measurements be what the owner/entrepreneur feels comfortable with. I, for one, feel comfortable with a startup company having a net worth at least 10% higher than the initial investment capital, consistently making a profit higher than current savings account interest, and having a cash flow that is capable of paying current due bills within the commonly extended 30-day grace period, thus gaining the ability to buy on credit.

Let's look at some of the mechanics of the 3 measurements, based on an example company.

The Balance Sheet

A balance sheet is an equation where everything financial about a company is either an asset or a liability. Some of it you may feel is a force fit and you might be right, but there is logic to the placement from the viewpoint of the company being a physical entity. Remember this is the status of the company, not you. So saying that an owner's equity balances off assets as a liability is really saying the company owes you money for the shares you own.

A balance sheet is an equation. Makes sense? Balance? Equilibrium? Yes it does. The equation is:

$$\text{Assets} = \text{Liabilities} + \text{Owners' equity}$$

Assets are everything the company owns converted to cash value. This includes property, automobiles, machines, stationery, any other physical entity that it has title to, and cash and accounts receivables (bills sent to customers not yet paid). Liabilities are everything the company owes to the outside world. Included here are bills it owes to vendors, taxes due for the current year, salaries committed to pay to its employees, bank loans, and any other items that it is required to pay. Owners' equity is the cash value of the worth of the company that the owners would get if everything would be liquidated. On Day 1, the owners' equity would be equal to the money they invested. If the company is profitable, the owners' equity increases in time. If not, it decreases.

The funny thing about balance sheets is that values for anything other than cash on hand and to be received, and specific bills and loans outstanding, is subject to interpretation concerning its objective values. Compare the various values you can appraise a used car for or the value for real estate properties and you get the picture. Therefore, we can say that balance sheets are "objectivisations" of many subjective appraisals of values, very much subject to interpretations. So when viewing a balance sheet, keep in mind what is absolute and what is subjective opinion. Cash in the bank account is fact. It is exact. The value of the factory building on Elm Street is subjective opinion. It is the best guess of 3 appraisers, averaged, saying what the building would sell for this day or week.

Let's look at a balance sheet and see how it's structured. For our example I've made up a fictitious entrepreneurial startup company that develops and sells engineering training software via a Web site. Figure 7-1 is its current balance sheet.

This is the common format for balance sheets. For simplicity purposes I've shown 2 quarters and labeled them "last quarter" and "current quarter." Balance sheets for a business should always have the preceding period shown for comparison purposes. Actually the more comparison columns vs. the current column shown gives the analyzer more faith in the accuracy and objectiveness of the current period report for the company because it portrays its history.

The top section is the Assets part of the equation. For our startup company, we show 4 major categories: cash, accounts receivable, inventory, and miscellaneous. We see that cash in our checking account and other accounts

Assets	Last Qtr	Current Qtr.
Cash	$205,967	$358,379
Accounts receivable	$1,202,300	$1,385,525
Inventory		
Engineering database	$700,000	$770,000
Video seminars	$600,000	$630,000
Technical courses	$3,131,974	$3,515,968
Misc.	$35,000	$35,000
Total assets	**$5,875,241**	**$6,694,872**
Liabilities and equity		
Current liabilities		
Accounts payable	$809,224	$919,553
Salaries & benefits payable	$261,300	$313,560
Taxes payable	$47,439	$54,868
Total liabilities	**$1,117,964**	**$1,287,981**
Equity		
Paid in capital	$2,868,881	$2,868,881
Surplus/deficit	$131,776	$152,412
Total equity	$4,757,277	$5,406,891
Total liabilities and equity	**$5,875,241**	**$6,694,872**

Assumptions for balance sheet:
Engineering database asset value = 2 x cost to produce.
Paid in capital = initial capitalization + increases or decrease in current period.
Surplus/deficit = retained earning or losses from previous periods + current periods.
Pre-venture capital investment: $37,000.

Figure 7-1. Sample balance sheet—start up engineering training company.

grew $152,412 since last quarter. This says we appear to have substantial cash available. The next line is accounts receivable. Here we see that the current total is $1,385,525. These are sales for which we haven't yet received payment. It is up 15% over the last period. This could be either good or bad. If it is a result of increased sales, then it's good. If it is a result of stable sales but customers taking longer to pay their bills, it is not good because we may not be receiving that cash to pay our own bills on time. The third line is the inventory of products and services, representing their cash values. Again we see that it has gone up over the previous quarter. This is okay if it is a planned increase to meet anticipated demand. However, if inventory is growing because sales are not what was anticipated, it is a negative. The fourth line is the miscellaneous category. This is the value of all the non-cash assets. It could be anything from automobiles to Xerox machines. Some companies show this line with all of its particulars on every balance sheet, and some only on the annual version. For illustrative purposes of balance sheets, I've left out the details. The total assets are the sum of all the major line items listed as assets. Our sample company has an asset base of $6,694,872 as of the current quarter report. It is a snapshot in time and is current as of today, only if the report is for today's values of these accounts.

Now we go to the other side of the equation: liabilities and equity. We start with current liabilities. Here we show the 3 major liabilities that this example company has: accounts payable, salaries, and benefits owed to employees, and current tax liabilities. These are all obligations that have to be met within the current period. For balance sheet purposes, accountants usually say within a year, but it can be fuzzy and open to interpretation, in other words, a chance to look as good as you can be. Accounts payable is fairly straight forward; it is what is owed to vendors from bills that have been received. Here we see the current period has accounts payable of $919,553, up approximately $110,000 over the last period. If it is due to increased volume due to increased sales or anticipated sales, then it is probably okay. If it is because the company cannot pay its bills on time due to a slow cash inflow of receivables, then there may be a problem and perhaps securing a line of credit loan is required.

The salary and benefits payroll line typically represents the quarterly or annual salaries and benefits that the company is committed to pay. The period, quarterly or annual, most often depends on the length of time between issuances of the balance sheet reports. Some companies do it another way. Salaries and benefits are shown only for the most current due payroll. Obviously, this would reduce the amount on this line. The theory is that the company owes only for work done, not for future work to be done. The only exception would be for those employees working under a contract for a longer period of time. Since most employees could be let go with a customary 2-week notice, there is technically no formal future work contract to list as a liability.

The tax line similarly shows what "known" tax liabilities currently exist. In the example, this is primarily sales tax, FICA, and quarterly income tax, all very calculable values.

The sum of the liabilities comes to $1,287,981. This is significantly smaller than the assets. Therefore, to balance the equations, the remainder has to be equity. Equity could be positive or negative. If the liabilities exceeded the assets, then the equity would be negative. In other words, the worth of the investment is less than its original input. The equity section of the balance sheet is slightly different than the typical assets and liabilities section. Here, owners try to hide as much as possible from the prying eyes of analysts. But the figures do compute. For the current quarter we see a paid in capital of $2,868,881. That's how much the entrepreneur and his or her fellow investors put into the company to launch it on its way. Any additional money they invest would also show up on this line. The next line, surplus or deficit, is the amount of earnings that have been put into equity that could be paid back to the owners in the future if they chose to have a dividend payout. In the example, the company now has $152,412 in that account, up from $131,776 during the last quarter. This means the accountants felt it was prudent to salt away this much in earnings for future use determinations. What is not shown on the balance sheet is the algebraic difference between total equity, which is simply assets minus liabilities, and the sum of paid in capital and surplus/deficit. That algebraic difference is virtually a meaningless number. It is sufficient to say that the paid in capital has grown in worth to the level of total capital; and by the way, we have reserved a surplus that could be paid to the owners if they so chose. Another way of looking at total equity is that if the owners chose to liquidate and could do so on a timely basis, this is what they would be left with. One way to liquidate would be to sell. So if they put the company up for sale, the minimum price they should accept would be the equity value.

Those of you who have had some accounting training may say I've taken some liberties with the line items making up assets, liabilities, and equity. I don't have everything that could exist. For example, I show no long-term liabilities and a true balance sheet would. That may be so. But our example company has no outstanding loans of more than a year; therefore, I chose to not show a line that has a zero value. What this balance sheet portrays is a simple balanced equation of this company showing the real information making up the firm's financial position. There could be more detail but would be of little additional value for instructional purposes.

The Profit or Loss Statement

P/Ls show how much money we made, not whether we have it in hand now or when it came in or whether it will be received. This may seem like a contradiction. How can you have made a profit and not have the money? Well, if we give customers time to pay their bills or if they pay on the installment plan, like paying off an automobile loan, then we've made the profit and, barring fraud and default, we will collect the money per an agreed-to schedule. This means the profit money isn't all here yet and therefore is not usable to pay our own bills and salaries.

In Chapter 6 we saw the P/L used as a tool to help make decisions on new product launch activities. Likewise, it could be used as a budgeting tool to set up the plan for spending corresponding to forecasted sales, and we will explore that use later in this chapter. But the original purpose for a P/L is an accounting one to show how a company is performing. Figure 7-2 shows the P/L for the entrepreneurial startup company that develops and sells engineering training software via a Web site.

The 1st and 2nd quarters listed in Figure 7-2 represent the same 2 quarters shown on the balance sheet example. The current quarter on the balance sheet corresponds to the 2nd quarter, and likewise the last quarter represents the 1st quarter. It may not look the same but the information relevant to both reports is there. It just takes a little deciphering to find. Let's look.

Starting with Assets. Cash on the balance sheet has no corresponding entry for the P/L. Cash is simply cash available, usually in the bank. Since it is not a period sale it doesn't enter into the P/L report. Accounts receivable on the balance sheet shows $1,385,525. This appears on the P/L as total revenue for the 2nd quarter. Here we are assuming that there are no accounts receivable from a previous time period being shown on the balance sheet, which for an example problem is logical. (It is logical because most companies work very hard to not let customers be in arrears more than 90 days.) None of the inventory valuations shown on the balance sheet are accounted for on the P/L because they are not involved in period sales or expenses. Neither do the misc. line of the balance sheet. This is the worth of other tangible assets such as computers, automobiles, machines, etc. and have no direct account impact on the sales or expenses.

Now we'll go to the liabilities and equity portion of the balance sheet. Only current liabilities need to be shown on the P/L. This is so because only those entries will affect the period with respect to profit or loss. On the P/L we see the total expenses being $1,233,113 for the 2nd quarter. On the balance sheet the total liabilities are $1,287,981 for the same time period. Why isn't it the same? The reason is the balance sheet shows a tax liability of $54,868, which isn't shown on the P/L. Why not? The taxes weren't paid during the 2nd quarter. They are a known liability but are not required to be entered as a 2nd quarter expense because they are theoretically not due, sort of like, a grace period for paying a life insurance premium. Could it have been shown on the P/L? Yes, but only if we are using a accrued accounting approach.

Accruals are obligations or sales that are credited exactly when they are committed, not when the transaction (passing of cash between the parties) occurs. The problem with accruals is that they paint an unrealistic picture of profits or losses with respect to cash flow. It would indicate far different portrayals of events than actual happenings. Consider the tax bill, if accrued for on the P/L, then our real cash would be much higher than the calculated cash because the P/L would have us believe it is already paid for. This leads us to the last of the 3 classic financial measurement tools, the cash flow statement.

	1st Quarter	2nd Quarter
Monthly Web site visitors	800,000 - 1,000,000	1,000,000 - 1,150,000
Revenue		
Web site advertising	$540,000	$645,000
Online technical courses		
Individuals	$337,500	$403,125
Corporations	$90,000	$90,000
Custom training courses	$170,000	$170,000
Book sales	$64,800	$77,400
Total revenue	**$1,202,300**	**$1,385,525**
Operating expenses		
Internet advertising	$432,000	$516,000
Magazine advertising	$75,000	$75,000
College publication ads	$12,000	$12,000
Search engine placement	$1,500	$1,500
Authors - engr. database	$30,000	$35,000
Video seminars	$24,000	$30,000
Online technical courses	$60,000	$72,000
Custom training courses	$119,997	$119,997
Legal fees	$4,500	$5,000
Accounting fees	$3,500	$4,000
Insurance	$1,200	$1,200
Phone service	$3,427	$4,456
Promotional items	$2,000	$3,000
Web site news feed	$9,000	$9,000
Web site hosting fee	$1,200	$1,500
Web site maintenance	$1,200	$1,200
Salaries	$201,000	$241,200
Benefits	$60,300	$72,360
Rent - office space	$15,000	$15,000
Utilities - office space	$1,200	$1,200
Misc.	$7,500	$7,500
Web site development	$5,000	$5,000
Total expenses	**$1,070,524**	**$1,233,113**
Net profit (loss)	**$131,776**	**$152,412**

<u>Notes</u>
Bounty for each book sold is 15%.
Advertising rates on other web sites are $40 CPM
Each magazine ad is $3000 to $5000, depending on size and publication.
Avg. technical course costs $6000 to produce.
Technical courses are sold for $375 per course.
Custom training courses cost on average $6000 to produce.
Custom training courses sell for $8500.

Figure 7-2. Sample profit or loss statement—start up engineering training company.

Cash-Flow Statement

The cash-flow statement is used to portray the liquidity of a company, meaning its ability to pay its bills with current, on-hand cash. This is extremely important. Many companies have great balance sheets and the P/Ls, both current period and pro-forma, show a very profitable company. And yet they fail or are taken over by shady venture capitalist sharks because they can't pay their bills. They have all their profits tied up in slow payers, or their assets are hard physical equipment and inventories. So they can't pay their bills. A cash-flow statement portrays such weaknesses.

Think of a cash-flow statement as a checking account statement. It has a beginning cash balance. It lists bills paid and similarly lists cash receipt for the accounting period. Then with an arithmetic summing we calculate an ending balance. The main feature is that we can see the path of cash balances after each transaction if we want to. This allows us to plan expenditures during the reporting period.

Cash on 10/31/02	$150.00
Pay for tires balancing, 11/2/02	−$50.00
Cash on 11/02/02	$150.00
Receive stock dividend on 11/3/02	$10.00
Cash on 11/03/02	$160.00

And so forth as transactions occur. What we see is the record keeping going on in the company's general ledger, which is the working detailed copy of the cash-flow statement. If a cash-flow statement had every transaction that occurred over the reporting period, it would be an enormously large document containing many pages. We reserve that for the general ledger. So the actual cash-flow statement is an abbreviated form of the general ledger.

Figure 7-3 shows the cash-flow statement for the example company. We can see the basic same sets of numbers we had with the balance sheet and P/L. And as with those other 2 documents, we find the data is the same but presented in a different fashion to portray the data that the analyst wants to see. We can see in Figure 7-3 that the cash, accounts receivable, and current liabilities are used in their summary format. We see 2 cash values, beginning and end of period. In fact, it is this information we want to make visible because it tells us if we're cash solvent or if we have positive cash flow. Obtaining the answer to that question is what this document expects to achieve.

We can make judgments as to future cash flows by observing the increases and decreases in sales and increases and decreases in accounts payable and salaries and benefits. These are important to knowing what kind of additional projects and obligations to pay should be taken on. Figure 7-3 shows that for the current quarter, the ratio of accounts receivable plus salaries and benefits to sales was 0.924, whereas the same ratio for the last quarter was 0.890. By evaluating these ratios it is apparent that the difference between obligations and future incoming cash is less favorable in the current quarter

	Last Qtr	Current Qtr.
Cash at beginning of period	$205,967	$358,379
Cash inflow		
Sales	$1,202,300	$1,385,525
Interest		$1,287
Other inputs	$20,636	
Cash outflow		
Accounts payable	$809,224	$966,992
Salaries and benefits	$261,300	$313,560
Cash at end of period	$358,379	$464,639

Notes:
1. Other inputs: additional cash from equity surplus
2. Tax liablity $47,439 from last quarter paid in current quarter
3. Bank checking account changed to interest bearing in current quarter

Figure 7-3. Sample cash-flow statement—start up engineering training company.

than the last quarter. This would indicate the company may have to take care to not incur new obligations until the ratio gets further away from unity. This measure of cash flow indicates that while cash is increasing, the obligations claimed against that cash flow is increasing at a greater rate. It is interesting to note that neither of the other measurements, balance sheet or P/L statement, show this in an easily discernible fashion. This example gives a vivid reason why all 3 measurements are needed to run the business.

CONSTRUCTING BUDGETS

Budgets are based on the pro-forma profit and loss statements (P/L) for the budget period. The pro-forma P/L, as we described, is a projection of income and expenses based on market conditions and the company's plans. Budgets are detailed breakdowns of accounts assigned to the various departments of the company. Measurements are a means by which we can determine whether budgets are being complied with or achieved; and they consist of financial and nonfinancial, direct and indirect methods.

The P/L statement, therefore the budget, is made up of 2 major components, commonly called *above the line* and *below the line*.

- *Above the line* represents all income minus expenses (called cost of goods sold) directly related to producing and selling the product or service, the results of which are the gross profit of the company or the department.
- *Below the line* represents the overhead costs associated with producing and selling the product, the total of which is subtracted from the gross profit to obtain the net profit (before federal and state income taxes).

- The mathematics and format are shown below.

    Sales – Cost of goods sold = Gross profit > above the line

    Gross profit – Operating expenses = Net profit > below the line

Budgeted sales are targets the company expects to achieve. The total sales budget is divided into the various departments having direct sales responsibilities, usually along product or service lines.

Cost of goods sold (COGS) budgets includes all direct labor required to make the product or perform the service. It also includes all materials required to make the product or do the service. If labor or material is used only for specifically delivering the sale to the customer, then it is included above the line. Any labor or materials used to support the overall capability of the company to make sales is included in overhead accounts, also known as operating expenses.

Typical COGS accounts found above the line are in the following list:

- Factory labor employed in making the product.
- Service labor used in delivering the service to a customer.
- Materials used directly in the factory that become part of the product, usually identified on the product's bill of materials (BOM).
- Materials purchased to be used to deliver a service provided by the company, and would not have been purchased if the service had not been provided.

Gross profit is the difference between the money received for the sale of the product or service minus the cost of goods sold. In a very small business where no additional costs are necessary for the company to exist or plan for future jobs, this is essentially the same as net profit. However, this condition virtually never exists. There is always a cost of doing business that exists regardless of the levels of sales achieved. We call this overhead or operating expenses.

Operating expenses are all expenses that indirectly support the company's efforts to make sales and deliver products and/or services to its customers. This means every expense not categorized as a COGS is an operating expense. There are many subcategories of operating expenses, ranging from advertisements to utilities costs. The number and types of operating expense categories depend on the nature of the business the company is engaged in. Typical operating expenses:

- Advertising
- Automobile operating expenses
- Bank loan principle and interest charges
- Dues for memberships and subscriptions
- Equipment purchases
- Insurances, health, liability, workers' compensation
- Licenses and permits
- Office supplies

- Payroll (all except direct labor in COGS) and benefits (all personnel)
- Postage and delivery
- Professional development of personnel
- Rent
- R&D
- Repairs
- Supplies for factory not directly associated with products
- Taxes, sales, and property
- Telephone
- Travel and living
- Uncategorized expenses
- Utilities

These are not the entire list of categories found under the operating expense heading. For example, if a company required certain employees to wear uniforms, such as medical professionals in doctor's office, then a category for uniforms would be listed as an account category.

Each line item of operating expenses represents the amount of money the company is planning to spend for that category. Like sales, these specific line items are divided between the various departments that will spend money for those purposes. Not all line item operating expense budgets are included in all department budgets. However, if a department spends money that is in an account for which it has no budget, it will still show up on any reports showing expenses vs. budgets.

Net profit is the deduction of operating expenses from the gross profit. Recall gross profit is simply the difference between sales receipts and expenses directly related to making those sales. To derive a truer value for profit, we need to include the overhead the company incurs to support the sale, for example, a company has an office that requires rent payment. That payment is a cost associated with all sales, therefore a portion of the rent cost should be added to the cost of making the sale. All operating expenses are overhead, so a portion of all of these costs should be prorated against the gross profit of the specific sales. However, not all sales require all the support of all operating expense activities. For example, if a company makes a product and also provides a distinctly nonrelated service, it would not be an accurate accounting of costs to prorate manufacturing support costs against the service. Cost such as loan costs for purchasing factory equipment should not be applied against the gross profit of the service. Similarly, automobile and vehicle charges used only for the delivery of services should not be a charge against the factory-produced product's sales. This means net profit is not easily determined except in the aggregate, unless we know what accounts and what costs against those accounts belong to each department of the company. Doing this separation is the task of budget preparation.

Budget preparation begins with business planning where the company determines what it will be able to sell in the coming budget period, usually a year (a detailed discussion of business planning will be presented in a later chapter). Based on the sales expected, the company determines the labor hours and materials necessary to deliver the sales. For labor hours, the company will include any productivity program results it expects to achieve. These are converted into dollar values. Materials costs are estimated based on market prices for supplies and any improvements that can be made to use less materials, cheaper materials, or negotiate more favorable prices. With sales, labor, and materials identified, we can create the above-the-line budgets.

The below-the-line budget is constructed by estimating what costs are necessary in the various operating expense categories to support the above-the-line revenues and costs. Many estimates are based on previous experiences prorated up or down based on the expected level of sales. If a company has no previous experience with a particular account, then it is necessary to do some research to determine a good first cut at a budget recognizing that it may have to be adjusted after some experience is gained. Also included are the fixed expenses such as rents, leases, and other items that will be spent regardless of the amount of sales generated. These are usually contractual so the exact amounts are known. An example of a typical small company budget is shown in Figure 7-4.

A full budget will show above-the-line as well as below-the-line accounts. It portrays what sales need to be achieved as well as the direct and indirect expenses the company plans to spend to achieve their profit goals. Note in Figure 7-4 that the budget is remarkably similar to a P/L statement. This shouldn't be a surprise because, in fact, a budget is a P/L statement; but in the budget we show the accounts by number so they can be tracked via a general ledger entry system to measure actual against plan. That means that every line, or account, has an identity number, as we see at the extreme left of the document. Also note, that we subdivide accountability to the various departments within the company. In Figure 7-4, the first column of figures to the left is the total budget, that is, the P/L pro-forma. Then, progressing to the right we've subdivided the total into the share of the budget that each department is responsible for. Notice we also show at the top the percentage of sales that each department is to achieve. You may ask, isn't the marketing department responsible for sales? Yes, it traditionally is, in that it has the lead responsibility for contacts with the customers. However, with the example company, an engineering development company, we find that there are product sales and service sales. So it is sometimes convenient to give the lead department in accomplishing the delivery of the sales, that is, the completion of orders, the accounting responsibility for the sales. In that situation, marketing becomes an agent for manufacturing for product sales. This is just one convention. Many companies still favor the traditional approach of assigning sales to marketing only giving that department budget responsibility. This is a doable arrangement if marketing has the ability to commit the other departments to do their bidding to complete sales per the marketing department's perceived customer needs. If not, then they have a difficult time being effectively responsible because they have no authority to make it happen. For this reason, we see a shift of budget responsibility toward the source of the actual work for accomplishing the completion and delivery of the product or service to the customer.

Sample Budget, Small Engineering Development Company

	Account category			2002 Budget by departments ($000)			
		Total budget	admin.	engineering	manufacturing		marketing
Account No.	Department number		A100	A200	A300		A400
	percent of sales	100%	1%	10%	65%		24%
	Income						
10010	Fees	$323.1		$230.8			$92.3
10020	Products	$1,500.0			$1,500.0		
10030	Reimbursed expenses	$115.4					$115.4
10040	Sales (misc.)	$23.1	$23.1				
10050	Services (Sales)	$346.2					$346.2
10060	Uncategorized Income	$3.1	$3.1				
11000	TOTAL income	$2,310.9	$26.2	$230.8	$1,500.0		$553.9
	Cost of goods sold						
20010	Production materials	$405.0			$405.0		
20020	Production development materials	$167.0		$127.0			$40.0
20030	Misc. materials	$11.5	$1.0	$2.5	$5.0		$3.0
20040	Professional services (DL)	$690.0	$5.0	$85.0	$500.0		$100.0
21000	TOTAL Cost of goods sold	$1,273.5	$6.0	$214.5	$910.0		$143.0
22000	Gross profit	$1,037.4	$20.2	$16.3	$590.0		$410.9
	Expense						
30010	Advertising	$15.0					$15.0
30020	Automobile	$45.0	$5.0	$5.0	$5.0		$30.0
30030	Bank service charges	$1.0	$1.0				
30040	Building materials	$2.0			$2.0		
30050	Contributions	$2.0	$2.0				
30060	Damages	$1.0			$1.0		
30070	Dues & subscriptions	$2.6	$0.1	$0.3	$1.6		$0.6
30080	Equipment purchases	$15.0		$5.0	$10.0		
30090	Equipment rental	$4.0	$1.0	$1.0	$1.0		$1.0
30100	Health insurance	$90.0	$0.9	$9.0	$58.5		$21.6
30110	Liability insurance	$38.0	$0.4	$3.8	$24.7		$9.1
30120	Workers comp. insurance	$30.0	$0.3	$3.0	$19.5		$7.2
30130	Interest expenses	$10.0	$10.0				
30140	Late fees	$0.1	$0.1				
30150	Licenses & permits	$1.5	$1.5				
30160	Misc. expenses	$5.0	$1.0	$1.0	$2.0		$1.0
30170	Office supplies	$12.0	$0.1	$1.2	$7.8		$2.9
30180	Payroll (mgmt & non-exempt)	$477.0	$477.0				
30190	Postage & delivery	$1.0	$1.0				
30200	Printing and reproduction	$2.5					$2.5
30210	Professional development	$15.1	$0.2	$1.5	$9.8		$3.6
30220	Professional fees	$15.0		$5.0	$3.0		$7.0
30230	Recruiting	$0.5	$0.5				
30240	Rent	$15.1	$0.2	$1.5	$9.8		$3.6
30250	Repairs	$2.0			$2.0		
30260	Taxes	$75.1	$0.8	$7.5	$48.8		$18.0
30270	Telephone	$12.0	$1.0	$3.0	$3.0		$5.0
30280	Travel & entertainment	$25.0	$2.0	$4.0	$4.0		$15.0
30290	Uncategorized expenses	$2.0	$2.0				
30300	Uniforms	$1.0			$1.0		
30310	Utilities	$12.0	$0.1	$1.2	$7.8		$2.9
30320	Write offs	$0.5	$0.5				
31000	TOTAL expenses	$929.9	$508.7	$53.0	$222.2		$146.0
32000	Net Income	$107.5	-$488.5	-$36.7	$367.8		$264.9
	COGS % of sales	55.1%	22.9%	92.9%	60.7%		25.8%
	Professional services (DL) % of sales	29.9%	19.1%	36.8%	33.3%		18.1%
	Gross income % of sales	44.9%	77.1%	7.1%	39.3%		74.2%
	TOTAL expenses % of sales	40.2%	1941.6%	23.0%	14.8%		26.4%
	Net income % of sales	4.7%	-1864.5%	-15.9%	24.5%		47.8%

Figure 7-4. Typical small company budget.

At the bottom of the budget sheet we see the percent ratios, which measure the proportions of expenses to sales. Obviously, we want these ratios to result in the largest net income as a percentage of sales. This is the return on our investment. More on these measurements later.

The budget of Figure 7-4 is not a very good operations tool because it portrays such a long period of time, usually a year. It's difficult to make corrections based on the inputs of data from the general ledger. As bills are received, invoices sent out, payments received, and payrolls made we would fast lose track of where we are. Are we doing well or not? With the 3 reports, balance sheets, P/L statements, and cash-flow statements, we could tell where we are on a daily basis if we had the ability to accumulate, sort, and evaluate data virtually as it happened. But this is very impractical. So we use convenient time periods, typically a month. The company is literally operating in the dark, not knowing exactly where it is with respect to the plan until the data is compiled. We gather all the data for a month, and usually within a few days after the conclusion of the month, we categorize the data into the various accounts for the period and then prepare our 3 reports. To do this in a way for comparison purposes, we need to know what should happen during that period of time. This is the budget again. Only this time we take the annual budget and divide it into 12 monthly periods, but not simply dividing by 12 because we need to reflect what we expect to happen. Figure 7-5 shows an example of taking the budget and dividing it into 12 periods, that is, months, for one department, manufacturing.

All months may be created equally, but what we do and spend is not. That depends on the business cycle our company is engaged in. Looking at Figure 7-5, note that the sales are not the same for each month. The period of April through August shows lower sales than the other months. There is also a corresponding change in materials and professional services Direct Labor (DL). These variations reflect what the management team believes will happen over the course of the budget year. Most accounts will vary in the same manner as the sales, since the money we spend should be just enough to achieve delivery of the sale to the customer. In the example, I've simplified the Sale and am showing it to have occurred in the same month as the work done to make it. In the real world, this may not be the case; there could be an offset greater than the period of the monthly budget, and we would have to account for it in the monthly budget breakdown. Also, you may ask, why isn't rent constant? It is, but the department is paying only its share of the rent each month, based on its monthly percentage of the company's total sales. This is a puristic approach, which is correct for the example, to make the point of shared expenses based on the use of the support facilities during the budget period. But it is perfectly acceptable to simply charge the same amount for base services such as rent each month. This simplifies the accounting work even though it isn't entirely fair. I suppose it's fine for most budgets as long as everyone being measured by their expenses and sales understands the rules.

Figure 7-5, the monthly budget by department, gives us the baseline to measure monthly results against. We can see if we are tracking according to plan or not. And if not, it gives the opportunity to investigate for root cause and take corrective actions. We do this by getting the details from the general ledger (more on the specifics of how, later) for each account that has a variance we're not satisfied with. We try to understand why we spent more

Sample Budget, Engineering Development Company
by Month
Manufacturing Department

Account No.	Account category	manufacturing Total	2002 budget ($000) percent of sales A300 65%	Jan.	Feb.	Mar	April	May	June	July	August	Sept.	Oct.	Nov.	Dec.
	Department number														
	Income														
10010	Fees														
10020	Products	$1,500.0		$150.0	$150.0	$150.0	$100.0	$100.0	$100.0	$100.0	$100.0	$100.0	$150.0	$150.0	$150.0
10030	Reimbursed expenses														
10040	Sales (misc.)														
10050	Services (Sales)														
10060	Uncategorized income														
11000	TOTAL Income	$1,500.0		$150.0	$150.0	$150.0	$100.0	$100.0	$100.0	$100.0	$100.0	$100.0	$150.0	$150.0	$150.0
	Cost of goods sold	$405.0		$40.5	$40.5	$40.5	$27.0	$27.0	$27.0	$27.0	$27.0	$27.0	$40.5	$40.5	$40.5
20010	Production materials	$5.0		$0.5	$0.5	$0.5	$0.3	$0.3	$0.3	$0.3	$0.3	$0.3	$0.5	$0.5	$0.5
20020	Production development materials														
20030	Misc. materials	$50.0		$5.0	$5.0	$5.0	$3.3	$3.3	$3.3	$3.3	$3.3	$3.3	$5.0	$5.0	$5.0
20040	Professional services (DL)	$910.0		$91.0	$91.0	$91.0	$60.7	$60.7	$60.7	$60.7	$60.7	$60.7	$91.0	$91.0	$91.0
21000	TOTAL Cost of goods sold														
22000	Gross profit	$590.0		$59.0	$59.0	$59.0	$39.3	$39.3	$39.3	$39.3	$39.3	$39.3	$59.0	$59.0	$59.0
	Expense														
30010	Advertising	$5.0		$0.5	$0.5	$0.5	$0.3	$0.3	$0.3	$0.3	$0.3	$0.3	$0.5	$0.5	$0.5
30020	Automobile														
30030	Bank service charges	$2.0		$0.2	$0.2	$0.2	$0.1	$0.1	$0.1	$0.1	$0.1	$0.1	$0.2	$0.2	$0.2
30040	Building materials														
30050	Contributions														
30060	Damages	$1.0		$0.1	$0.1	$0.1	$0.1	$0.1	$0.1	$0.1	$0.1	$0.1	$0.1	$0.1	$0.1
30070	Dues & subscriptions	$1.6		$0.2	$0.2	$0.2	$0.1	$0.1	$0.1	$0.1	$0.1	$0.1	$0.2	$0.2	$0.2
30080	Equipment purchases	$10.0		$1.0	$1.0	$1.0	$0.7	$0.7	$0.7	$0.7	$0.7	$0.7	$1.0	$1.0	$1.0
30090	Equipment rental	$1.0		$0.1	$0.1	$0.1	$0.1	$0.1	$0.1	$0.1	$0.1	$0.1	$0.1	$0.1	$0.1
30100	Health insurance	$58.5		$5.9	$5.9	$5.9	$3.9	$3.9	$3.9	$3.9	$3.9	$3.9	$5.9	$5.9	$5.9
30110	Liability insurance	$24.7		$2.5	$2.5	$2.5	$1.6	$1.6	$1.6	$1.6	$1.6	$1.6	$2.5	$2.5	$2.5
30120	Workers comp. insurance	$19.5		$2.0	$2.0	$2.0	$1.3	$1.3	$1.3	$1.3	$1.3	$1.3	$2.0	$2.0	$2.0
30130	Interest expenses														
30140	Late fees														
30150	Licenses & permits	$2.0		$0.2	$0.2	$0.2	$0.1	$0.1	$0.1	$0.1	$0.1	$0.1	$0.2	$0.2	$0.2
30160	Misc. expenses	$7.8		$0.8	$0.8	$0.8	$0.5	$0.5	$0.5	$0.5	$0.5	$0.5	$0.8	$0.8	$0.8
30170	Office supplies														
30180	Payroll (mgmt & non-exempt)														
30190	Postage & delivery														
30200	Printing and reproduction	$9.8		$1.0	$1.0	$1.0	$0.7	$0.7	$0.7	$0.7	$0.7	$0.7	$1.0	$1.0	$1.0
30210	Professional development	$3.0		$0.3	$0.3	$0.3	$0.2	$0.2	$0.2	$0.2	$0.2	$0.2	$0.3	$0.3	$0.3
30220	Professional fees														
30230	Recruiting														
30240	Rent	$9.8		$1.0	$1.0	$1.0	$0.7	$0.7	$0.7	$0.7	$0.7	$0.7	$1.0	$1.0	$1.0
30250	Repairs	$2.0		$0.2	$0.2	$0.2	$0.1	$0.1	$0.1	$0.1	$0.1	$0.1	$0.2	$0.2	$0.2
30260	Taxes	$48.8		$4.9	$4.9	$4.9	$3.3	$3.3	$3.3	$3.3	$3.3	$3.3	$4.9	$4.9	$4.9
30270	Telephone	$3.0		$0.3	$0.3	$0.3	$0.2	$0.2	$0.2	$0.2	$0.2	$0.2	$0.3	$0.3	$0.3
30280	Travel & entertainment	$4.0		$0.4	$0.4	$0.4	$0.3	$0.3	$0.3	$0.3	$0.3	$0.3	$0.4	$0.4	$0.4
30290	Uncategorized expenses														
30300	Uniforms	$1.0		$0.1	$0.1	$0.1	$0.1	$0.1	$0.1	$0.1	$0.1	$0.1	$0.1	$0.1	$0.1
30310	Utilities	$7.8		$0.8	$0.8	$0.8	$0.5	$0.5	$0.5	$0.5	$0.5	$0.5	$0.8	$0.8	$0.8
30320	Writeoffs														
31000	TOTAL expenses	$222.2		$22.2	$22.2	$22.2	$14.8	$14.8	$14.8	$14.8	$14.8	$14.8	$22.2	$22.2	$22.2
32000	Net Income	$367.8		$36.8	$36.8	$36.8	$24.5	$24.5	$24.5	$24.5	$24.5	$24.5	$36.8	$36.8	$36.8
	COGS % of sales	60.7%		60.7%	60.7%	60.7%	60.7%	60.7%	60.7%	60.7%	60.7%	60.7%	60.7%	60.7%	60.7%
	Professional services (DL) % of sales	33.3%		33.3%	33.3%	33.3%	33.3%	33.3%	33.3%	33.3%	33.3%	33.3%	33.3%	33.3%	33.3%
	Gross income % of sales	39.3%		39.3%	39.3%	39.3%	39.3%	39.3%	39.3%	39.3%	39.3%	39.3%	39.3%	39.3%	39.3%
	TOTAL expenses % of sales	14.8%		14.8%	14.8%	14.8%	14.8%	14.8%	14.8%	14.8%	14.8%	14.8%	14.8%	14.8%	14.8%
	Net income % of sales	24.5%		24.5%	24.5%	24.5%	24.5%	24.5%	24.5%	24.5%	24.5%	24.5%	24.5%	24.5%	24.5%

Figure 7-5. Typical small company budget by month for the manufacturing department.

or less than the plan and then make a judgment as to whether it was justified. If not, we put controls in place to prevent reoccurrence and simultaneously make changes to operating procedures to make the operation perform as planned.

In the case of sales being less than planned, we do basically the same thing. We had a plan to sell X volume and only did Y. We would have to understand why, then take corrective actions. If it's due to poor salesmanship, that would be fixed. If it's due to our product or service not being suitable for the customer, we'd have a more basic problem and would be forced to go through an analysis such as described in the new product introduction chapter. Sales misses are always more traumatic than other deficiencies because the entire budget is based on specific sales levels being achieved. This means companies react more rapidly to these variations than to any other. The budget process and the measurement reviews that result from it are vital to the company's success.

The company has now created a total overall budget. For the budget to be a useful measurement tool to evaluate the performance of the various departments, it is necessary to divide up all accounts into the departments that have responsibilities for managing those accounts, as discussed above. This is done first by allocating responsibility for the above-the-line items by giving sales budgets to the departments directly involved in selling, then giving direct labor budgets to those components directly involved in manufacturing the products or delivering the services. Materials budgets, likewise, are given to those involved in actually buying and using the materials.

Below-the-line budgets require more analysis. If an operating expense item is directly related to a specific department, the entire budget is allocated to that department. If an expense item is shared by 2 or more departments and the amounts can be easily discerned, then it is split accordingly and given to those departments. This may or may not be easy to do. For example, a budget for computers may be capable of splitting if all employees get the same terminals. Therefore, if 10 terminals are required overall, and 1 department has 6 workstations while another has 2 and a third has 2, then a 60, 20, 20 percentage split is appropriate. Likewise, salary and benefit expenses can be split, depending on the number of employees and how much they individually earn allocated to each department. The hard part comes when such things as rent, telephone, and utilities must be split up. There are several ways of doing this.

- By percent area of total space occupied
- By percentage of historical use for a service
- By prorating by population of people using the service
- By any other way management feels gives a true representation of cost allocations

Sometimes management doesn't feel an allocation would serve any useful purpose and perhaps just clouds the ability to measure performance. When that happens, the cost is put into an administrative account. Even salaries can sometimes go into this account if enough people are used in more than one department, and the amount of time they spend there varies considerably throughout the year. This is common for matrix organizations. The administrative account is a catchall that belongs to the most senior member of management, usually the president in a small company, and contains discretionary accounts that the president feels he or she needs to control and perhaps keep confidential. Also, other accounts (it's not worth the bother to figure out where they belong) would be included in the administrative account. The key to a successful administrative account is to ensure that accounts that could be used for important measurements for the departments are excluded from it.

BUDGET-RELATED MEASUREMENTS

The budget is not complete until we have a way of keeping score with it. The simplest way to do this is to create account and department numbers so that charges as they occur can be put in the proper "bucket." This means charging to the proper account and department so we can make periodic comparisons between actual spending occurrences and the budget.

The numbering system for all accounts is arbitrary. Whatever strikes the budget-makers' fancy is suitable. One bit of advice is to leave at least 10 digits between each account. This allows for additional accounts to be placed between existing accounts at a later date. Often it turns out we would like a more detailed breakdown of costs for analysis purposes.

Example: If we have several types of automobile expenses and they're all lumped into one account, we would have difficulty determining which type of expense is causing a potential problem. Perhaps we use automobiles for deliveries and also for sales calls. Suppose we are over-budget for automobile expenses by 50%. Is it due to deliveries being up, which would be good? Or is it because sales calls are up, which could be either good or bad, depending on the number of sales generated compared with the sales budget? To tell where the automobile expenses are over-budget, the company may elect to divide the account into 2 accounts. If the original account number was 30020, we may reserve that one for sales' use of automobiles and create a 30021 for deliveries' use of automobiles. But we couldn't do that if there was no digit room between automobile expenses and the next account, bank service charges, or the previous account, say advertising.

Now, how do we keep books on expenses to compare to budgets? We use a "general ledger" concept. Every time a bill is received and an expense occurs for such things as salaries, it is coded with the account number and the department number. For example, suppose we make a salary payroll disbursement to all salaried employees, and suppose we charge salaries to the administrative account. Then the accounting would be something like this:

We would make an entry in the general ledger for $10,000 charged against account 30180 department A100, where 30180 represents the salary operating expense account and A100 is the administrative department's identification number. In simplified form, the general ledger account entry would look something like this:

> A100 30180 11/23/01 $10,000.00   sal.

From left to right we have department I.D., account number, date of transaction, amount of transaction, and the reason for the transaction. Of course this is only one transaction. In reality, there would be hundreds each month for even a small company. So usually these general ledger entries are made daily and summed perhaps daily if there are lots of them, but usually weekly. Most of the general ledger entry work is now done by entering data into packaged computer programs such as Peachtree and competitor equivalents.

At the end of each reporting period, the sum for all accounts for each department is totaled and presented on a report with a comparison against the budget for that period. This is used as a management report for evaluating how well the company did for that period. Typically these reports, called monthly financials, are generated for each department and summed for managers responsible for more than one department. Most of these accounting programs have the ability to tailor reports to suit the specific needs of individual companies.

The purpose of any accounting system is to evaluate the financial performance of a company as compared with its plan. The first thing a manager does when he or she receives the monthly financial reports is to compare bottom line results with the budget (plan). Obviously we want to be on budget or better than budget. But just achieving that is not enough. We need to know why we are where we are so we won't be surprised next month or the month after that. We need to understand why the numbers came out the way they did so we can make corrections, if necessary, to achieve the year-end goals. After we look at the overall results, we need to scan back up the page to look for accounts that are at variance to the budget. As a general rule of thumb we should understand every account number that has a variance of greater than 10%, plus or minus. Under-spending should not be automatically looked at as being a positive. There may be some dire reasons for this happening. For example, we may be slow in initiating a project, and the delay may cause us to miss sales opportunities we were counting on. So the message is to understand why things are not per budget. If an under-spending has occurred, but everything else associated with it is satisfactory, then perhaps the under-spending is due to better-than-expected performance, and that's good. But you wouldn't know this unless the variance was investigated. It is very important to instill the discipline of thoroughly investigating monthly financial report variances every month and as a high priority management task. You'd be surprised how much can be learned by just perusing the general ledger and asking questions as to why an expense had occurred, especially if it's at variance to the budget. Often we find the budget is flawed and was based on overly optimistic assumptions and the expenses reported are realistic. By knowing this we can take corrective actions before the company has committed too many resources to a faulty plan. Similarly, over-expenditures may indicate an under-trained workforce that if not corrected, could cause the company to miss commitments, thus disappointing customers with all the detrimental effects that that has on future company growth. So we see that using the monthly financial report as a measurement tool is a significant segment in managing the business and needs to be encouraged.

Based on the P/L structure of the budget, over the years managers have developed some interesting ratios that can be used as measurements of economic performance. There are 5 of them of note in general use:

Cost of Goods Sold as a Percentage of Sales

This measures the direct expenses of making the product or delivering the service against the sale price for doing so. It is usually expressed as a percentage. Here we want the percentage to be as low as possible so the gross profit will be as high as possible.

Direct Labor as a Percentage of Sales

This is the labor component of the COGS/Sales ratio. Companies that are very heavily dependent on labor, usually service companies, often use this measurement instead of the total COGS measurement. It often is used to set productivity improvement goals for products and services where prices are very stable and there is little potential for increases.

Gross Income as a Percentage of Sales

This is the opposite concept of the COGS/Sales ratio. Gross income is the difference between income received and direct expenses necessary to achieve it. The ratio shows the absolute highest profit we can achieve if overhead was zero. It tells us what the upper limit of expenses can be before we actually lose money on the product or service. It is very useful in setting prices if it can be based on a known overhead cost.

Operating Expenses as a Percentage of Sales

This is the way we obtain a value on a macro basis for the overhead percentages we should use for new products or services. This is the below-the-line costs ratio to sales.

Net Income as a Percentage of Sales

This is a method of expressing return on investment. It is the profit number by which we measure the worth of the company. Companies desiring investment capital via venture or angel investors or by selling stock are interested in keeping this number as high as possible. A benchmark is to compare this number against what an investor could get by investing in good commercial paper such as bonds and treasuries, as well as in other companies. It is also a way for the company to determine in a very macro way whether its owners would have been better off investing in stocks and bonds or simply putting the funds into a bank savings or money market account. The old axiom that a company's net income has to be higher than bank interest always applies.

These are the common financial measurements associated with budgets. Measurements by themselves mean nothing. It is the actions taken as a result of the measurements and the succeeding measurements over time that tell the story. We should use measurements to point the way to root causes of problems, then solve those problems, then the succeeding measurements at a later date will tell us if the problem has been resolved effectively.

I have already mentioned the use of analysis to understand reasons for variances and then correcting the root cause. I cannot overemphasize this philosophy. Most companies failing to do a good job in finding root causes of problems fail because they do not do an adequate job in this arena. Using a financial measurement to discover a problem in an operating expense line should lead to an investigation for root cause. At the point of being redundant on this point for the need for adequate analysis, let me give another example.

Example: If we find that sales are on budget but that direct labor is 10% over budget, we suspect that the productivity is not at the level it should be. We would investigate how jobs are being done and compare that against how they were planned to be done. If we find that the method is not being followed correctly, we need to find out why. Perhaps the training of the operators was inadequate. If that's the case then we have to delve deeper and understand how we go about training the operators and find out where we're going wrong. Perhaps we're assuming the operators have a base knowledge level that they really don't. If this is the root cause, then we fix it by instituting remedial training. As you can see, an economic variance can lead to a complex decision tree–type investigation to get to the root cause. That path has to be followed; otherwise the variance will never be corrected and the company's financial viability will eventually be detrimentally affected.

I believe so strongly in the need for proper analysis that I have employed specific report forms to be used for the analysis. This way I can ensure that the level of the analysis from month to month is consistent. An excellent way to track changes made as a result of an analysis is by doing a re-forecast of the next few reporting periods. A good way to do this is through a monthly 3-month rolling forecast, as shown in Figure 7-6.

A rolling 3-month forecast means we do a forecast for the same month each month for 3 months, much like standing in a queue and moving up one

	Account Category ($000)	Next Month: Budget	Next Month: Forecast	2nd Month: Budget	2nd Month: Forecast	3rd Month: Budget	3rd Month: Forecast	Reason for variance more than 10% Use back of page if more space required
Account No.	**Income**							
10010	Fees							
10020	Products							
10030	Reimbursed expenses							
10040	Sales (misc.)							
10050	Services (Sales)							
10060	Uncategorized income							
11000	TOTAL income							
	Cost of goods sold							
20010	Production materials							
20020	Product development mat'ls							
20030	Misc. materials							
20040	Professional services (DL)							
21000	TOTAL Cost of goods sold							
22000	Gross profit							
	Expense							
30010	Advertising							
30020	Automobile							
30030	Bank service charges							
30040	Building materials							
30050	Contributions							
30060	Damages							
30070	Dues & subscriptions							
30080	Equipment purchases							
30090	Equipment rental							
30100	Health insurance							
30110	Liability insurance							
30120	Workers comp. insurance							
30130	Interest expenses							
30140	Late fees							
30150	Licenses & permits							
30160	Misc. expenses							
30170	Office supplies							
30180	Payroll (mgmt & non-exempt)							
30190	Postage & delivery							
30200	Printing and reproduction							
30210	Professional development							
30220	Professional fees							
30230	Recruiting							
30240	Rent							
30250	Repairs							
30260	Taxes							
30270	Telephone							
30280	Travel & entertainment							
30290	Uncategorized expenses							
30300	Uniforms							
30310	Utilities							
30320	Writeoffs							
31000	TOTAL expenses							
32000	Net income							
	COGS % of sales							
	Professional services (DL) % of sales							
	Gross income % of sales							
	TOTAL expenses % of sales							
	Net income % of sales							

Figure 7-6. Short range forecast form.

customer at a time to purchase a commodity in a store. This is an excellent way of documenting changes due to variances in the business situation. The form requires the budget and its related "new" forecast to be input next to it for the next 3 months. It requires an explanation for variances by account where they occur. Forcing this forecast to be done every month makes managers and those responsible for achieving company goals to be cognizant that their actions will always be reflected in financial results. This is a very effective way of achieving the awareness of preplanned actions vs. simply reacting to circumstances.

For a short-range forecast to be made, a starting point is needed that shows the baseline. This is the budget initially, but quickly becomes the results from the first reporting period. An excellent way of recording those

Financial Potpourri • 179

Year:_____		Month:_____			Completed By:_____
Department Name:_____					Date:_____
Department Number:_____	Budget	Forecast	Actual	Difference Act./Fcst	Reason for variance more than 10% Use back of page if more space required

Account No.	Account Category ($000)	Budget	Forecast	Actual	Difference Act./Fcst	Reason for variance
	Income					
10010	Fees					
10020	Products					
10030	Reimbursed expenses					
10040	Sales (misc.)					
10050	Services (sales)					
10060	Uncategorized income					
11000	TOTAL income					
	Cost of goods sold					
20010	Production materials					
20020	Production development mat'ls					
20030	Misc. materials					
20040	Professional services (DL)					
21000	TOTAL Cost of goods sold					
22000	Gross profit					
	Expense					
30010	Advertising					
30020	Automobile					
30030	Bank service charges					
30040	Building materials					
30050	Contributions					
30060	Damages					
30070	Dues & subscriptions					
30080	Equipment purchases					
30090	Equipment rental					
30100	Health insurance					
30110	Liability insurance					
30120	Workers comp. insurance					
30130	Interest expenses					
30140	Late fees					
30150	Licenses & permits					
30160	Misc. expenses					
30170	Office supplies					
30180	Payroll (mgmt & nonexempt)					
30190	Postage & delivery					
30200	Printing and reproduction					
30210	Professional development					
30220	Professional fees					
30230	Recruiting					
30240	Rent					
30250	Repairs					
30260	Taxes					
30270	Telephone					
30280	Travel & entertainment					
30290	Uncategorized expenses					
30300	Uniforms					
30310	Utilities					
30320	Write offs					
31000	TOTAL expenses					
32000	Net income					

	Budget	Forecast	Actual	Difference
COGS % of sales				
Professional services (DL) % of sales				
Gross income % of sales				
TOTAL expenses % of sales				
Net income % of sales				

Figure 7-7. Review of operations form.

results and understanding the reasons why they happened a certain way is through the use of a "Review of Operations" report (Figure 7-7).

This form is similar in setup as the short-range forecast form, except that here we're dealing with the month that just concluded. We list the budget for the month, list the most current forecast from the latest short-range forecast, and show the results for the month in question under the actual column. We then show the variances between forecast and actual, with reasons for variances, if any. This then leads to an investigation and corrective actions, as discussed previously. The results of the review of operations becomes the basis for the next short-range forecast. This may seem like a lot of work, but it is very necessary. Managing the company's assets in achieving the goals is the only way a company can stay solvent. Those that take this task seriously and understand that it is an operations responsibility, with the aid of the finance department and not vice versa, will have successful companies virtually guaranteed. As for those that don't, statistics show those companies fail more often than we would believe to be the case.

One other point on measurements: Financial measurements are macros in that they indicate a problem, which requires investigation to find the real cause. Most companies supplement financial measurements with nonfinancial measurements aimed at known or suspected problem areas. Such measurements as defects per million occurrences or hours of labor expended per product produced are common quality and productivity measurements that tell quantitatively how the company is doing. When they are used in conjunction with financial measurements, it is possible to tell the incremental worth of a percentage improvement of productivity and quality. This becomes a very powerful tool in forecasting financial results and as a planning device for understanding the worth of improvement programs.

SOME USEFUL MANAGEMENT TECHNIQUES FOR MAINTAINING FINANCIAL VIABILITY

A company full of promise but unable to pay its bills will inevitably fail. Therefore, the entrepreneur needs to be ever-mindful of her or his company's cash position. This means it is prudent to know what's happening virtually day by day so there are no unpleasant surprises with respect to cash flow. It is important to be able to anticipate current cash needs and where the money will come from to satisfy those needs. While this is true for any company, it is especially true for new startups that do not have extensive cash reserves. Large companies have cash in bank accounts and readily redeemable financial paper. Small companies enjoy this backup to a much lesser degree. Startups are usually always scrambling to make payments and payrolls. So it becomes even more important for startup managers to have tight controls on cash.

Even with all the financial controls we see in place for most companies, the entrepreneur needs more. The entrepreneur is trying to grow the fledgling business and at the same time remain solvent such that he or she maintains control and doesn't barter away the future ownership in return for survival cash. This is a tough chore and there is no simple all-encompassing equation to make it happen. But one thing that will aid in this endeavor is good cash control over and beyond what we get from everything discussed up to this point. Let's see what else can be done.

Needs vs. Receipts Check List

If I know what I have to pay today and can plan where the money will come from, I am 90% home in achieving the goal for the day. Remember, the goal for the entrepreneur is to remain solvent every day so the entrepreneur can grow the enterprise to the point where that happens virtually automatically. As a manager, I want to know what bills are really due today those must be paid. Then, I want to know what receipts I expect to get today to add to my checking account for paying bills. This can be done by making up a simple check sheet based on the P/L statement format. A sample is shown in Figure 7-8.

Engineering Development Co.
Mgr's. Estimate Cash Position Sheet

Date	5/16/01			Projection For Next Payroll (5/25/01)			
Starting Balance	$4,658						
Adds		date here estimate	date here actual	Subtracts		date spent planned	date spent actual
Tropical Mfg.final payment	$270	5/14	5/14	Sales tax	$1,000	5/18	
Cadwell Partners final payment	$1,480	5/18	5/16	Bayman Metal, 1st Pay	$1,800	5/14	5/14
Cingul Tampa (6) finals	$12,478	5/18		Short Loan Payback-Joe	$4,200	5/15	5/16
Cingul Naples/Brad.(2) finals	$6,602	5/21		Lawyer	$2,800	5/16	5/16
Crown final	$2,400	5/21		Short Loan Payback-Dan	$4,615	5/25	
City of Payola 1/4 payment	$927	5/15	5/14	payroll 5/25/02	$34,000	5/24	
Steve Inc. final (a)	$2,917	5/25		Bayman Metal, 2nd Pay	$1,800	5/16	5/16
Brian Ind. deposit	$401	5/15	5/16	Ground detectors (600)	$3,200	5/25	
Hood Ind. Add'tl Deposit (a)	$1,000	5/18	5/16	King Electronics Supplies	$2,000	5/25	
Dipple Co. final	$896	5/25	5/15	Phil Transportation	$3,500	5/25	5/25
Walker Mfg.final	$128	5/15	5/15	Minute Man Digital	$2,500	5/25	
Ladelly Instruments, final	$2,414	5/18	5/14	Mill Wire Co.	$1,000	5/25	
Oak Park Co. final (a)	$1,000	5/24		Industrial Tech. Weld	$1,700	5/25	
Roschman Stores final	$2,085	5/18	5/17	Test Lab	$4,000	5/25	
LS School Deposit (a)	$5,000	5/21		Insurance	$800	5/25	
Lipan Fan, final	$2,700	5/25		American Mach. Repair	$500	5/25	
Higgins Camera, final (a)	$1,800	5/18		telephone	$350	5/25	
Messer Auto., final	$900	5/18		Vehicle payments	$850	5/25	
Pelton Trans., final	$600	5/25		Utilities	$1,500	5/25	
Taff Ind., final	$600	5/21		Plastec	$500	5/25	
Salter Co, deposit	$3,600	5/25		Rent	$2,000	5/15	5/15
Summer Run Hotel, final (a)	$3,300	5/25					
Sova University, deposit (c)	$3,980	5/14	5/14				
Kupple Co., final	$2,258	5/18	5/16				
Fink Distributors, deposit (c)	$2,190	5/16	5/16				
Schelb Mfg., deposit (c)	$880	5/16	5/16				
Ganger Int'l., deposit (c)	$545	5/17	5/15				
Iavino Co., deposit (c)	$586	5/21	5/16				
Brian Ind., addendum	$1,067	5/21					
Rodriguez Export, final	$620	5/24					
Bradly Distributors, final	$1,727	5/25					

Collectables - no date for receipt (A/R's beyond 90 days)	
Bay Mfg.	$46,000
Leyot Co., 2nd payment	$12,000
Silber Trans., Credit	$1,000
Alon Co., final	$3,700
Total	$62,700

new deposits (b) $15,833 5/25

Total Adds $87,842

Total Subtracts $74,615

Balance $13,227

(a) At risk of not arriving as scheduled
(b) Unknown - estimate based on current activities.
(c) New jobs logged against new deposit estimate of $25,000 for period.
    Amount shown as Total Adds is remainder of new deposits expected.

Figure 7-8. Manager's cash flow check sheet.

This is a simple Excel spreadsheet that the manager uses to keep track of pending receivables and payables for the immediate period. I recommend making the time period coincide with payroll, because payroll is usually the biggest expense a startup company has on an ongoing basis. If a payroll is disbursed every other week, then the cash-flow check sheet would be made up the day of the previous payroll disbursement and updated daily by the manager. Some say this is redundant to what the accounting clerks are doing but it's not. The accounting clerks are not involved with day-to-day operations in a decision-making manner. They assist the manager with inputs of financial data from which decisions can be made. By the manager keeping this simple check sheet, he or she knows what is happening as close to real time as possible. This way, decisions become as close to real time as possible, thus affording the opportunity for quick responses to customer needs, as well as monitoring the solvency of the company.

Here is how the check system works. At the end of each period, the manager, with inputs from finance and operations, makes up the check sheet. She or he lists the current checkbook balance, then lists the expected customer payments for the new period as far as they are known. If the customer has already placed an order but the deposit or first payment has not been received, it is shown on the list under the Adds column as a deposit to be received. We see on Figure 7-8:

"Brian Ind. Deposit $401 date here estimate 5/15"

For orders that are in the production queue, they are listed as final payment with a date coinciding with the shipment and lead time for receiving the payment. Many companies will not make final shipments until the customer makes final payment. So it is reasonable to be able to predict with a high degree of certainty when the payment will be received. This is reinforced by the fact that payment is made upon completion of manufacturing, so the schedule is virtually all in the hands of the entrepreneur's company. A typical entry shown on Figure 7-8 would be:

"Cadwell Partners final payment $1,480 date here estimate 5/18"

In addition to deposits for already booked orders, we also include an estimate of new orders we expect to receive during the monitoring period. For the period, according to note (c), the manager selected $25,000 for anticipated new business. However, we see a total of $15,833 listed for new deposits shown under the Adds column of Figure 7-8. This is the remainder of the $25,000 of the original estimate at the beginning of the period. Note (c) explains the arithmetic and the job deposits, such as Sova University on Figure 7-8.

The tally of beginning of period balance, deposits known to be in transit, anticipated additional deposits, and planned final payments makes up the anticipated Adds or receipts. This is the total amount of cash the company possesses to run the business for this period.

On the right-hand side of Figure 7-8, the Subtract items are listed with their due dates. A typical entry would be:

"sales tax $1,000 date spend planned 5/18"

These are the bills and the payroll that must be paid during this period. They are the key items that have to be paid now. There may be and probably are more bills that the company must honor, but not at this time. For example, a vendor may have supplied transistors for circuit boards on the 15th of the month, payable in 30 days. That means the bill payable on June 15 would not appear on this edition of the check sheet because it is a transaction that must be executed at a time beyond the immediate concern. Of course the manager needs to be cognizant of the future needs in making payment plans, but that is secondary to getting through this period.

The arithmetic of the Adds/Subtracts culminates in a balance at the end of the period. This is the amount of money available to start the next period. Whether that is sufficient depends on the receivables/payables activities into the future. If the P/L statement, Cash-Flow statement, Balance sheet, and the measurements and forecasts are being done with the attention they require, there will be few surprises. If not, there will be some rocky roads to travel.

The last item on the check sheet to be aware of is what I call the advertisement section. Those familiar with the commercial web sites on the Internet could properly call this the paid for banner section. It is a reminder to the manager that expecting everything to go in accordance to plan is naive. Here we see a list of collectables that are more than 90 days old. Their total is quite large, approximately two thirds of the entire Adds for the period. These are funds owed to the company that may never be collected, a serious problem for a startup. It is shown here to remind the manager of the need to make decisions that keep the cash inflow leverage positive for the company and to not take risks with doing work and not being paid for it. It also points out that these are opportunities for the manager to make her burden lighter by making those overdue collections. So it is both a warning to husband cash and an opportunity to obtain more cash. Also, for a very new startup the banner section may want to list the amount of cash reserves available in case the 2-week plan really goes astray.

I've explained the mechanism of the cash check sheet. Its use is straightforward. The manager conducts business as usual but adds keeping an eye on the inflow of cash to her responsibilities. Knowing the status of the inflow, the manager can make judgments as to how to manage the production schedule to optimize the final payments cash flow. Also, based on the cash flow, the manager can regulate the flow of cash, leaving the company as much cash as possible for current bill payments. Probably most important for future success is the management of cash so that payrolls can be met, thus keeping the startup's employees motivated. This balancing can be exciting, exhilarating, and terrifying the latter especially if the manager sees a period where cash will be tight, even though the medium- to long-term future is very bright and positive. What do we do when that happens? Let's take a look.

Managing Cash Flow Beyond Payables and Receivables

Sometimes no matter how well you pay attention to the details, the incoming cash temporarily doesn't meet the outflow requirements for a specific period of time. It's easier to become cash negative than cash positive. The former is entirely under your control. You have to take the initiative to make a transaction. Although becoming cash positive requires you to take an action, a corresponding action is needed on someone else's part. If the company sells a product, it is due a cash payment. But it doesn't receive the payment until the customer takes action to pay. With an expense, you or your agents have purchased some goods or service that now places an obligation to pay when the bill becomes due. We can say that it is simply putting the shoe on the other foot. I can chose not to pay just as a customer can chose not to pay me. True enough. However, in the case of not paying your legitimate bills you are taking a chance in damaging your financial credit rating. The customer not paying his bills is chancing damaging his credit rating, but that's not a concern of yours for the viability of your company. So, responsible entrepreneurs should feel an obligation to pay all bills on time, regardless of the actions of his debtors. With this sense of responsibility in place, it's easy to see how it's relatively easy to create obligations to pay vs. more difficulty in collecting funds due to the company.

Often a company may have to buy equipment and supplies well ahead of receiving payment for those good. The saying "you have to spend money to make company" is certainly true. But if we're depending on cash flow to pay for all of these expenditures, we're liable to run out and not have paid all the current due bills. If I have to spend $100 this month to bill $1000 for next month, I still will do that. The problem is where the money comes from to write that $100 check. It will come out of cash reserves, such as the starting balance shown in Figure 7-8, the manager's estimated cash position sheet, or from another source. Those other sources will be loans of some sort from a variety of sources. Here are a few:

Sell Shares of the Company

This is selling equity in the company to others. It changes the percentage of ownership among all of the investors. The manner and amount of equity that can be sold in this manner is regulated by state and federal law to some extent, usually through the tax codes and the securities and exchanges codes. Realistically, this is a "crap shoot" for a startup company. More funds can be raised for company reserves but this will not happen immediately and probably will not affect the current situation. So think of equity selling as a way to bolster future strategic cash reserves and probably not an answer to today's immediate problem.

Raising funds through equity selling requires finding interested investors, and once investors are found, setting up a contractual situation between existing investors and potential new investors or even among existing investors, if they are putting more money into the company. Determining how much a share is worth is definitely subjective and speculative. If people want to buy in, then the value of share prices can be set higher by the owner. If the owner needs funds to operate the company, then the percentage of ownership decision advantage goes to the buyer. As a general rule, the more desperate the entrepreneur gets to raise money, the more ownership percentage he will have to give up. As an aside, this is how venture capitalists tend to legally "steal" a company from the entrepreneurs. They are standing with the money but will give it only on terms advantageous for them.

Obtain a Bank Loan

Theoretically, this gets the entrepreneur money without giving up control of the company. All the entrepreneur has to do is pay back the loan plus interest at the contractual times. In practice it doesn't work quite that way. First, to get the loan, the bank will want some collateral, which is essentially giving hostages that can only be repatriated when the loan and all of the conditions attached to it are satisfied. Second, the bank may insist on being able to examine the financial records of the company periodically to make a judgment as to whether its investment is still safe. In other words, is the entrepreneur running the company in a fiscally responsible way that will ensure the loan obligations are met?

Notice I didn't say that the company is managing in a fiscally responsible way to ensure growth and reach the company's market potential. Those items may or may not coincide with the bank's interest. The entrepreneur may want to take more risks than the bank is willing to tolerate in order to achieve his or her ultimate goal. The bank could be very unhappy about that and in some cases even recall the loan, thereby really causing a financial calamity for the startup company. Yes, many loan contracts have provisions for the bank to make the loan fully due and payable at any time at the bank's choice. What this all means is that once a bank loans a startup funds for operations, it in essence becomes a partner in running the business. Sometimes this is good because the bank's representatives may have excellent business sense and be a real asset to the new company. Sometimes the reverse is true: The bank hinders proper operation because it does not understand the particular mechanisms necessary to make the entrepreneur's startup successful. Whatever the case, the bank becomes a partner, for better or worse.

Bank loans come in all sorts of varieties. They range from a one-time straight cash dispersement to a line of credit that the company can draw from when needed. Payment plans also vary considerably. Many lines of credit are revolvers in that they have an upper limit and the company can use the funds up to that limit. When payments are made, the amount of credit available increases because the difference between the upper limit and the amount of already committed increases. Sometime lines of credit have no due date on the entire principle due. All that is necessary is the monthly interest payments be made on time.

Whatever the type of loan being considered, keep in mind that the bank will want some type of collateral. It can be assets of the business, if it has anything worth enough to be pledged. If need be, the bank may want the entrepreneur's home or outside investments such as stocks and bonds pledged as collateral. Also, be aware that banks are very conservative and often do not loan money to startups. They are more interested in doing business with well-established business. It is not unusual to find a very attractive new venture turned down, while a steady-state, no excitement, no-real-growth-potential business will be approved because it has survived and has virtually no chance of defaulting on the loan. The better established a company is, the better the deal the company gets from the bank. In fact, it is often contemptuously said that banks loan money only to those that do not need it.

Credit Cards

This is an easy way to get quick cash, but it is very expensive. Credit cards usually charge 3 to 4 times, and perhaps even more, the Federal Reserve's prime rate. Credit cards are also only good for limited amounts of money, which is based on the entrepreneur's personal credit cards extensions of credit. When borrowing from your own credit card, the risk is on you personally. You are responsible for paying back the loan, not the company you're supporting. If the company is owned 100% by the entrepreneur, this is a moot point. However, when partners are involved, the personal risk is dependent on the relationships between the owners. There is always the chance that the credit card debt will have to be made good by the card holder, not the company. In defense of credit cards, they are extremely easy to access, and repayment requirements are very minimal, usually the interest on the outstanding debt. Also, no approval requirements are needed before the money is available. Once you have the credit card, you can apply it toward anything you want, up to the credit limit of the card. Business history is full of "lore" telling the story about how a now very successful company was kept alive in its formative years through the use of funds obtained from the entrepreneur's personal credit card. Credit cards represent a good emergency cash flow reserve and can be very useful if used sparingly. Keep in mind that a 20% annual interest fee is not unheard of.

Factoring

This is selling or loaning receivables for cash. Here's how it works. You have invoices outstanding to customers who, however reliable for paying, are slow payers. They typically take more than 60 days to pay. With a cash-flow problem you would like to see that money arrives as soon as possible. To help in this area, there are financial companies that will buy your receivables at some discounted rate. Say you have an invoice for $100 due in 30 days. The factor will give you $90 for it now. You get the immediate cash relief and the factor realizes a $10 profit when the customer pays.

There are all sorts of variations on this scenario, usually evolving about percentages charged, such as how the client qualifies for factoring, based on their credit worthiness. This is a very individualized arrangement relationship with many variables. For example, some quirks in the scheme require the seller of the receivable to make good the amount received plus interest if the customer fails to pay. It goes on and on.

Using factoring will get the company virtually instant cash, without having to take on a bank as a silent partner, but it is expensive. The example is 10% a month! However, factoring is a good way to get funds if the entrepreneur is very careful not to get entangled with high interest (called discount) rates and penalty clauses.

Some more progressive banks are entering the factoring business. They tie it to a checking account (which becomes the collateral). They deposit the receivable in the checking account deducting usually 1 to 3% and collect the money from the debtor. If they collect the money within the agreed-to time period, usually a sliding scale from 30 to 90 days, they refund some of the interest charged. They do take control of the receivables by having your company require your customers to send payments made to your company to a P.O. box that is controlled by the bank. And they have the right to intercept and take their share before it goes into the company's checking account. So the fact that you're using a factoring service is transparent to the customer. Transparency may or may not be in place with non-bank factoring companies.

Factoring is a viable way of maintaining the cash-flow liquidity, but like all the other methods of gaining financing it has to be explored thoroughly and all of the ramifications thoroughly understood beforehand.

Expediting Collection

This is the most direct method of getting needed cash. To become proficient in expediting collections, the person in the company managing the finances must understand that this is a prime responsibility, as important as raising operating funds. Here you're dealing directly with the customer and asking him to pay. How to go after slow payers will depend on circumstance. But in every case it is important to stress, diplomatically of course, that your company has fulfilled its obligations of the contract and now you expect the customer to fulfill his.

The first item of business for expediting collections is to make sure the customer has been billed. This may be so obvious you might ask, why mention it? Unbelievable as it may seem, because it is a leading cause of late payments. We can't expect a customer to pay a bill until he receives it; therefore, make sure it is actually sent to him. Many times errors occur in billing, especially when the business control system is not a computer-based Manufacturing Resources Planning or Enterprise Resources Planning (MRP/ERP) integrated one, and that's usually the case for a startup company. What happens is that an order is completed and shipped and the documentation to finance to generate an invoice is overlooked or misplaced. Rather than having a check system in place to ensure that all facets of the company are in communications with each other, the startup company strives to clear up glitches in its production cycle and develop the products or services it started the business to capitalize on. The company's focus is primarily on its product stream, not on bill collecting. So it happens that bills sometimes aren't processed. The cure to this is embarrassingly simple: Make sure you understand the 7 steps of the manufacturing system and comply with them. We will investigate the 7 steps and what it means to the startup company, in fact any company, in another chapter.

When an invoice is sent, it needs to be correct. Again this seems obvious. But incorrect invoices invite delays by customers in paying their bills. They have a legitimate claim to delay paying until they agree the bill is accurate. It's unrealistic to expect the client to pay questionable amounts, either underbilled or overbilled. By doing so, the client is complicating her general ledger by requiring correction entries. This means their reports are going to be incorrect and so forth. No good comes from inaccurate invoices: Besides creating documentation complications, it also creates a loss of confidence in the capability of the company with the customer. It causes questions such as: "If they can't get the billing correct, what can we expect for the quality of their product?" These are not the questions a startup company wants to have brought up. So, make sure each and every invoice is correct before issuing it.

With correct documentation on the invoice, we can expect customers to pay on time. Without correct documentation it requires negotiation and explanation of the invoice to the customer before he will pay. In this situation, the customer has the upper hand. He can demand all sorts of backup proof that the bill is correct and that could further expose the startup company's billing woes as the company tries to expedite payments. The old carpenter's rule applies, "Measure twice, cut once," modified to "calculate and document twice, invoice once."

When we're sure our invoicing is correct but the customer still doesn't pay per the contract, what do we do? Here are some tips.

For clients that are only a few days behind on payments, a simple reminder to their buyer may be all that's necessary. If that doesn't yield any results, then the next call goes to the accounting department. Always ask for payment, stating that it's overdue, then ask when your company may expect to receive payment. Always get a date. Do not be put off by vague promises. If the client is evasive, go up the management chain. Eventually you will get a commitment, if the client is a responsible company. If not, then the lesson learned is to be wary when doing business with this company and require larger up-front payments.

Usually, when customers say they will pay by a certain date, they do. However, occasionally a customer will renege. When that happens, first confront the customer with the facts, specifically, that they haven't lived up to their agreement. Then ask them what their resolution is. If it is satisfactory, then agree; if not, propose your own. You can suggest, perhaps, a partial payment to show good faith. Usually these lack-of-payment situations will resolve without further incident. Sometimes they do not.

When a customer continues to be in arrears, it becomes necessary to take other steps. A first step could be holding back deliveries. If your product is something the customer needs to maintain his business, this will normally get a positive response. When payment does come, release the material, but not until the customer's check has cleared your bank. Then in the future, until the customer demonstrates a good payment record, require full payments before work starts; later on you can permit partial payments before work starts. Also require payment for the remainder of the amount due before making deliveries.

These may seem like draconian measures and certainly not customer-friendly. But remember, business is a relationship between a seller and buyer, both having obligated themselves to provide worth to each other. There is a need to be civil with customers, but not to let them take advantage of your company. Your company is no good to them if it cannot stay in business. And maintaining good cash flow by receiving what is legitimately due to you is a way of staying in business so you can service the customer. It is also ethical and proper for clients to pay their bills on time. By going after notoriously slow payers, you are being ethical and forcing your customers to reciprocate in kind.

Last Resorts

Sometimes, no matter what you do to make collections, the customer still doesn't pay or set up a plan to make good on his bills. This will require either drastic action or none at all.

There is legal recourse available to the company. The client can be sued for the amount of the bill, plus damages suffered trying to collect, plus legal costs to pursue the errant customer. This takes time (upward of a year or even longer), lots of effort to provide the lawyers with the necessary documentation, and in general a lot of patience and strong resolve to see it to its end. Legal recourse is usually not pursued unless the amount owed is substantial. It must be well over $5000 to make it worthwhile. Why that number? Because below $5000, most jurisdictions allow disputes to be resolved in small claims courts. Here simple presentations before a judge are made and a ruling handed down quickly. Also the legal system allows for appeals. The loser, either plaintiff or defendant, can petition for an appeal. It takes quite a while for the system to even decide if the appeal will be heard. Then once that's done and an appeal is allowed, the clock starts over again. With the system stacked against expeditious resolution, legal claims against an errant customer need to be for large amounts to make it worthwhile. And then on top of that is the cost of a legal suit. At a minimum, the law firm will receive 30% of the award the system delivers for the company. So legal suits are really a last-resort attempt to collect a large amount due. Otherwise it is simpler to threaten a law suit and go through the motions to "scare" the client into settling with your company at some percentage of the money owed. Lacking that, it may be cheaper overall from the viewpoint of fees and time spent on the case instead of on the true needs of the business to write off the debt as uncollectible. I am reminded of some sage advice and I pass it on: "For the most part, law suits are for the benefit of the lawyers, not the clients."

There is an intermediate method of getting some money from "deadbeats." Sell the account receivable to a collection agency. This is the ultimate factoring situation. Except in this case the company is permanently ridding itself of the account receivable. They actually sell it to the collection agency at a very low percentage of its true worth, which is probably significantly less than 50%. The collection agency then owns the account receivable and can do what ever they chose to get some payment from the customer. The company is completely divorced from this process and has no say in its final outcome.

Legal recourse and turning over accounts receivable to collection agencies are taken only when the company has exhausted every other means of being paid. They are truly adversarial in nature and most often end up destroying any commercial relationship between the company and the client.

KEEPING FOCUSED ON CONTROLLING THE OPERATION

As we can see, many management tasks need to be done to keep the finances of a company under control. In fact, this expands to keeping all of the activities of a company under control. New entrepreneurs can surely get discouraged if they stop to think about all they have to do just to maintain control of the company, aside from thinking about the products or services for which they started the company. Most entrepreneurs find this administrative control aspect to be daunting and not what they envisioned running their own company would be like. Quite simply, they don't like doing these things. They want to spend their time "playing" with the technology or creative ideas that initiated the business in the first place. But the successful entrepreneur knows his or her business is more likely to fail because of lack of attention to the administrative control functions than for any other reason. So the entrepreneur reluctantly emerges himself or herself into the myriad of details necessary to remain financially and operationally valid.

The good news is that it's not necessary to memorize all the administrative tasks that are necessary to do. The bad news is that they all still have to be done. Most administrative tasks are repetitive, complying with some sort of time cycle. This means we have daily, weekly, monthly, quarterly, semiannual, and annual tasks that have to be done. So knowing what has to be done and when can be scheduled and put on the business calendar. Once we put the standard tasks on the calendar, we can check them off as they come due and are completed. Our administrative control tasks reduce down to 2 items:
- Maintain an accurate calendar of standard events.
- Assign each task to the proper individual or department to do.

Let's look at both tasks.

Maintain an Accurate Calendar of Standard Events

We need to first take an inventory of standard activities that we think we need to do to manage the business. Some companies say anything that improves effectiveness of how they do their business ought to be included on the list. Others say only activities that are required by regulations and contracts ought to be on the list. I'm a supporter of the former. I believe any task that is necessary for maintaining the continuity of the business system needs to be on the list. This means the items related to the classic 7 steps of the manufacturing system are "musts." While any activity that indirectly supports the 7 steps are "wants." The 7 steps will be discussed in detail in a later chapter. However, for purposes of our current discussion, we should understand that all businesses comply with the 7 steps, whether they realize it or not. And those that do so with a clear knowledge that they do, tend to be more successful. The 7 steps are listed again for reference in Figure 7-9.

Every one of the steps will have some standard activities that are done on a periodic basis. Some have additional steps that occur when a project is initiated. An example would be a product design review under Step 2. This would not be listed on the calendar checklist because it is not periodically repeatable. It is random, therefore it's a special event that management would schedule as part of the process of doing a specific project. However, virtually every financial chore the company must conform with is a periodic event and would be on the calendar checklist.

To create the inventory we go through each of the seven steps and list the repeatable tasks that occur for our business that makes it possible for us to complete the step. We then make a judgment as to the time of the repeat task cycle; e.g., weekly, quarterly, etc. Most of the time, the cycle is obvious. Examples: doing the short range forecast under step 6. is monthly, and the schedule for the factory is updated daily under step 5. We then put all of

1. Obtain product specification.

2. Design a method for producing the product, including the design and purchase of equipment and/or processes to produce, if required.

3. Schedule to produce.

4. Purchase raw materials in accordance with the schedule.

5. Produce in the factory.

6. Monitor results for technical compliance and cost control.

7. Ship the completed product to the customer.

Figure 7-9. The 7 steps of the manufacturing system.

	Task	Daily	Weekly	Monthly	Quarterly	Semi-Annual	Annual
1	Read/answer mail, e-mail, fax	x					
2	Return/make telephone calls	x					
3	Implement schedule of the day	x					
4	Review results for next day's schedule	x					
5	Make next days schedule based on results of day and master schedule	x					
6	Send out bills for completed work	x					
7	Collect A/R $'s for completed work	x					
8	Record hours worked for individual employees	x					
9	Update completed job hours; actual compared to planned	x					
10	Update quality records	x					
11	Make sales visits		x				
12	Update master schedule		x				
13	Pay bills		x				
14	Make purchases per master schedule		x				
15	Make out payroll		x				
16	Update productivity records		x				
17	Update quality trend charts		x				
18	Make equipment purchases per business plan			x			
19	Review financial and operational results			x			
20	Do rolling 3 mo. forecast of financials and H/C needs			x			
21	Review status of goals achievement			x			
22	Develop and implement corrective action plans for items 16-21			x			
23	Review of operations, actual vs. planned, as compared to business plan				x		
24	Review capital equipment needs and update business plan				x		
25	Correct current budgets for adequacy as required					x	
26	Update 3-year business plan						x
27	Create operating budget and plan for next year period.						x

Figure 7-10. Periodic management activities.

these activities on an actual calendar, preferably a computer based one, and print it daily weekly, etc. what ever the company needs. Microsoft Outlook or Entourage are excellent PC- and MAC-based software programs that can issue daily calendar reminders. With the items on a calendar, we are reminded in a timely fashion as to what tasks have to be accomplished. This way we've institutionalized the administrative tasks. They will not be forgotten, and the probability of them being done is very high. In this manner the company can be assured that it is "minding the store" while at the same time doing the creative work it set out to do in the first place. A typical check sheet to prepare a computer calendarized database is shown in Figure 7-10.

Assign Each Task to the Proper Individual or Department to Do

Like all schedules, and the calendarized database for standard tasks is one, the activities need to be assigned to departments and eventually individuals in those departments to accomplish. This is a management task that cannot be taken lightly. It is important that each standard event be assigned to the person with the responsibility and authority to carry out the actions required. This way, not only will these reporting tasks be done, but those doing the reports have the authority to carry out the actions that will be required. Otherwise we simply have a reporting system of dubious accuracy and have imposed an unnecessary step between cause and corrective actions.

It is very import to get down to root responsibility levels when assigning standard repeatable tasks. One common mistake is to assign all of the financial standard repeatable tasks to the finance department. This can lead to wonderful reports but poor follow-up on corrective actions. For example, a short-range forecast and review of operations should be done by operations managers, not accountants. A short-range forecast portrays what will most likely happen in the next 1 to 3 months based on a review of what has just happened during the past month. Who better than the person who is responsible for making it happen to be assigned to make the forecast? Some would argue that this person is unfamiliar with financial reports, therefore it's better for the accountant to do the report. That's a null argument. He or she had better become well-versed in financial reports. After all, the department's performance is going to be judged significantly by the financial results, so the department manager can't afford to be a spectator. He or she has to be the prime participant. Likewise, how can an accountant make up an operations forecast without any control of what actions to take and probably only a scant knowledge of the activity? The point is, the person with the responsibility and authority is always best suited to do the reports. That person must understand the nuances of the reports, otherwise he or she is unsuited for the position.

FINANCIAL MEASUREMENTS AND THE BUSINESS PLAN

One final thought about finance documentation. It becomes the basis for all business decisions. We start out with an idea for a business or for a business expansion, and then after assuring ourselves that the idea has currency, we evaluate it for financial benefits for the company. And as seen in this chapter, we have lots of means of measuring the worth of various plans, which we do periodically. However, we should take special notice of the business plan, this annual event for providing direction to the company. There are no tricks or other techniques pertaining to finance that only apply to the business plan. But the business plan allows us to use all of the financial techniques to forecast the company's future based on the current assessment of the world and the pertinent markets. The exact structure of the business plan, which is much like the NPI process on a grander scale, will be dissected in a later chapter. But I bring it up here to assure you that all of the financial techniques have been discovered and will be employed. The numbers may be more on a macro scale, but like any other analysis, whether it's doable or not depends on the details.

SUMMARY

Volumes have been written about financial management. The purpose of this chapter has been to introduce the entrepreneur to the operational characteristics of financial management of a business. I wanted to show that operational people, such as most entrepreneurs, can use financial measurements to benefit their businesses in a prime way and not as an unpleasant adjunct or necessary evil to being successful. Understanding finance principles aids and abets the fulfillment of the dream of bringing a good idea to fruition. I tied the operational nature of a startup or small business to the "great" tools that are traditional and not-so-traditional uses of financial theory for the use of the entrepreneur.

The financial tools and an understanding of how to use them gives the entrepreneur the ability to navigate his or her idea to business success. Without these tools and without a dedicated commitment to use them, the chance of bringing that outstanding idea to business success is much reduced. Understanding and using the financial tools to gain and control capital will enable entrepreneurs to realize their dreams and really commercialize her or his idea.

Chapter 8

E-Commerce and the Virtual Corporation: Opportunities for the Entrepreneur

Entrepreneurship is about business. Therefore, being successful in business is all about being good at transactions. E-commerce is a new way of doing business, which we shall see in this chapter. But before we can do this and in order to present a clear picture, we should describe virtual corporations and their common identity—supply chains. To start, we look at transactions, the heart of all business activities. Here is a definition of transactions.

Transactions:
Finding customers; understanding their needs; defining how your company can satisfy those needs; making a proposal to the client. Then convincing the customer to agree with you, take you up on your proposal, and enter a contractual relationship with your company.

Today, more often than ever, transactions are becoming more complicated. We see more establishments of longer-term relationships between clients and vendors once an agreement is reached, stating that the vendor can be a useful adjunct to the customer's success. This is happening for 2 reasons:

- The demise of the larger corporation, thus less vertical integration of manufacturing
- The ability to merge production schedules via electronic media to make them virtually seamless

These trends offer tremendous opportunity for the entrepreneur. The dismantling of vertical integration means companies no longer make everything about their products. In fact, many companies became assemblers rather than combinations of fabricators of subcomponents and assembly. This means companies buy more components than ever before, both off-the-shelf generics and items made to their special specifications. With the ability to merge schedules between sellers and buyers electronically, it makes it almost as easy to order parts from suppliers as it would from internal sources. The dismantling of vertically integrated companies gives smaller companies vast opportunities to partner with assemblers to be part of their "virtual" factory, that is, part of their integrated schedule but not under their financial-ownership umbrella. These relationships offer the opportunity for the small startup to act as if it were an established company with all the management trimmings in place. In this chapter, we will discuss this phenomenal opportunity and learn how it works, pointing out both pitfalls and rose gardens to be aware of. It is mainly a factor of understanding and applying what we call the supply chain concept, which also entails a good grasp of pragmatic marketing and salesmanship.

IMPERATIVE OF BEING WANTED

"Everybody Loves Raymond" is the title of a long-running television sitcom. The basic premise is that Raymond is a good guy who is always linked with family and friends. That's what the entrepreneur wants to emulate. However, in the entrepreneur's case, she wants it to be her product or service primarily, not necessarily her, too. How does the entrepreneur make this happen? Here are the 4 "must" prerequisites:

- Have a product that is useful to the customer.
- Offer quality equal to or better than the customer's needs.
- Make the product available when the customer wants it.
- Offer a price the customer is comfortable with.

Have a Product that is Useful to the Customer

In the best scenario this means being the provider of a generic product that the customer needs to include in his assembly operations when there is no need for special design to make it suitable for the customer's needs. For example, a provider of standard fastener hardware could fall into this category. In the past, large companies frequently made all of their standard, as well as special, hardware because they were totally vertically integrated. The philosophy at the time was to not rely on any outside sources for their ultimate success. This philosophy is now considered too expensive based on the value received. So most large companies will buy commodity-type supplies from smaller companies that specialize in such items, hence can manufacture them more efficiently.

The next best scenario to being useful to customers is to be a provider of products that are specifically designed for the customer's use and to have the technology to make the product the customer needs while the client doesn't. Ideally, the customer doesn't want to acquire the capability either.

Going down to lesser scenarios, the client can make his own parts specific to his design but chooses not to. In this case, the vendor can supply parts as if he were part of the client's supply chain, but the link is weak because the customer can choose to break it any time he wants. It's a matter of turning on his own machines. Further down the security chain, it could be that the customer cannot make enough product volume within his own facilities and hires a company to be a supplemental producer. In this case, the contracts are usually short-term and many times a "one shot deal." This last scenario is one that a customer would use to handle peak loads, and is not very attractive to the entrepreneur.

In all cases, the customer will evaluate the potential vendor's technical capability. The vendor must be judged capable for the customer's needs to even be considered as an outsource supplier. If the vendor passes this hurdle, the next step is to demonstrate adequate capacity to deliver required quantity within the time constraints imposed by the client. If both hurdles are successfully navigated, then the vendor will be listed as a qualified supplier. Qualified suppliers are traditionally given first preference to bid on jobs.

Offer Quality Equal to or Better than the Customer's Needs

In addition to being technically qualified, the vendor needs to have in-house quality-assurance capabilities that the customer is comfortable with. In other words, the products supplied by the vendor have to be transparent, quality-wise, to the customer's own. They have to be as good as he would have done himself, if they were made by him.

Notice, I didn't say zero defects or any other standard. Standards will vary from client to client. Quality is truly in the eye of the beholder. The client needs a level of quality that will satisfy his customers and nothing more. However, as the supply chain becomes more sophisticated, we find that quality requirements are being nudged upward, primarily because the customers know that things like "six sigma" espousal of zero defects exist, whether that's realistic or not for the specific application. So in many cases, a vendor trying to gain a long-term contract with a client will want to ensure that his quality-assurance programs are state of the art for his industry.

Make the Product Available When the Customer Wants It

There is absolutely no wiggle room here. If an entrepreneur wants to join a supply chain as an outlet for his products, he must comply with the schedules the supply chain demands of him. It doesn't matter if the supply chain is only 2 companies, the vendor entrepreneur and the client, or composed of several companies all feeding into a chain for the ultimate production of a complex entity, such as an airplane. Those who input to the supply chain need to be able to produce at the rate and quantity required. If they can't, they fail the "must" test and cannot participate.

Offer a Price the Customer is Comfortable With

This doesn't mean the lowest price, but one that is reasonable and allows all parties to the supply chain to make an acceptable profit. In long-term relationships, we need to have healthy customers and vendors. If either is pushed to the point where there is not enough profit margin in the transaction, then they will drop out. We are finding that large companies ganging up on their smaller suppliers to push purchasing prices down only works if at the same time the larger company is helping the smaller company become more efficient, for example, the startup being funded by the entrepreneur. By becoming more efficient, the vendor can lower his prices and still enjoy the same profit margin. Becoming more proficient in lowering the cost of goods sold and operating expenses can lead to lower selling prices and still get the same net profit. This is a win-win situation that vendor–customer relationships should strive for.

In true supply-chain relationships, profit margins are agreed to for all parties, and all members of the supply chain pledge their resources to help other members in reducing their internal costs; accordingly, overall costs of the assembled product are as low as possible, while at the same time, all members of the supply chain are experiencing profits per the agreements.

THE SUPPLY CHAIN EVOLUTION

The supply chain is a phenomena of the demise of the vertically integrated company. There is no doubt that the most efficient transformation of raw materials to finished product is the integrated factory. Integrated factories control their destiny entirely by making everything they need, hence not being dependent on anyone else. The self-sufficient organization is one that makes all of its own decisions based on its own best interests. It owns the sources of raw materials. It owns the various factories necessary to make all of the component parts. It possesses its own transportation system to ship the product to market. And in many cases, it has its own retail outlets to sell the products to the ultimate customer. So we see the integrated company can set its own profit margins between departments; in fact, it need not even consider interdepartmental profits. It can be as efficient as it deems necessary and be at a quality level it feels comfortable, at least until it reaches the ultimate customer.

As you can imagine, supporting such an empire can become very bureaucratic and resistant to change. In the beginning and for some time thereafter, the vertically integrated company is super-efficient. But it reaches a point where this is no longer true. The inward focus inevitably causes an arrogance and contempt of new ideas because it has enjoyed the fruit of original creativity from long ago that has never been successfully challenged, until now. Being so large, the vertically integrated company can offer lots of false security from the realities of the real world. These monoliths often miss market opportunities and fail to exploit new technology they may have developed because they tend to be inward-focused and see no need to aggressively compete. Their entire universe is within their control, or so they think. They are busy managing the various factions and processes, some in conflict with each other, to the point they virtually withdraw from the real competitive world. They often live on for decades on pure inertia. Its like stopping a giant supertanker. It goes and goes and goes until all of its energy is finally dissipated. In this case, the energy can be thought of as "clipping coupons" or getting "royalties." The inward-looking company doesn't offer many new innovative products to the market place. It primarily lives on products that are eventually yielding to the product life cycle curve. And nowadays the cycle time is shrinking ever faster.

So vertically integrated companies, once super-efficient, are almost all gone. Does that mean the efficiency of an industrialized society is gone too? Or at least retrograded? No. The supply chain has taken its place and demands the same system efficiency that integration can provide. How does this work? Let's take a look.

The supply chain can be thought of as a vertically integrated company that has been broken up; the analogy of AT&T and its breakup comes to mind. The monolith was broken into many segments by court decree. Yet the telephone system still runs and probably more efficiently. A supply chain can be thought of as a series of departments, or even plants, that are now independent companies. They band together in contractual relationships to be both vendor and customer to other members. So Lucent Technologies with its Bell Laboratory becomes the supplier to the regional Bell telephone companies, thus creating a chain from raw materials to finished product. This was the original supply chain after the breakup. But like all supply chains, the members have to maintain quality, technology, and capacity competitiveness to be able to (and want to) stay together. We all know this hasn't been the case with the breakup of AT&T. The original supply chain probably doesn't even exist anymore, and we have successful interlopers such as Northern Telecom that have pushed aside some of the founding members and have replaced them or have even become such dominating forces to have been able to establish their own supply chain. Think of it. The scion of vertical integration was first forced into a supply-chain relationship, and then that relationship was fractured by the entry of outsiders and the displacement of weak sisters to form more efficient supply chains. The bottom line is that the much-vaunted efficiency of the vertically integrated business has been bettered by a loose federation of smaller but more focused, thus more efficient, companies.

A very spectacular example of a successful supply chain is the Airbus and its ability to compete with Boeing. Airbus is not one company like Boeing, but a series of smaller European companies, each having a narrow scope of capabilities. When these capabilities are integrated through a schedule and agreed-to profit margins, we see the creation of a linked chain of companies capable of competing against the giant Boeing. Figure 8-1 shows this comparison between a supply chain and an integrated company.

Just like in an integrated company, each link in the chain, like the various departments, are experts in that particular aspect of the job. The difference is that the supply chain is like an agricultural cooperative, a group of companies that have banded together to compete more efficiently and with more clout. In this case, the supply chain is a group of dissimilar companies banded together to produce a product that none of them can produce by themselves. Unlike the vertically integrated company, the members of the supply chain can kick out poor performers and their links are taken over by better performers. There is no guarantee of membership except performing at the level that meets the needs of the supply chain.

Since supply chains are independent companies, it takes a considerable amount of patience and skill to set up the parameters they will operate

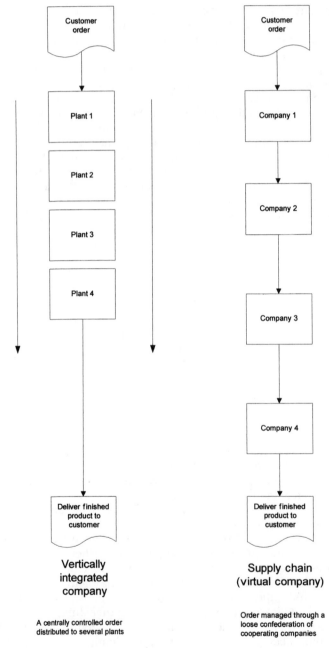

Figure 8-1. Supply chain vs. integrated company.

under. These are usually design procedures and schedule documentation, quality systems, and profit margins. Companies help each other achieve and maintain the required standards. It is not unusual for supply-chain members to share expertise among the entire chain. Remember, they are not competing against each other but most likely against another supply chain. So if a company has a strong technology base in, say, shop operations management, it will share it with its supply-chain team members because it is to its own advantage for all supply chain members to be strong in manufacturing. It lowers costs, increases productivity, and gives all the members a better chance in winning contracts over the competing supply chain.

HOW TO JOIN A SUPPLY CHAIN

An entrepreneur joins a supply chain when and if the entrepreneur's product or service is beneficial to the members. But the net gain must be greater than the initial investment the supply chain is going to have to make to bring the newcomer up to standard. It would be very unusual for an entrepreneur to have the operating systems in place that represent the degree of maturity of those being employed by the supply chain. In fact, it's quite possible that the entrepreneur's company has voids in its system, such as an integrated purchasing systems that would have to be completely filled by supply chain members' "generosity." I highlight generosity because it really isn't so. It is driven by the net gain calculation, making the supply chain wanting to recruit the entrepreneur's company. And, most likely, the cost of implementing the system into the entrepreneur's company would be deducted from her share of the profits, probably over an agreed-to period of time—like a young lawyer or physician being made a partner, thus having to buy into the business to compensate older partners' loss of equity.

But this really is a good deal for the entrepreneur. As we've discussed previously, having a good idea is not sufficient for business success. You also need the structure to manage the 7 steps of the manufacturing system optimally to succeed. By joining a supply chain, the small startup all of a sudden has access to resources for business systems and technology development that weren't available before. Employing these resources, such as readily accessible experts and CIM systems, allows the startup to reach operational maturity much faster than by going it alone.

So if joining a supply chain is such a good thing for entrepreneurs, why isn't it virtually common knowledge and happening all the time? Well, it happens more than you think. Virtually all spin-offs of university-developed technologies have access to university-sponsored business-development groups. These groups, while not nearly as strong as large commercial supply chains, do have access to markets through the university's networks. They also have access to MEPs—manufacturing effectiveness programs. MEPs are not for-profit corporations established to promote local area producers by making available practical expertise in manufacturing and business systems for relatively low costs. These combinations of MEPs and networks tend to form very loose supply chains, which more often than we realize become real supply chains. Although it is less likely that a brash new startup will get to join an established supply chain, it is highly probable that it could audition itself to other companies via MEP sponsorship. In this manner, the MEP is the sponsor of the supply chain and is the provider of much of the technical expertise normally associated with mature supply chains.

Even startups that do not have links to university beginnings can form bonds in supply chain–like arrangements through another type of network, the regional manufacturers association. The vitality and strength of these associations vary considerably; therefore, the ability for them to nurture startups is all over the place. But assuming the entrepreneur is starting up business in an area where there is a good manufacturers association, by joining, she or he may find there is a large variety of training available that could bootstrap the startup's business skills. Also, the association may have a cadre of consultants willing to work for lower fees because they don't have to market services. The association does it for them. (Don't get turned off by the title "manufacturers" association. Most accept service providers as members too. A good supply chain needs service groups too, for example, an architectural engineering firm to design structures as part of the foundation to hold the pump, etc.) Supply chains are formed through alliances with the various companies done via networking. Regional manufacturing associations are excellent ways to network. This doesn't happen automatically. It requires "getting involved" in the programs, which means attending and participating and taking advantage of relatively low-cost training programs, and finding ways to do mutual business with each other and gradually spinning outside the circle to find work that several association members would team together to do. This is the beginning of a new supply chain.

But how does the entrepreneur's company get good enough to even attract the types of companies found in regional manufacturers associations? The nominal size of members ranges from 5 to 500 people and upward of $50,000,000 annual revenue. They attract attention by making it known they have great technology and large potential. Then they work to establish a mentoring relationship, not in a formal sense, but more or less in a camaraderie manner. They also take part in same events that the companies they'd like to link with do. All this leads to becoming a known quantity. When this happens, the entrepreneur has set the stage to do "soft" benchmarking. For each of the 7 steps of the manufacturing system, the entrepreneur wants to find a local company that is not a competitor that the entrepreneur can query about how the company handles order entry, for example, or sets up an inventory control procedure. The entrepreneur is interested in being at least as effective as those she or he would like to team with. The best way to learn what it takes to be on par with future team members is to benchmark with them, ask them how they do specific functions and then see if the technology is sufficiently free of product specificity that it can be used within the startup company. Knowing what is needed and might work is a great first step. But unless there is excellent to expert-level capabilities within the startup company, the entrepreneur will need help in launching the upgrades. Here we go back to the association. Sometimes there are training programs available that bridge the gap between how something being done now and how it needs to be done. Project management training is a common one where significant skills can be learned and directly applied. Other times, there are needs that require an on-site expert to lead the company with implementation skills. Implementing MRP II as part of CIM is an example where bootstrapping won't work in a practical sense. You need to hire experts to do it. Again, the association may be able to help. Some of their cadre of consultants and trainers may have the expert skills the entrepreneur seeks. If that's the case, the association can enter into a contract with the entrepreneur to do the work. And since the association is a not-for-profit corporation, the cost will be significantly lower.

There are ways for the startup to gain the benefits of supply-chain membership and not be a full-fledged member of a supply chain, as explained above. However, getting involved with a successful supply chain is something every new provider of products and services should strive for. Remember the key element is a successful audition, being able to show that the product or service fits within the product or service domain of the supply chain and that it is superior to any currently available to the other member companies. The goal is to show supply-chain members that they will be more profitable with the entrepreneur's company being brought onboard, than without them. Whatever way a startup joins a supply chain, the goal of doing so as a means of selling its products or service ought to be high on the priority list. Teaming with compatible and noncompeting companies servicing the same customers brings greater strength to all.

INTRODUCING YOUR NEW PRODUCT TO THE WORLD USING A SUPPLY CHAIN

The new product introduction discussion in Chapter 6 went over the details of that entire process. What we will do now is focus solely on how the supply-chain concept aids and modifies that procedure.

The supply chain is set up to foster the development and sale of products that meet the needs of the customer by essentially doing a group Quality Functional Deployment exercise. This means instead of a single company investing an effort to see if their products or services are good fits to meet the needs of the customer, all the companies combined in the supply chain consortium are jointly doing the same thing. They are evaluating whether the supply chain has the expertise and capacity to support the customer effectively. How this happens is an exercise in group dynamics. A lead company, probably one with close relationships to the client, will emerge and coordinate the effort. Here in lies the opportunity for the entrepreneurial startup.

The entrepreneurial startup, having been accepted into the supply chain, now uses that same chain as an extended marketing arm. The company now has the ability to match its product or service offering into a bigger matrix than ever before. While once thinking of itself as a provider of a product with no real stake as to how it is to be ultimately used by the end purchaser, it now knows precisely what the end use will be. This gives the company an opportunity to introduce a new product specifically tailored for the supply chain's end use.

> **Example:** A producer of heat sinks to remove excess heat from electronic components acting as a sole and unaffiliated producer is forced to make generic parts suitable for many different applications. As part of a supply chain, it can introduce better, probably more efficient, products specifically tailored to the end users need. This means a specific heat sink for a specific chip supplier for a specific laptop computer can be developed at low risk to the entrepreneur. As part of the supply chain, the entrepreneur could issue the new product with a virtual guaranteed market. This means that all of the design and manufacturing engineering decisions are not going to have the same rate of iterations as before for a general product offering that marketing and sales are going to find customers for. Instead, marketing and sales will link up with the other members of the supply chain and obtain a much more refined set of specifications to give design and manufacturing engineering.

Working within a supply chain reduces the risk of new product introduction to the individual company by being able to share it with other members of the supply chain. Since the goal is to service the customer per information gained via the QFD analysis, the entrepreneur develops the new product without worrying about competition from within the supply chain. This means what they develop will be used as long as it meets the cost goals and performance requirements of the rest of the supply chain. It's akin to being a department or division within the old vertically integrated company, where the degree of guaranteed market (via other departments and divisions) is very high.

Using the supply chain takes great marketing risks out of the individual company's domain. This, in turn, often liberates the internal creativity of the design group because they know specifically what the target end product will be and therefore can come up with unique and very efficient designs, once again showing that the supply chain can compete more effectively than individual competitors. The supply chain preserves the high resources levels of the vertically integrated company, while at the same time allowing significantly more creativity by its member companies.

BEING ABLE TO HANDLE TRANSACTIONS VIA E-COMMERCE

Understanding how supply chains work and structuring your company to do so is vital for success. As we've seen, being part of a supply chain is a credible goal for the startup entrepreneur, since it makes success more probable than failure. The supply chain virtually guarantees a market at a fair profit. With profits, the young company can nurture itself to maturity and sustained growth. Now the entrepreneur, having gained entry to a supply chain, must be able to fit in as if his company is just another cog in the system. To do so, transactions have to be fast, accurate, and complete. We now need to understand how transactions occur and why the supply chain demands they be electronically linked, company to company.

The problem of accurate order entry has plagued Manufacturing Resources Planning Systems since their inception. It is an axiom among product or service providers that they are only as good as the understanding they have about their customer's true needs. In a supply chain, this is even more important because of the merging of customer vendor relationships. Business has employed huge reservoirs of resources in solving this problem and it has always come down to the fact that to be successful, they had to depend on the sense of responsibility and devotion to duty of their order-takers to accurately enter data into the fulfillment system. Recently we have seen the introduction of e-commerce via the Internet for a host of things from obtaining mortgages to relationship counseling. It is apparent that e-commerce makes data entry more reliable. Add to that the accuracy of bar code data entry, and we may have a way at last to approach 100% accuracy in order entry. This being the case, the Manufacturing Resources Planning System is on the verge of finally integrating and automating product and service fulfillment at 100% accuracy.

Order Entry and Fulfillment

Order entry and fulfillment is at the heart of any business transaction. Customers place orders for goods and services and vendors fulfill them, for a profit. For a supply chain this must be done from one member company to the others with the same information along the entire path. The more efficient the system of order entry and fulfillment, the lower the internal cost to the supply chain becomes and the higher the profit margin to the members becomes. Contrary to popular thought (and a reason supply chain membership is so desirous), long-range profits do not lie in creating an advantage for the buyer or seller over the other but for an efficient system so they each gain. The goal is to create a balanced system that both buyer and seller agree is fair and that requires knowing precisely what the customer wants and precisely how the vendor can fulfill that want. In this case, the customer is the end recipient of all the member companies' vendor efforts.

The quest for the optimal order entry and fulfillment system focuses businesses energies on striving for error-free fast transactions systems. Electronics systems based on common databases offer the best possibilities for achieving optimum systems. Early electronics-based systems, commonly known as Materials Requirements Planning (MRP) systems, made significant progress toward achieving optimum order entry and fulfillment. They were stand-alone internal company systems that used the computer's capability of accurately handling vast quantities of information.

The first step for an MRP system was to get the order into the vendor's business system (the MRP system). From then on, the entire fulfillment process marched "lock step" through the various stages of providing the product with unerring accuracy. It worked exceedingly well as long as human intervention was eliminated. This of course meant it could never be so. There were too many situations where the smooth order entry information flow was interrupted due to it being incomplete or wrong because of data-entry errors. The entry of data was (and in most cases still is) a task assigned to people. It was typically output from the customer's database reduced to hard copy, which then had to be manually entered into the vendor's database. Now, with the introduction of true e-commerce, it is possible to eliminate this manual transfer and make it part of a continuous superhighway system, where order entry is transparent to the source of entry. Even more important, e-commerce offers the first true potential for error-free transmission of the customer's desires, which is the order entry into the fulfillment chain. Let's explore an e-commerce system and see how the efficient flow of information happens and why it can provide for a super-efficient order entry and fulfillment system. To do so, we need to start by building the foundation based on the fundamentals.

Manufacturing Resources Planning Philosophy

The place to start is with the MRP/Master Schedule. This is a computer-driven and monitored system that keeps track of large numbers of business transactions virtually error-proof. Here we see all the functions of a product or service provider company laid out in a manner showing the integration of each of the steps with each other and the status of jobs monitored by the computer system (Figure 4-3). The interesting thing to note is that all the steps are laid out in accordance with the 7 steps of the manufacturing system (Figure 7-9), which in actuality represents the entire business system.

The entrepreneur startup company must be able to integrate its MRP II system into the supply chain's master system. In fact, even more basic than that, the startup company must commit to have an MRP II system as soon as possible. And if it needs help to get one up and running, it is prudent that the startup company inform the supply-chain members and ask for assistance. Retrogressing a bit, not having an MRP II at the start of the supply-chain relationship is okay. Procrastinating and not putting one into operation is not okay. Remember, the startup company was invited to join the supply chain because of its potential to assist the overall well-being of the entire commune. Living up to that potential is critical.

Another interesting observation implied by Figure 4-3 is that once an order is entered into the system, the information is transferred electronically from one operation to the next. It is the same basic information from operation to succeeding operations with additional information added to represent the work that has been completed. This is either value added, making the product, or design or administrative information needed to further clarify what has to be done or was done. The significance of this is that we can store all information pertaining to the order in one common database. This simple fact eliminates the need for entry and re-entry of information as the process of making the product or performing the service goes from operation to operation. As long as the information was originally entered correctly, it stays correct forever. This is a tremendous advantage in assuring that customers get what they really want. With common information available at all operations we approach a state of communications excellence. This means opportunities for mistakes decrease to much lower levels.

With manufacturing resources planning systems driven by computers utilizing common databases, the world of commerce has evolved from a philosophy of *acceptable imprecision* to one of *available precision*. In practice this means it is possible for every job going through a flow type production system (commonly known as mass production) can be individualized and different than its predecessor or successor. This is accomplished simply by entering new data into the process equipment on a real-time basis as the product passes through. This increases the potential for highly individualized products being available at or near the price of mass flow items.

The prime need is for the correct information to be entered in the system at the very beginning with order entry. Order entry is the first step in the "7 steps of the manufacturing system" titled *"Obtain Product Specification"*. Figure 4-3 shows two entries of data. Either forecast or customer order. Forecast is easy for the company to handle, because it takes on the role of both customer and vendor, and pleasing oneself is irrelevant. Customer order is an outside party placing a requirement on the organization to perform a service or make a product. The degree of success in performing the work will depend on two things:

- How well the company understands the requirements to be met, and
- How well the company can execute against the requirements.

Understanding is the key for order entry. Execution on that understanding is the key for fulfillment. Understanding means many things. But in the context of a business transaction it means understanding the full extent of the customer's needs and translating those needs into language and experience bases that the fulfillment process recognizes and can respond to.

The fulfillment process is an internal activity of the vendor until the end, when the results need to be given to the customer. The vendor must deliver the product or service to the customer in a way that the customer recognizes as achieving her/his needs. Those are the same needs as originally put forth at order entry. Here again we see the need for perfect communications. The vendor needs to receive the message absolutely clearly, 100% accurately, from the customer. Then the vendor has to fulfill the requisition in accordance with that instruction set. Finally, the vendor must deliver a finished product and communicate back that 100% of the original request has been accomplished. It's logical that original information must be correct and not corrupted during any portion of the process for this scenario to be successful. Electronically controlled systems can do this virtually 100% of the time, regardless of the complexities of products involve; manual systems cannot.

Unfortunately, order entry is the weak link. In most systems it uses electronic data entry as an aid, not as a prime driver. This is a point where a supply chain can break down.

The Generic Problem with Order Entry Systems

The internal workings of MRP, now called MRP II (Manufacturing Resources Planning) because the system expanded to integrate labor and facilities needs with materials needs, is fully defined and is a mature technology. The weak link is order entry at the start and as a result, fulfillment accuracy at the end. To satisfy the start activity, most manufacturers and providers of service have established an elaborate system of manual checks and verifications at the order-taking ends of the spectrum. This means large staffs and complex systems. In some businesses, these systems far outweigh in complexity the rest of the entire MRP II system because so much is at stake to get the order right. An order entry needs to have the following as a generic minimum:

- A complete description of the product (or service) to be provided
  - Details as to functionality
  - Details as to size and weight, and ultimate geometric configuration
  - Details as to materials of which it is to be constructed
  - For services, a detailed description of what constitutes delivery of the service
- A complete description of the quality standards the product has to meet
  - Details of variances acceptable in functionality
  - Details of how functionality will be measured
  - Details of how variances outside the range of acceptability may be repaired and what further tests are then required
  - Warranty guarantees and length of time
- A complete agreement on price and delivery dates
  - Details on how variance in delivery schedule will affect payment
  - Details on payment schedules

Based on these generic needs, businesses create specifics for the their products and this can be very elaborate. Here again is where a supply-chain concept can simplify the process. The entire confederation of companies works on the specifications in such a manner that there will be no surprises at any stage and for any component of the subassembly. All challenges are known and all contingencies are pre-planned. To help this effort, member companies in the supply chain offering products often develop catalogues describing offerings along with commitments to maintain certain levels of quality conformance, usually to an outside code-setting authority (for example, UL certification for electrical equipment safety). The process involves a certain set of wall building against the customer (not the members of the supply chain; they already have contractual relations to share potential liability) to protect the vendor from potential future liability. This means the process is full of forms and disclaimers that become part of the order. So it is no surprise that internal policies within the supply chain require meticulous attention to these details at the stage of taking an order from a customer. Think of the forms required to be executed (excluding government-inspired registration) to buy an automobile. This takes time, and every entry that has to be manually entered is a chance for an error to enter the fulfillment stream.

The generic list shows 10 specific categories to contend with. Let's assume each category has only 1 entry to make in the order-entry process (an oversimplification, but sufficient for this example). Now let's further assume we have a very efficient order-entry work force and they have a 99% accuracy level. That means that for each opportunity to enter data for an order for each of the 10 items, the chance of error is only 1%. Since there are 10 different sets of information to enter, there are 10 opportunities to not have an error (0.99 raised to the 10th power). If we multiply 0.99 by 0.99 ten times, we will get an answer for total system accuracy. Probability theory would show that in this case, our system's probability of *not* making a mistake is 90.4%. Conversely, we have just under a 10% chance of disappointing a customer by making a mistake during order entry, thus dooming the fulfillment process to making a product that is not quite what the customer wanted. Is it any wonder then, if we special order an automobile we find it arrives at the dealer with the wrong stereo system option? Or, if not that, something else equally as aggravating?

With a 10% error in our example, it's easy to see why the order-entry staff and its trouble-shooter, the customer service representatives, make up such a significant part of the white-collar work force in industry today. Their presence is everywhere, from credit card companies to rust belt providers of industrial goods. They represent a significant segment of the overhead cost associated with doing business today. This certainly is not optimal and results in higher costs for both the producer and the consumer. So what do we do about this? We need to find a way to improve the input of order entry into the MRP II system so it is accurate 100% of the time. We need to eliminate the manual data transfer from one system, the external customer, to another system, the multi-linked systems along the entire supply chain. The best way to do this is to make it one system. A way to do this is through an e-commerce link. Let's see how that can work and what it will take to make it work.

An E-Commerce Order-Entry System

E-commerce is an opportunity to simplify order entry for the entrepreneur as part of a supply chain and even as an individual competitor. This is how. As I've said, the weak link in the order-entry and fulfillment system is the manual order entry from a customer into the supply chain's business system. The ideal way to overcome this is for the customer to enter the data for her or his purchase and not the vendor. This is very appealing and has been tried, but with only minimal success for 2 interrelated reasons. However, with supply chains, what I will now explain is only pertinent to the initial receipt of the order. Once the order is received into the supply chain, problems described below no longer apply.

- The customer doesn't have access to the vendor's system.
- The customer doesn't understand the vendor's system.

Let's look at the first problem. The initial thought about a decade ago was for vendors to allow customers to have live terminals under their control and usually at their business location, tied directly into the vendor's system. This in effect made the vendor part of the customer's supply chain, albeit not an internal company member. This hasn't been too successful because few companies want to expose all of their business systems to clients. To set up systems that are limited in scope, that is, restricting the files the customer is allowed to access, is difficult to accomplish from a technical viewpoint. It is also unpalatable from a customer-relations viewpoint. There will always be that element of distrust between the parties that the vendor is trying to hide something. Also, a lot of companies make or supply many different products and services to a large variety of customers, and unless they are linked together as a true supply chain, there is only limited commonality. This means that a lot of products on a vendor's database are irrelevant to specific customers. The vendor would be pressed to purge the database to only customer-specific relevant information. It would be an extreme strain on resources to provide and maintain these external terminals with editing capability so that only the items of the mutual relationships are hooked to the system. Further complicating the situation is the desire for the customer to be online on a real-time basis. The desire to be real time is a selling feature of the link so that the customer can self-check the status of her or his orders.

The first problem manifests itself in another way. Sometimes customer vendor relationships reverse, depending on the product. A paper company may supply computer printout paper to a computer producer and at the same time may be a customer for those computers. For this reason, each party may need to keep the other at a certain distance to protect their competitive structures, and quite possibly for antitrust reasons. The paper company, if it had access to the computer company's business system, may discern some information that would give it an advantage when it came to understanding the next bid for computer equipment (such as its overhead structure pertaining to the quote). Also, neither the computer company nor the paper company would want the other to know what it is charging their respective competitors for similar products. This becomes another reason the customer and vendor may not want to share a single database.

The second problem may illustrate more succinctly why sharing databases has had only minimal success. The customer may feel inadequate in learning the vendor's system so that he or she can enter an order. So many different order-entry systems are in existence that it is virtually impossible for customers to be competent in all that are being used by their vendors. The customer really doesn't want to be responsible for errors, especially if there is a fear of inadequacy in understanding the vendor's system. In fact, the customer may be so intimidated that he or she may refrain from placing the order and going elsewhere. Another reason the customer may not want to enter orders on the vendor's system is that they may not know all the items they need to specify in order for enough information to be transferred into the fulfillment system. They may not feel confident that they know enough about the technology to be purchasing the correct parts and/or assemblies. In the area of services, they may not be aware of what set of services the vendor can provide that are relevant to their situation.

So it would appear that order entry by the customer is a nonstarter, especially in non–supply line companies' relationships, and as just presented, it certainly is. We can overcome these barriers through a true e-commerce system. If we have a system that can eliminate order-entry errors and be relatively fast, this would greatly simplify the process and eliminate a large portion of the overhead expenses (order-entry clerks and the customer service trouble-shooters). E-commerce can do this via the Internet. It eliminates the need to have dedicated terminals between vendor and customer so all of the security reasons disappear. The vendor's catalogue can be found on its Web site, along with an electronic-based prompting system to step customers through proper feature selection for their needs. This takes away a good part of the fear expressed by the customer that he may order the wrong thing, providing, of course, that the information on the Web site is properly designed for clarity and ease of querying. There can also be a segment of the Web site dedicated to consultation and questions, once the customer goes through the various prompts and still isn't sure if the application would suffice. If the customer takes advantage of this "chat" service, at least we know we have a customer with a strong interest in the product. If after going through all of the prompts and there is still no commitment to order entry, there may still be a way to satisfy the need. The customer, by answering the prompts, has given the vendor a description of the problem in language and facts that are meaningful to the vendor. The vendor then can modify the designs based on the best match between existing designs and variations the customer may desire (this is a group technology approach, which is well defined in the literature and easily added to the software of the Web site) and then present this solution to the customer. There may be several iterations of this process, but it is being monitored by the prompt database so changes are being constantly upgraded. When the customer makes the final decision, the latest update becomes the order-entry information and it is 100% accurate.

The key is to educate the customer in e-commerce, but not in the traditional way. We want to teach by doing but not spend time or energy explaining how the system works. We only want the customer to be trained to meticulously follow the prompts. When a customer orders a product or service, he or she must be stepped through the process of transmitting all of the information to the vendor. The allegory is ordering dinner in a restaurant. The dinner patron is asked by the waiter what her choice is, based on the menu and the recited specials. The waiter prompts responses as necessary. But even in this system, there are mistakes. Sometimes the salad arrives with the dressing applied instead of on the side as had been requested. To be successful for order-entry purposes, prompting must be done with a 100% verification check after every step of the process. Also it would help if the order-entry process compares responses with expected combinations and questions items that are out of the ordinary. A statistical relationship would have to be built into the system to define what is typical and query those that are not. An example would be availabilities of certain options. For example, an automobile being bought for rental car fleet use would likely have an automatic transmission and not a standard 4-speed manual shift.

Introducing bar coding to the process promises to further improve accuracy of order entry. Not all of a customer's products will be listed on their Web pages unless they have relatively few offerings or a very comprehensive Web site (such as some of the book sellers, but even they do not cover the entire spectrum of publishing). To further close the loop of obtaining 100% correct order entry, it would be beneficial if all of a product's information could be linked to a product code. Fortunately this already exists for many, in fact, most products. It is the Uniform Products Code, which classifies products by various industrial categories and uses. We know it as the bar code we see on large varieties of products. There are many different types of bar codes but they are all "codes" and readily available to vendors. Such systems would need to have as much of the vendor's product features for order entry included in a bar code printed in their catalogues and other sales literature. If this is the case, the customer simply has to swipe the code, read it into the vendor's Web site, and he or she would be automatically ordering everything encoded. Since bar coding and capabilities for reading bar codes into computer databases are readily available, it is conceivable that the customer can enter the bar code swipe in his internal database system and electronically transfer it to the Web site. This would be truly automated order entry in that the customer doesn't have to transfer any catalogue information to the prompting of the electronic monitor in the Web page, hence greatly minimizing customer input error. Of course, if the customer needs features not included in the product for which he or she entered the bar code, the customer would have to revert back to the prompting algorithm. Once that scenario is completed, the vendor, if she chooses, could produce a new bar code (it's really a Group Technology Classification and Characteristics Code) for the next iteration of sales literature. Several companies are experimenting with bar code systems such as these.

The bar code entry has another outstanding attribute. It would ensure very fast and accurate downloading to the rest of the fulfillment system. All of its Bill of Materials structure would preexist and therefore eliminate the need for any engineering inputs or purchasing inputs. It would create an automatic entry into the entire manufacturing system (Figure 4-3), including the all-important placement on the production schedule. This would shorten the portal-to-portal cycle time to a great extent. The literature is full of articles on flexible automation where an order is automatically processed without human intervention once the order is understood. By using an automated bar-coded order-entry system, we would be even closer to a fully responsive computer-driven system than ever before. We would also have a system several orders of magnitude more accurate than before. This system would be ideal when there is a match between customer desires and available products. Since this will probably continue to represent the vast majority of all products, we will have an order-entry process that is accurate and capable of eliminating the drudgery of order entry as it presently exists. This frees up resources to deal effectively with items that truly need to be "different."

SOME CONCERNS ABOUT E-COMMERCE ORDER ENTRY SYSTEMS

The technical problems of an e-commerce system enhanced with bar-coded entries are ones of capacity and training. They are certainly surmountable. The needs for computer database capacity having orders of magnitude increases are already solved. The next generation of computers and linked servers now coming into the market have ample capacity to service the needs. So the need is for trained personnel to manage the system. Information technology (IT) personnel are quite capable of handling these servers and ports. The problem will be with the operations people. Most of these people will come from sales and unfortunately their record of computer integrated systems competency is not impressive. Most computer integrated experiences in industry lie within the manufacturing, engineering, and finance functions. This will have to grow to include sales and marketing. The hardware and software capacity increases will be handled as a matter of course; as companies foresee needs, they will purchase it. It remains to be seen, however, if order-entry database management can be improved as quickly as needed.

The real problem is a culture change. Vendors will have to product code their offerings in a way that describes the product and does so in a way that all of the information the fulfillment process requires is included. To do so has been the goal and dream of MRP II systems for years. What it does mean for businesses is that MRP II will have to grow to include Sales systems, particularly how products are offered for sale to customers and support provided. Sales persons will have to be computer literate and learn the nuances of the internal workings of their companies. They will more than ever need to be experts on how to convert commodity-type products into unique applications. At the same time, e-commerce will more than likely eliminate the need for sales positions selling commodities by handling virtually all transactions through the software programs. This will likely put an emotional strain on the sales force and perhaps cause a "ludite"-like reaction with the associated tendencies to resist change.

The prime concern we need to be aware of is whether the e-commerce system philosophy can overcome resistance to change. We will have to be successful in solving the nontechnical problems for e-commerce to be an all-inclusive way to do business. If we can, there will be sufficient momentum, so the incentives to have trained personnel (primarily sales, but also technical) to manage and maintain it will flourish; and a paradigm shift in how business is conducted will occur.

The opportunity for the entrepreneur to be in the forefront of this change is large. Entrepreneurs by nature are risk-takers and willing to try new promising techniques. This being the case, entrepreneurial startups are more likely to embrace e-commerce at the initiation of such technology, thus having a significant advantage, as described above, when it reaches true commercial viability. The same can be said for supply chains, although they are perhaps slightly more conservative in their approach. The supply chain can use its mutual business information system to advantage to add barcode capability. This way, the probability of input error is greatly reduced. So the entrepreneur as a member of a supply chain can greatly increase the probability of a successful outcome.

SUMMARY

The concept of mutual support is demonstrated by selling via the supply chain. An entrepreneur, tying in with a supply chain either as a purely commercial supplier or through a more esoteric connection as an outcome of being a member of a network, gains a true mentoring advantage. This mentoring advantage is not totally altruistic. Entrepreneurs are mentored by more mature companies because they see some long-term mutual advantage: you scratch my back, I'll scratch yours.

Joining a supply chain is definitely an advantage to gaining better access to the market and for the entrepreneur startup being able to reach a mature status much faster. Coupling that with the use of e-commerce offers even more ability for a startup to grow to the level that they can become self-sustaining. E-commerce is a new way for transactions to occur and will become commonplace. It's only a matter of time. The entrepreneurial startup, offered an opportunity through linkage to a supply chain to implement e-commerce, will find it is developing a beneficial maturity and discipline in doing business. In my opinion, this has no downside.

Chapter 9

The Business Plan for the Entrepreneur

So far I have described the techniques and methods of organizing a new company and defining how to establish its products in the marketplace. We have seen numerous procedures for organization, developing products by finding financing, managing operations and its project subsets, and philosophies to use these tools. This is all fine and very necessary. But why do it? Why bother? Because the entrepreneur has a strong desire to build a successful business and make money. Is this sufficient, and is using the tools effectively all we have to do to succeed? Probably not—besides an idea, albeit a well-developed one with significant financial and product analysis, we need to have a sustaining vision. Do we have an inkling of an idea of where we should be 1, 2, 3 years from now in terms of bottom-line profits? Possibly yes but probably not. We may have a goal in mind and even some financial numbers, but nothing firm or even focused.

This is where strategic planning enters the picture. A business needs a plan in a strategic sense. Otherwise, any road traveled is as good as another and only happenstance will get us to a desirable location. We want to employ all the very important tools, procedures, and philosophies we've learned, but we want to do it in a cohesive, coordinated way so that the road we choose is the correct one for our competitive situation. We need to create a business plan to grow and sustain the new business.

Strategic planning gives us focus. It transfers an idea into pragmatic reality as played against the business environment within which we need to compete. The strategic planning process will produce a business plan for the company, whether it be a new venture or an established firm. It is equally valid for product-producing or service companies. This document will set the stage for all functional planning; it should be the basis for the next 2 year's operating budgets and provide direction and guidance as far out as 5 years for all aspects of the business. Simply stated, a business plan ties all the pieces together. We will now see how a business plan is developed and how it uses many of the techniques previously discussed in this book.

Throughout this discussion, we will use the terms *strategic planning* and *business plan* virtually interchangeably. This is okay but the purist would say we are technically incorrect. Just for the record then, here is my definition for each.

Strategic planning:
The process of evaluating the environment a company chooses to compete in by identifying the constraints and creating a path for the company to traverse to meet its goals.

> Business plan:
> A method of achieving a set of established goals within the confines dictated by the strategic situation.

So we see the definitions are quite similar. Perhaps we can say that strategic planning is more cerebral while business planning is tactical and pragmatic. I prefer the term *business plan* because to me it implies doing something, executing, while *strategic plan* suggests a think-tank atmosphere, which is far from the reality of the entrepreneurial startup company. The business plan is structured within the following broad categories, which we will examine and illustrate with examples:

- Executive summary
- Strategies
- Programs
- Resources required
- Financials
- Contingency plan

The purpose of the business plan is to provide a unifying document that coordinates and communicates all of a function's responsibilities in carrying out the stated purpose of the business. Business plans are normally constituted for more than 1 year, with the first year being very precise, and the subsequent years being less precise due to more market unknowns. The business plan ought to have a 3-year horizon with at least 6 months of the initial year fixed.

We will describe the various parts of the business plan in accordance with the format of the finished document. The sequence of events for preparing the business plan is shown in Figure 9-1.

However, the best way to understand the makeup of a business plan is in the order of presentation, that is, by the table of contents, which is not the same as the order of its development. I will describe the business plan content in the order of presentation of the finished document and illustrate with examples as we go along. Then, after concluding the explanation of the business plan's content, I will use the chronological sequence to describe how it's done. This way, an abstract communications device will be given some real-world meaning before we try our hand at developing a business plan.

THE CONTENT OF THE BUSINESS PLAN

It is important that we have an understanding of the content involved, so when looking at a business plan you'll know what to expect and what to search for to get a complete document. One more thing before we start. Business plans have 2 major purposes; therefore, they should be written to satisfy both purposes, which include the following:

1. *Implementation directives and guidelines*—basic instructions by the most senior management as to what the overall needs of the company are, the schedule for completing the business plan, and who is assigned to do what.
2. *Vision development*—what the company stands for
3. *Mission statement*—how the company intends to carry out the intent of the vision
4. *Objectives*—broad-based statements of intent that support the meaning of the vision and mission statement
5. *Situation analysis, market assessment, market segmentation*—a pragmatic assessment of the current "world" state of being that sets the stage for how the company will and can compete through implementing its vision, mission, and objectives
6. *Sales Plan*—a setting of desires, usually numerical ($ and units), to be achieved during the time period of the business plan (typically 3 years) that are compatible with the evaluation of the "world" from Step 4. How to achieve the desires are spelled out in detail.
7. *Production plan*—refining the numerical data of the sales plan into units of production, which are further broken down into materials and labor, by total and per time period. Often a capital-equipment purchase plan is appended to the production plan. Comparisons are made between existing labor and forecasted needs, with plans to accomplish.
8. *General and administrative plan*—staffing for the company, along with necessary training and recruiting plans. All other plans not specifically covered elsewhere.
9. *Financial plan*—budgets, pro-forma P/L, and balance-sheet development per the inputs of all the plans. Also a plan for raising capital and obtaining credit to fund the business plan.
10. *Measureable goals*—after all plans are completed, the set of goals that are developed to specify what needs to be accomplished and by when. (This becomes the driver for all projects.)
11. *Contingency plan*—the "what if?" scenarios if the projections, guesses, and hunches associated with any of the above steps do not happen as forecasted. These are the backup plans, usually done in concept format only.

Figure 9-1. The chronological sequence of business plan development.

- Define the strategies employed that lead to the company accomplishing its goals, and
- Create a sales document defining who and what the company is to prospective financiers and customers.

All well-written business plans satisfy both purposes.

Executive Summary

The executive summary contains a vision, a mission statement, a situation analysis, ongoing objectives, and period goals. The mission statement and ongoing objectives should precede the business plan formulation process. Situation analysis and period goals are a result of the business plan process.

Executive summaries can be short and simply a bullet-point summary of the strategic or business plan, or not be a traditional summary at all but take up the majority of the document's pages and fully describe the programs to be employed. It depends on the style of the company. However, whatever style is employed, it must contain the elements mentioned previously.

We covered visions, mission statements, objectives, and goals extensively in Chapter 4. What we haven't specifically covered is the situation analysis concept. So with this explanation of the executive summary we'll review visions, mission statements, situation analysis, ongoing objectives, and period goals from the viewpoint of the executive summary's content and reason for inclusion in the business plan; the majority of the discussion will be devoted to the situation analysis.

Vision

Vision is what the company stands for. It is important to have a focus for the business plan that gives an overriding reason for why decision choices have been made.

Mission Statement

The mission statement is management's tool for describing the philosophy of how the business is to be operated and what the expected outcomes are to be. Contents of the mission statement are described as the key statements of intent (usually 3–6) that delineate the company's reason for being and what it wants to accomplish on an ongoing basis.

Situation Analysis

The situation analysis describes the current business "state of being" in an exact definition with facts as possible. It contain 6 major categories:

- Environmental influences
- Market assessment
- Market (customer) segmentation
- Competition
- Self-evaluation
- Strategic issues

Environmental Influences

In this category, we describe the climate in which the company must operate. Therefore, its content is very open-ended. Topics frequently included:

- *Demographics*
  This is a description of the company's products' and services' end-user population, and where they're located. Included would be their standard-of-living quotient, hence what they can afford.

Example: For an electronic component company, an end user could be the owner of a notebook-type personal computer, nominally earning a $50K annual salary. This would set the price for the product and all of its component parts (the electronic components company's contribution). Location is also important because it may ultimately determine where factories must be located.

- *Government influences*
  Government influences define what laws and regulations must be complied with now and in the future. They will set the stage for understanding the company's potential cost exposures and plans to contain them.

Example 1: Known or anticipated changes in environmental protection requirements forcing capital expenditures. Both manufacturing and engineering would need to plan, for example, for the impact on capital budget and on profit margin.

Example 2: Potential changes in the tax codes, and the effects on profits.

Example 3: Foreign government policies affecting a company's ability to manufacture locally or even sell in that country.

- *Economic considerations*
  The situation analysis should define the economic climate the company will face during the forecast period. This includes, but is not limited to, the state of economic growth or recession and forecasts of rate of change in either direction. Also included would be currency exchange rates prognosis, economic evaluations of major customers (their ability and need to buy), and shifts of economic strengths of regional markets.

Market Assessment

The market assessment is an evaluation of the opportunities to sell products or services by category. This information will be instrumental in creating the build/service provide/sales plan mix, which in turn will create the investment and manpower plans.

Market (Customer) Segmentation

Closely related to market assessment, this section defines the strategies and plans for selling each of the company's product lines. Using a quality functional deployment technique, the company looks at product lines and major customers and defines the selling characteristics necessary to reach that customer or group of customers. This section would outline the marketing strategies to be employed for each identified segment. Items such as specification tolerances and quality requirements are also highlighted, even though these may be issues needing to be addressed by other than the marketing function.

Competition

This is an identification and analysis of competitor strengths and weaknesses that the company needs to contend with. It is important that this be as objective an assessment as possible because marketing plans will have to take these factors into account. It is also important to identify all competitors so that the company encounters no surprises. Too often, foreign competitors are ignored, which could leave a big hole in the company's sales strategy.

Self-Evaluation

This is perhaps the most difficult portion of the situation analysis to accomplish. The need to be objective in describing weaknesses as well as strengths is not an easy thing to do. It is necessary to look at the business as if we were considering it as a potential acquisition and doing a due diligence investigation. The purpose of this evaluation is to optimally pursue the objectives based on a true assessment of capabilities. A true assessment will allow generation of the proper programs for improvement couched in the context of what really has to be done.

Strategic Issues

This section must deal with the weaknesses discussed in the self-evaluation section, as well as those items that will drive the very nature of the business. Obviously, weaknesses must be assessed for proper solutions. But it has to be done in a manner compatible with the external forces that will affect the business. Here, trends in the product marketplace need to be addressed and assessed for the ability to not only survive but prosper. The changing nature of customer requirements directs the type of products that must be supplied, which in turn creates significant issues of internal change that must be coped with. These include organizational structures, personnel requirements, design and manufacturing capabilities, distribution channels, and all other issues regarding the continuous satisfactory health of the company. The real strategic issues must be defined so the business plan can be a concise and focused working document with resources optimally deployed.

Ongoing Objectives

This section of the executive summary contains the company's ongoing objectives and each function's (for example, marketing, manufacturing, engineering, etc.) derivative of those objectives. They are broad-based statements of intent that support the meaning of the vision and mission statement, but usually are focused on performance. Keep in mind that since objectives are broad-based statements of intent, they are not directly measurable. They serve the purpose of creating direction and thus focusing goals for the business plan period. Objectives are ongoing in that they remain in force year in and year out.

Typical ongoing objectives include:

Objective 1.: Decrease Direct Labor (DL) as a percentage of sales (improve productivity).
Objective 2.: Create additional capacity to increase sales and decrease outside vendor costs.
Objective 3.: Manage creation and implementation of new products.
Objective 4.: Reduce operating expenses as a percentage of sales.
Objective 5.: Increase the value of the company to the shareholders.
Objective 6.: Increase market share.

Period Goals

Goals for the business plan are specific, measurable statements of intent bounded by time. They never extend beyond the time horizon of the business plan. While goals are normally function-specific, a few goals transcend function and are listed separately under this broader headings. Typical cross-functional headings in this category include

- Profitability
- Sales
- Margins
- Financial Working Capital

In addition, we would have the major goals for each function listed under the appropriate heading. The CEO, president, and senior staff decide the question of what constitutes a major goal.

Goals change with the evolving nature of business dynamics, that is, changes in the business climate as derived from the situation analysis. Goals must always support the ongoing objectives. Goals are accomplished through the development of supporting programs or projects (Chapters 3 and 4).

Strategies

Strategies for each function are developed concurrently. The idea is to create mutually supportable strategies to support the ongoing objectives and the mission statement. From these strategies, the specific goals are developed. The process is iterative but normally follows the pattern set forth below.

Sales Plan

This is considered the starting point of the business plan. It is based on work done to produce the situation analysis that sets the stage for what is possible to sell. The sales plan should create the matrix of products or services to be sold during the business plan period by month, quarter, and year. Also included in the sales plan are the promotional, advertising, and distribution plans, along with associated expenses. Strategies to be employed to execute the plan are outlined in broad-based concepts, which become the salient points of the strategies section of the business plan. Some typical examples of strategies derived from the sales plan:

- Distribute the product through existing distributors and develop new distributors for territory X,Y,Z.
- Based on demographics, expand market force in region 3.
- Price products and services to take advantage of anticipated industry-wide production equipment shortages.

Production Plan

The labor hours and materials matrix is produced on the basis of the sequencing, quantities, and types of product or services to be produced or delivered and using data from the sales plan. This is compared with available resources and a production strategy is developed. As in the sales plan, the salient points of the strategies are placed in the strategies section of the business plan.

Some examples of strategies derived from the production plan:

- Achieve a smooth production flow throughout the manufacturing cycle by developing a flexible schedule that compensates for unplanned production constraints.
- Foster a culture of quality performance throughout the manufacturing process.
- Introduce new manufacturing processes to lower production costs.

General and Administration Plan

This plan represents the "all other" category lumped together. The most prominent component is the personnel staffing plan, which needs to be driven from the sales and production plans. Levels of staffing for all functions of the company must be derived from the expected activity load to support these plans. Also included here is the information systems (IS) plans to support development and maintenance of necessary communications and computer systems.

Some examples of strategies derived from the general and administrative plan:

- Provide appropriate levels of automation to reduce administrative costs.
- Use standard software and systems that are supported by major companies and are proven by the experience of multiple users.
- Conduct systematic training and qualification programs for all employees.
- Staff all departments to meet production and service needs, taking into account the productivity improvement objectives.

Finance Plan

Based on the sales, production, and general and administration plans, finance develops funding strategies to provide the necessary cash flows for successful implementation of these plans and to ensure that the bottom-line net profit is per the original goals. As would be expected, the finance plan is not developed until at least all the other plans have gone through a first-pass evaluation.

Some examples of strategies derived from the finance plan:

- Secure lines of credit at lowest possible costs to support the company's strategies.
- Create budgets for all functions to comply with company strategies.
- Derive financial measurement systems that optimize the operating components abilities to manage in a fiduciary responsible manner.

Programs

Programs consist of measurable goals to enable the strategies to be achieved. Let me repeat again, in the parlance of the objectives and goals theory, that they are measurable statements of intent bounded by time to accomplish them within. For each major plan described in the strategies portion of the business plan, we would have precise supportive programs goals. These can be singular or complex, depending on the strategy. For ease of measurement and control, they are usually defined along functional operating units, such as engineering, manufacturing, and sales and typically along the functional lines of authority represented by the company's senior staff. However, it makes sense in some programs to transcend traditional functional area responsibilities, such as the Capital Expenditure Program, and these programs are treated as separate entities. Programs should have supporting major projects included in the business plan. This leads credence toward how the program may be accomplished and data for required resources.

Resources Required

This section is a description and refinement of major resources required to make the business plan succeed. For most companies offering a product or service, it typically contains 3 sections:

- Employment summary
- Capital plan summary
- Market and facilities expansion summary

This section of the plan may be omitted and its content included in the respective plans. Small companies often chose this option to minimize the size of the finished document.

Employment Summary

This section has a recap of employees by category and justification for the plan level. A typical breakdown is by direct labor, indirect labor, manufacturing and engineering overhead staff, marketing and sales staff, and administration. For business plan purposes, the administration category is a "catch-all," a place to home-base general management and office personnel, and the finance staff.

Capital Plan Summary

The capital plan summary gives reasons for selection of capital projects, along with a detailed explanation of the capital plan by individual planned expenditures. Also, a summary of costs, savings, and financial justification/payback period is often included.

Market and Facilities Expansion Summary

A section covering the topic of growth, or in some case contraction, is often found in a business plan as a separate, indexable section of the plan. Contraction or growth is a significant challenge for any company and often requires a place in the plan for specific information detailing the resources (human, capital, and materials) necessary to carry it out.

Details of these activities are like all other activities but in many ways not like all other activities. Programs all require resources to carry them to a successful conclusion. So do expansion or contraction programs. However, expansion and contraction programs are not typical ongoing activities. Growth and contraction are usually special-emphasis activities that cannot happen all the time, as much as we would like the positive aspect to do so. Expansions as well as contractions take considerable resources to accomplish. So like any good strategist, the business leader must manage the resource and only allow real expansion and contraction to occur at times of true opportunity. In this manner, they are akin to project management and follow all the rules for such, as explained previously.

As with other activities, we would include broad-based strategies (objectives) through detailed program statements (goals and projects) in the respective proper sections of the business plan.

Financials

This section contains the traditional financial spreadsheets with adequate details to support operational and financial planning. The type of data found here would be similar to that presented in our review of financial measurements but for the entire period of the business plan. P/Ls, balance sheets, and cash-flow statements are the type of documents populating this section of the business plan. I should point out that the data presented and its format are arbitrarily selected, usually to meet some selling or regulatory needs. There is no one correct way to present financial data in a business plan. However, sticking with generally accepted accounting practices is a safe way to go.

Contingency planning

This is the "what if" scenario and recovery plans to be instituted if the situation analysis does not work out as anticipated. The more "iffy" the situation analysis is, the more thought and work is required to support the contingency plans. Contingency plans ought to be conservative but with a flair for the dramatic because they are recovery activities for plans that didn't work. Therefore, the need to be conservative is ever-present. We certainly can't be wild and speculative. We need to be focused and bold but working off a firm recovery base platform.

The rules of recovery are simple. The first step is to stop the bleeding. In other words, understand what is causing the harm and stop doing it. The second step is to understand why the discovered cause is causing harm and plan how to remove the toxic aspects without destroying the rest of it. And the third step is to implement and monitor the fix.

Contingency planning, however, starts before the harm has begun. It says let's imagine that the plan is no good and fails. Then we can speculate how to stop it, how to understand the cause of the poor results, and finally delineate a plan we will engage if the harm does become reality.

In the business plan we list contingencies as simple statements and then say what we will do to prevent or at least minimize detrimental effects. Here are some examples.

Example:

Possible detrimental effect: The business cannot operate for a month because of fire damage.

Contingency plan: Actively get an insurance claim for loss, adjudicated so money is available to rebuild. Invoke business loss insurance requests for funds.

Example:

Possible detrimental effect: Sales at 78% of expected value.

Contingency plan: Initiate price rebate program to stimulate sales if sales fall below 80% expectations level. Eliminate Saturday overtime except for emergencies if sales dip below 80% expectations.

Note there are trigger points for actions to commence. This is a most frequently used methodology for treating contingencies.

SOME FINAL WORDS ON THE STRUCTURE OF A BUSINESS PLAN

All 6 items of a business plan are interrelated. The *executive summary* tells the story of the company and gives its aspirations as well as its potential problem areas, and describes how it will achieve its vision, mission, objectives, and goals. *Strategies* show the way the objectives and goals will be achieved. *Programs* are the structures for achieving the goals. *Resources required* spells out the materials and people needs to be successful in achieving the goals. The *financials* give us a plan for having sufficient funds to work the programs and to measure progress. And finally *contingency planning* gives us a head start in recovering if our initial strategies and programs were not successful. The flow of information cascades from ideas to actions, with measurements for corrective actions as needed. A business plan is a plan; it's as simple as that.

As simple as a business plan is in concept, it is always a mystery to me why some companies, even relatively large ones, do not take the time to develop a business plan. This affliction seems to affect newer companies, usually less than 10 years old, that are in upturning markets. Their experiences have been everything rosy and peachy. Business is good, in fact it comes to them with little effort to encourage buyers for their products or services, sort of like fishing in a brook teeming with very hungry fish. Anything that resembles bait and has a hook will do. The fishing is good. When that's happening, who has to bother to plan for when the fishing isn't so good? Frequently, companies that are growing rapidly don't feel they need to plan for their future. They think the good times will continue to roll along. Then it happens. All of a sudden they find their competition is taking business away from them. Or probably more accurately, the pasture land where customers have been grazing has shrunk considerably and the competitor is

ALLOW YOURSELF TO THINK OUTSIDE OF THE BOX

" No! - I can't be bothered to see any crazy salesman. We've got a battle to fight! "

Figure 9-2. Failure to anticipate ahead.

now eating in the same plot of land that they are, but he's much better at it because his company has a plan that understands the customers, their needs, the other competitors, and all about the future of the business. And he has a contingency plan that allows him to absorb competitors' feeding territory if he needs it to prosper. Too bad for the fair-weather competitors; they have no plans, they're not prepared, they lose.

And don't think that this is an affliction of only small and medium-sized companies. Here's an example.

Example: During a consulting assignment, I came in contact with a $1.5 billion dollar company in the medical supplier field that had no business plan. They started to drift during a recent recession, as explained above. It appeared to me that the leadership of the company was made up of entrepreneurs who were very adept at making deals to acquire new businesses but didn't understand how to integrate them and make them profitable. The only dictates they gave their various divisions were bottom-line profit goals, and nothing else. Therefore, one of their larger divisions took some very short-term actions that minimized costs way below what they should have been to achieve stable growth. Consequently, after meeting their profit goal for the period, it became apparent they were not able to sustain the goal for future periods. The answer to the problem was to develop a business plan. But this did not happen until the senior leadership of the division was fired and replaced. The new leadership did develop a business plan and found that they needed to structure it in accordance with the levels previously explained. They went right down the path I've outlined, and after many years in their particular market, they found they were lacking in understanding their market and how to respond to it when it was softening. They also found that since the fishing used to be so good, they didn't have the organizational strengths to carry out effective manufacturing and marketing when it was more than simply trolling for business and delivering as best they could, not necessarily what the customer needed. As a result, their business plan had the embarrassing need of requiring basic skills training.

The moral of this story is planning is always necessary; it's not a nicety to do because the theorists say it's required. Planning allows a company to see the entire picture and to be able to play to its own strengths, not being buffeted by the whims of the business cycle.

AN EXAMPLE OF THE DEVELOPMENT OF A BUSINESS PLAN

Now that we have explored the content of a business plan, let's go through the process of developing one for a hypothetical company.

First we'll set the parameters for our hypothetical company. Let's call it the ABC Maintenance Company. ABC does maintenance of a technical nature for commercial and residential property. It has all of the expertise, licenses, and building and construction trades necessary to maintain buildings. Its owner is an entrepreneur engineer who has a P.E. license and 10 years of experience in HVAC. He has decided there is a "fortune" to be made by being a technical resource to maintain structures and systems commonly found in buildings. He is convinced that a "one-stop" company, able to take responsibility for and capable of doing all repairs, would be welcomed by the building owner community.

The company is now 1 year old and the entrepreneur realizes he needs a business plan. First year's sales were $450,000, and the entrepreneur thinks his net profit was $16,000. He uses a P/L bookkeeping system set up by his accountant. He has a 2-person sales department headed up by his wife. He is the engineer and he has a licensed plumber and a licensed electrician who work for him on a contract-by-contract basis. His only full-time employees besides his wife and a sales order clerk is a handyman skilled in masonry and carpentry trades. This is indeed a small company startup. Let's say we have been hired to help the entrepreneur develop a business plan because we are consultants who do that for small businesses. We have the assignment; now let's do it.

Setting the Schedule

The first thing we need to do is establish a schedule. This may seem trivial but is essential. As we've seen, order of sequence is necessary for successful development of the plan. Figure 9-3 shows the traditional time model for preparing a business plan.

While the model shows an infrastructure for a well-established company, it is certainly suitable for any company. The reason this is so is because even in a startup company, these functions are being performed, albeit many functions are being done by the same person. The schedule defines the emphasis for development that the multi-hat wearer needs to focus on at any one time. For example, the owner/CEO will have to focus on mission statements and objectives and then later in the cycle, gives himself directions as the Operations Manager to develop the production plan. Actually, a benefit of a business plan development with only a few participants is a more effective communications of intent and directions to each function. However, fewer participants means more work per individual.

I can't over-emphasize the need to conduct the schedule in the sequence as shown. Notice how most steps require a predecessor step, and for good reason. Market analysis is necessary before a sales plan can be drawn up. To do a sales plan without an understanding of the market situation would be more like writing fiction than preparing a useful document. And a production plan too, would be fiction if based on a faulty sales input. Coordination of information is also critical for developing a good business plan. We want to make sure that the same basic information is used by each

(time to accomplish, 4–6 weeks, some steps concurrently)

Topic	Responsibility (traditional model)	Cycle Time (predecessor in parentheses)
1. Mission statement	CEO, President, Board of Directors–prime Senior Staff–inputs	1-2 days
2. Objectives	CEO, President – prime Senior Staff–inputs	1-2 days (1)
3. Situation analysis, market assessment, market segmentation	President, VP Mktg – prime Senior Staff–inputs	1 week
4. Sales plan	VP Mktg – prime Pres., Sr. Staff–inputs	1 week (3)
5. Production plan	VP Mfg., VP Engrg.–prime Pres., Sr. Staff–inputs	1 week (4)
6. General and admin. plan	VP.H.R., VP.MIS – prime Pres., Sr. Staff–inputs	2 days (3,4,5)
7. Financial plan	CFO – prime Pres., Sr. Staff – inputs	concurrent with 4,5,6 plus 2-3 days
8. Programs to support strategies (goals), including resources	Senior Staff – prime President–inputs	concurrent with 4,5,6,7
9. Contingency planning	Senior Staff – prime President–inputs	1-2 days (7)
10. Executive summary	President, CFO – prime Senior Staff – inputs	concurrent with 1 thru 9 plus 1-2 days
11. Approval	CEO, Board of Directors	1 day (10)

Figure 9-3. Typical order of accomplishment of the business plan with chronological cycle time.

function providing inputs to the various segments of the business plan. This ensures that all the data will match and be relevant. Relevancy is important because the business plan must be realistic and very pragmatic. To try to implement a business plan based on faulty information will sour the process for the organization, to say nothing about the harm it could do to the business. One sure way to eliminate the possibility of this happening is to adhere to the schedule sequence.

Schedule connotes sequence and time. The times shown are for established companies with full complements of staff. However, much to the surprise of startup companies, their cycle time is much the same. This is due to the multi-hat nature of startups. Individuals are doing lots of different jobs. They need to attend to many things, which means the ability to allocate time to do business planning competes with other tasks and there is less of a capability to pass off work to subordinates. In fact, subordinates may not even exist. The end result is that the time schedule for preparing the business plan is about the same for all companies.

Example of a Mission Statement

The mission statement sets the tone for the entire business plan. In one or two sentences it describes what the purpose of the company is and what it strives to achieve. The sample company, ABC Maintenance Company, does technical maintenance for commercial and residential property. It is not a full janitorial company in that it doesn't do cleaning, except when cleaning is a component of a maintenance service such as cleaning clothes dryer ducts as fire-prevention and energy efficiency–improvement maintenance activities. The company's mission statement should reflect what it does and how it intends to do it. The following is a suitable mission statement.

> "Establish ABC Maintenance Company as the preferred provider of high-quality commercial and residential building maintenance and related services in the Florida Southeast Region."

Notice that the statement says "the preferred provider," not "the low-cost provider." This sets the stage for all of the planning to follow. The emphasis will be on quality, not on methods of providing low-cost services. This would imply that the company will set its prices to ensure that correct tools are used and time to do the work is compatible with ensuring the highest-quality performance. No rushing to squeeze more work into a tight schedule and no skimping on having the proper tools, including diagnostic equipment. This emphasis will definitely shift the company away from a goal of gaining a dominant market share at the expense of superior quality. Instead, the emphasis will be on taking jobs where quality of performance is important, thereby being able to charge premium prices.

So we see mission statements set the direction of the entire business plan. In the mission statement, if the word "dominant" was substituted "preferred," and "high quality" was eliminated, then the message would be quite different:

> "Establish ABC Maintenance Company as the dominant provider of commercial and residential building maintenance and related services in the Florida Southeast Region."

This implies an entirely different course for the development of the business plan. It would emphasize profitability via market share growth, that is, a high volume of business with probably lower profit margins, similar to a supermarket being compared with a gourmet food shop.

In the next example we'll see how the mission statement will begin to give directional focus to the business plan development.

Examples of Objectives

Objectives, as we've seen, are ongoing statements of intent that tend to be generalized rather than specific. They define the mission statement in specific, usually functional categories. Looking at the mission statement for ABC Maintenance company we would want to break it down into functional area statements of intent. The mission statement is repeated below for reference. Let's see what we can do with it.

> "Establish ABC Maintenance Company as the preferred provider of high-quality commercial and residential building maintenance and related services in the Florida Southeast Region."

First we see "preferred provider of high-quality—building maintenance and related services"; this means we need to have an ongoing objective related to quality. Since we know we want to continue to improve and objectives are infinite in time rather than finite and specific, perhaps an appropriate objective would be:

> "Achieve 6 sigma quality standards in all departments."

Another objective related to delivery of quality services could be:

> "Maintain service innovation leadership to provide best value for customers."

We also know that to be a preferred provider, we would need to be giving customers services using leading-edge methods and technologies. So another objective to focus the company in this area could be:

> "Introduce new technologies to become a "world class" maintenance and related services provider as recognized by the industry."

So that ABC Maintenance Company can continue to be the preferred high-quality provider, we need to be productive, have customers, and be cash-flow positive. This leads to several more objectives for Operations, Finance, and Sales.

Operations-oriented: "Improve productivity to levels that are compatible with maintaining a positive cash flow," and "Provide a positive work experience for employees."

Finance-oriented: "Maintain a positive cash flow to negate the need to use line of credit financing," and "Improve profit margins by reducing break-even costs to be at the rate of best in industry standards."

Sales-oriented: "Improve market share to become a top-tier maintenance provider in the geographic area we service."

Notice we do not say gain share to become a leader in sales. Our objective is to become a known factor in the market so it's possible to become a preferred provider. Another objective along this line would have to do with what we charge so that we can become a player in this market, and it would probably be stated as:

"Create pricing structures to stimulate sales and increase profitability."

Using this approach of first making sure that the intent of the mission statement is represented by one or more objective, then attending to the ongoing viability of the business with the objective oriented respectively to operations, finance, and sales, we can create objectives for any type of business. To recap, the objectives for the ABC Maintenance Company are as follows:

ABC Maintenance Company Objectives

Mission-oriented
- Achieve six Sigma quality standards in all departments.
- Maintain service innovation leadership to provide best value for customers.
- Introduce new technologies to become a "world-class" maintenance and related services provider as recognized by the industry.

Operations-oriented
- Improve productivity to levels that are compatible with maintaining a positive cash flow.
- Provide a positive work experience for employees.

Finance-oriented
- Maintain a positive cash flow to negate the need to use line-of-credit financing.
- Improve profit margins by reducing break-even costs to be at the rate of best in industry standards.

Sales-oriented
- Improve market share to become a top-tier maintenance provider in the geographic area we service.
- Create pricing structures to stimulate sales and increase profitability.

We now have an intent, purpose, and focus on what we want to do. The next step is to understand the opportunities and constraints we need to contend with so we do specific planning. That leads to the next phase of business planning.

Example of a Situation Analysis, Market Assessment, Market Segmentation

We have 9 very nice-sounding objectives with noble intent. What do we do with them? How do we go from generalized intentions to specific actions? Well, the road leads through the valley of geo-political situations that we must understand. Then knowing what we must contend with, we seek to understand our own strengths and weaknesses with respect with what we've found the lay of the land to be. With this intelligence we can assess our ability to sell our services and products. And if necessary, we can cut out a parcel of sales territory for our own, one where we feel reasonably sure we can be successful. How do we do this?

We start with the situation analysis. Previously in this chapter we said that a full situation analysis would contain the following items:

- Environmental influences
- Market assessment
- Market (customer) segmentation
- Competition
- Self-evaluation
- Strategic issues

How we put them together in our narrative depends on the type of business we're in and how important each one is. The important factor is to make sure that each is covered. By covered I mean each is considered for its impact on our particular business for the period of time of the business plan. Note that the title of this subset of the chapter says "example of a situation analysis, market assessment, and market segmentation." This says that I've mixed up my outline parent-child relationship. While a little confusing, it's still okay. I will cover everything under the heading of situation analysis. What I'm doing is lumping all but market assessment and market segmentation under the heading of situation analysis, then upgrading market assessment and market segmentation to equal status in the outline to situation analysis. I'm doing this because small business startups need to pay special attention to market assessment and market segmentation. And the other topics traditionally under situation analysis are not as important, but still must be addressed to make sure there are no hidden surprises. Again, this may be confusing, but it points out the lack of rigidity that has to be reckoned with when doing business plans. While the general model is a very good guideline, it is not necessarily a hard-and-fast rule of science that needs to be followed.

Okay, now that the reasons for deviation from the general structure have been explained (I hope—but really it's not that important), let's get back to the ABC Maintenance Company. Here we have a new company that needs to understand its market so it can decide how to approach selling what it believes to be a unique new concept of maintenance. So what we've chosen to do is combine environmental influences, competition, self-evaluation, and strategic issues under the topic situation analysis and have separate investigations of market assessment and market segmentation.

Looking at the basic facts given for our example, here's what the situation analysis for the ABC Maintenance Company could look like.

Situation Analysis

The maintenance and related industry is a service provider to residential, retail, commercial, and new construction real estate. It is bounded by contracts to perform maintenance services ranging from one-time only to retainer agreements. The industry is considered low tech; even though as our company perceives it to be so, this is not true. This makes barriers to entry low, other than recruitment of labor. The industry is highly fragmented with no company having a dominant market share of greater than 5% in its regional area. Fees commanded for services are relatively low compared with other service industries such as electrical and plumbing contracting. This is due to the perception that the skills level to perform efficient maintenance services are low.

To be successful, a maintenance service company needs to be agile in performing jobs in the shortest period of time while satisfying the customer. Due to the relatively low fees the industry commands, excellent management of company resources is essential for survival and growth. It is imperative that companies in this industry practice sound fiscal and personnel management.

Management emphasis must be on:

- Obtaining jobs to maintain full use of staff and equipment. Whatever segment of the business (HVAC, painting, electrical repair, plumbing) gains entry to the customer, it needs to be the lead-in to selling other services.
- Accurately estimating time to do jobs. Needs to be based on recognized standards.
- Accurately estimating costs to do jobs, based on recognized standards.
- Maintaining an efficient scheduling regime. The methodology must be archival for comparison purposes and to measure effects of improvement programs.
  - Assignments of staff
  - Dispatching staff and equipment
- Doing jobs efficiently. Constantly comparing against industry standards and variances examined.
- Maintaining sufficient quality levels to gain
  - additional work from the client
  - recommendations to other potential clients

- Reconciling of estimates to actuals for costs and time to perform jobs. Understand reasons for variances and taking immediate corrective actions.
- Tightly controlling materials usage to minimize expenses.
- Recruiting and training of staff in proper methods to minimize staff turnover.

The standards of success from the client's viewpoint will be primarily subjective; that is, is the system still running? The client is not interested in knowing what technical feats had to be achieved to keep the air conditioners functioning, and the company will only be as good as its last job done for that client.

Since this is considered a low-tech industry, recruitment of staff and maintaining that staff is essential for success. Generally, the jobs in this sector are considered entry-level except for the skilled or licensed technicians. Therefore, a greater-than-typical effort must be placed on evaluating new employees and identifying those with leadership and future management potential. These individuals need to be nurtured so that they become the core of the workforce and the eventual department managers. The situation requires an emphasis on "bootstrap" education to train this cadre, realizing that the start level for the education process is going to be no better and perhaps lower than the normal industrial standard. Emphasis on training and education needs to be higher for tech-system-based industries and must be recognized and planned for. The additional problem will be convincing employees that they are not glorified janitors, but trained maintenance professionals.

Does the previous narrative cover everything that should be in this particular situation analysis? Let's look.

Environmental influences

The environment denotes the climate of doing business, which includes the perceptions of what the clients think about the services or products offered. There's no doubt that the narrative covers this topic. It is full of references and statements about perception. The emphasis of need to overcome lack of understanding on the part of the clients as to what constitutes maintenance is a telling factor in setting the environment of how the eventual marketing of the company must be played out. It will be necessary to sell on the professionalism of the staff and differentiate from normal janitorial work, such as the mental image of a whacko techno-nerd incapable of doing real jobs we often see of residential "supers," which will have to be overcome.

Competition

This is mentioned in the barriers to entry and the lack of market share dominance by any one company. By implication, there are many companies in the field competing, even though many do not have the skills of ABC Maintenance Company.

Self-evaluation

The section spelling out the need for higher-caliber people and training go to the heart of self-evaluation. It is implied that the company has tremendous capability but will need to sustain it through training and recruiting.

Strategic issues

All the bulleted items are strategic issues. They instruct what will be important in running the company. The entire flavor of the report is strategic. The market breakdown by share, perception, and the need to upgrade the profession in the minds of the client are all in essence strategic issues, as well as other tactical issues.

So we see the situation analysis does cover everything that it needs to for this particular business plan. Now let's move on to market assessment and segmentation.

Making Use of Competitive Information

Competitive information is the beginning of the market assessment and segmentation task. Understanding what your competitors are capable of doing allows your company to plan effective defensive strategies of how to answer clients' questions when you are compared with other similar service or product providers. It also give you ammunition to take the offensive by knowing how to blunt the competitors' strategies when going to market. Competitive information allows a company to compare its products and services to see how it measures up to other offerings in its market segmentation. When competitor information is available, it must be evaluated *objectively* against your similar offerings to discern differences, good and bad. The last thing you want to do is to become defensive and try to defend your product or service against a competitor's. What you want to do is learn what competitors are doing better than you so you can improve your product. You also want to learn what you're doing better than the competitors so you can devise ways to prevent them from catching up with you.

When doing the Business Planning segment—situation analysis, market assessment, and market segmentation—it is prudent to assign staff members to carefully evaluate competitive information. They should answer the following checklist questions in as objective a manner as possible. In essence they will be doing a benchmarking evaluation. The following is a list of instructions I gave to the imaginary staff of the ABC Maintenance Company for evaluating competitive information. You should do the same with your colleagues when you have an opportunity to construct a business plan.

- List the major differences between the competitor's products or services and our own at entry into the market.
- Try to understand why the competitor is doing work similar to our company but in a different manner. List your suppositions for each difference noted.

- Rate each difference as better or worse than ours. Break list down to
  - Better than ours
  - Worse than ours
  - About the same as ours
  - They do something we don't do.
- For each item on each list, describe why you put it on the list:
  - Use summary statements only.
  - Describe what we should do about it.
- List lessons learned from your evaluation that could help us compete better against this competitor.
  - Rate in priority order for inputs as improvement goals into the Sales, Production, General and Administrative, and Finance plans.
  - Remember, do this for competitors weaknesses as well as strengths. It is particularly important to widen the performance margin between a competitor and us in order to maintain our advantage.
- Think about how you will become an advocate for overcoming the challenges posed by competitors. These become the basis of goals for our company to pursue to take advantage of what you have learned from your analysis. But they will only become active goals if you are a successful advocate within the company. The following questions will help you formulate your reasoning. Remember, you know what the competitive advantages or disadvantages are from your study. Now you have to sell your associates on the need and value of taking an action based on the study you've done.
  - Why do it?
  - How do we do it?
  - What resources are required?
  - When must it be done?
- List recommendations in priority order.

This checklist, while general in nature and applicable to any study to find reasons for differences, can make a significant difference if you are in a situation where achieving specifications in a bids situation is important. Understanding the competitor's strengths and weaknesses will give your company significant information on how to structure a bid. Remember, a lot can be deduced about a competitor if you know what his service or product is, how it works, how it compares with yours. Knowing this difference gives insight on how the competitor goes to market, what he uses for selling, and what he wants to hide.

Always try to learn as much about your competitors before you develop the market assessment and segmentation strategy. We can't do this here to any great degree with our sample company, ABC Maintenance, because it would require setting up other sample companies for very little instructional gain. However, we will assume that the competition information is known and has been factored into the market assessment and into the segmentations decision we shall soon see.

The following is our market assessment and segmentation development plan strategy. We will use it for the specific example of ABC Maintenance Company and recognize that it is only specific for that example. The strategy, however, may be used as a guideline for any company.

Market Assessment Development Plan

- Understand the economic factors and trends of the market area.
- Determine how those trends affect ABC Maintenance Company's service offerings.

Category	% of Total (based on 2002 sales of $450,000)
HVAC	36% ($162,000)
Plumbing	27% ($121,500)
Electrical	18% ($81,000)
Carpentry	12% ($54,000)
Masonry	4% ($18,000)
Painting	3% ($13,500)

- For each category, evaluate growth potential (plus or minus or stay the same), where growth will change (what segment of the economic factors and trends; this can be geographic, business type, or both) in sales volume, for example, dollar amount of sales and number of orders.
- Determine which service categories will prosper, stay the same, or decline. For each do the following:
  - Use 0–10 point score for probability of assessment happening: 10 a certainty, 0 absolutely have no clue.
  - Multiply probability by % of Total business = relative worth to pursue
  - Prioritize order for major groupings as growth, stay the same, or decline.

Segmentation Development Plan

- Based on assessment, choose priority of services to concentrate on.
  Note: Unless a service is to be discontinued, minimal work must be done for each service.
- List buy, no buy factors for customers in each service area.
- Determine if the company can positively affect these decisions.
  - If yes, what are they?
  - If no, why not?
  - This becomes the basis of the successful sales plan.
- Determine whether the company has the resources to carry out all the "yes" items and list by service segment.
- Resources will probably be finite to do all the "yes" items for all the service segments.
- Make list of priorities for applying resources based on assessment rankings.

- This list will be a matrix of services vs. market segment.
- It will show what the sales plan should concentrate on to get the best return.

For the example, I chose my home locale to get demographic and trend information. What I used is real data, primarily available through an Internet search in the course of one afternoon. I mention this because it is easy for any company to obtain this type of data and there is no excuse not to.

Assessment

Based on the assessment development plan, here are the results of the work.

Example: Market assessment for ABC Maintenance Company
- Economic factors and trends of the market area

    The tri-county Area, Broward, Miami-Dade, and Palm Beach, continue to grow in population and economically. The population is growing at an average rate of 3% annually. The growth is about equal in residential and commercial. This means the growth rate of the company should grow at 3% just to maintain par and not fall behind in market share. Using the year 2002 as a base, a no-growth scenario would indicate total sales to be:

No growth–adjusted for population expansion

Year	Sales ($000)	Net profit ($000)
2002	$450.0	$16.0
2003	$463.5	$16.5
2004	$477.4	$17.0
2005	$491.7	$17.5

- Effects on the company

    Based on experiences through the first 7 months of 2003, the company will do better than simply keeping pace with the local population growth. Growth due to aggressive selling and building on new opportunities for masonry and painting has resulted in actual sales of $312,493 for the period. This has resulted in net profits of $12,500 (4.0%). Prorating the first 7 months' performance for all of 2003, it is forecasted that the year will generate $535,700 in sales and a net profit of $21,428. The sales growth is projected to expand by 19% over 2002, while net profits are expected to grow over the year 2002 to 4% vs. the 3.55% as calculated from the table above. This gain is due to customer acceptance of our masonry and painting services for major projects. It is expected that the next 2 years will primarily be consolidating these gains and creating a firmer market share for them. However, a growth over population expansion of 16% for sales and 4.5% net profit is entirely reasonable through focus on masonry and painting (where growth is most probable), improved management procedures, and tighter fiscal controls. The projections for the remainder of 2003, 2004, and 2005 are as shown:

16% growth, adjusted for population expansion

Year	Sales ($000)	Net profit ($000)
2002	$450.0	$16.0
2003	$535.7	$21.4
2004	$621.4	$28.0
2005	$720.8	$32.4

Additional growth could be achieved through incorporating other new services into the company's repertoire, such electronic diagnostics and maintenance for computer systems.

Notice the example is in narrative form. But this is what the results would be after having gone through the assessment development plan. You can assume that following the steps of the assessment development plan yielded more positive results for the masonry and painting than for the remainder of the business segments. Some people like to show the results of the assessment and segmentation steps as well as the narrative. I do not. I think it clutters the business plan with unnecessary data that in no way helps in implementing the action steps that we develop later on. However, it is a good idea to keep all the data used to come up with the final narrative. This allows for future "how did I do that and get that answer?" inquiries, and to validate future calculations, if needed. Let's move on to market segmentation, which is the next logical step after assessment.

Market Segmentation

In the segmentation step we get the opportunity to look at all the services or products the company offers and see how they will grow within their market niche or segment. For real companies we would go through the segmentation development plan meticulously to find out how much share of the total each contributes to sales and net profit. For our example company, we cannot do that in any practical manner. So we will assume that masonry and painting are the growth stars and the rest are somewhat behind them. A market segmentation section of the business plan would look like the following:

Example: Market Segmentation Assessment for the ABC Maintenance Company

All facets of the company's offerings have not experienced the same growth rates. It is expected that the growth of segments for the remainder of 2003, 2004, and 2005 will be as portrayed in the following chart:

Category	Share of total (based on 2003 sales projection)		(%) of forecast rate relative growth	2004/2005 expected sales as % total	Sales ($000)	
	sales (%)	sales ($000)			2004	2005
HVAC	34.4%	$184.3	3%	35%	$217.5	$252.3
Plumbing	26.3%	$140.9	5%	26%	$161.6	$187.4
Electrical	18.5%	$99.1	10%	16%	$99.4	$115.3
Carpentry	11.9%	$63.7	6%	10%	$62.1	$72.1
Masonry	4.9%	$26.3	30%	7%	$43.5	$50.5
Painting	4.0%	$21.4	40%	6%	$37.3	$43.2
Total	100%	$535.7		100%	$621.4	$720.8

While all categories of company offerings will experience growth and require sales efforts, the priority order for allocating resources is shown below. The top 2 growth groups are considered the growth opportunities areas based on economic projections for the tri-county area. Categories listed under the 2 priorities should be given specific attention for maximizing sales efforts. Base on available information, they are the areas where the company can gain market share through either current in-house strengths or from lack of competition in that very few maintenance firms offer full lines of services. These services should be able to piggyback on other services that have already gained entry to the client's facilities.

1st Priority	2nd Priority
Painting	Masonry

Notice that the segmentation and assessment examples are replete with numbers, percentages, and dollars. This is a common way of objectifying a decision process that has a significant amount of subjectiveness about it. While the numbers may be taken out to 1 or 2 decimal places, the important information to gain from the data is the relative magnitude compared with other categories. For example, in the segmentation we see that there is a relatively large positive change in growth rate for masonry and painting. This would lead to recognizing that something good is happening and should be explored and exploited for the benefit of the business. Without this segmentation analysis, not much attention would be paid to these 2 market categories because they are small in sales dollar magnitude. But the market segmentation analysis shows that these 2 service categories have significant potential that should be pursued.

As would be expected, and is, the next item in the development of the business plan is the creation of the sales plan. Let's take a look at the way ABC Maintenance Company develops its sales plan.

Example of a Sales Plan

The sales plan defines what the company intends to sell and the magnitude of sales, along with the methodology of how the sales will be achieved. It is the precursor of the production, general and administrative, and finance plan. The sales plan requires considerable interaction between sales, operations, and finance. Like any other plan, it needs a strategy for planning the outcome desired and tactics to follow to implement the intent of the strategy.

The example of the sales plan is shown in 2 phases: a planning stage we call a development strategy and the resultant sales plan. By expressing the sales plan in 2 steps, we see the action plan ready to be implemented and its outcome, the sales plan document for the business plan, a sort of before-and-after sequence.

Sales Plan Development Strategy

The sales plan development strategy is set forth below.

- Determine sales volume (dollar and number of service units performed for each service).
  - Units performed are determined by totaling hour quotes divided by number of hours to perform a specific service.

> **Example:** Painting year estimate is $21,400. The charged labor rate is $28/hr. (This the selling labor rate, which includes labor, overhead, materials, and profit.) So there are 1070 labor hours. It takes 8 hours to paint the average maintenance area job. Dividing 1070 by 8 equals 134 units of work to be performed during the course of the forecast segment. This number is used by the sales department to judge how well they are doing in selling this service in accordance with the plan.

- Show trends for previous years to use as a base line. 2002—$450,000
- Based on market assessment and segmentation determine, % change from trends, service by service
  - By weighted averages, total company growth potential by summing all of the services.
  - Factor in desired growth targets for specific services.
- Determine number of units of services to be performed.
- Convert to $ sales volume.
- Determine sales strategy for each service offered and in each market segment for 2003, 2004, and 2005. As a base line for sales strategy, find reasons the sales broke down the way they did in 2002.
- Determine how the company matched its abilities with potential customers, real and perceived.

> **Example:** This is not done formally through any simplified Quality Functional Deployment (QFD) technique, but rather through the quoting and proposal process and by observation and discussions with clients. All sales prospecting is based on providing basic maintenance services.

- Define how the company sells its services.

> **Example:** Leads are generated primarily through referrals. However, the company does maintain Yellow Page advertisements. The vice president of the company (the entrepreneur's wife) acts as the sales manager and is the primary seller of all services.

We can see that the sales strategy is to obtain referrals. Our example company is developing a plan to visit the prospective clients and to solicit a request for proposal from the client. Then a proposal will be written and in all cases the goal is to gain a long-term contract for services from the client. This is a simple sales strategy and could be very effective, depending on the tenacity and investigative capabilities of the sales team to find and not let go of the client until a contract is gained or another vendor has succeeded in direct competition. One more item needs to be considered: how to compare effectiveness of the approach from year to year. We would ask the following questions in the sales plan:

- Are the 2002 facts relevant for today's situation?
- What needs to be changed to be more effective?

And then:

- Prepare a list of strategies to be implemented.

Sales Plan

Here's an example of using the sales development plan and creating the sales plan for the business plan.

> **Example:** Sales Plan ABC Maintenance Company
> Service offerings are divided into 6 distinct categories, each with different methods of cataloguing units of service. However, after 1 year of recordkeeping, the selling price of the quoted job is made up of 52% direct labor, 48% materials, overhead, and margin. Therefore, it is possible to define a selling labor price. Since only 1 year of records are available, the selling rate is set at twice the labor cost rate. The sales goals are arrived at by growth forecasts from the Market Assessment and Segmentation review.
> Figures are shown below for years 2003, 2004, and 2005.

Service categories	2003 $000	2003 Labor rate per hour	2003 Service hours	2004 $000	2004 Labor rate per hour	2004 Service hours*	2005 $000	2005 Labor rate per hours	2005 Service hours
HVAC	$184.3	$30.00	6143	$217.5	$33.00	6591	$252.3	$36.30	6950
Plumbing	$140.9	$30.00	4697	$161.6	$33.00	4897	$187.4	$36.30	5163
Electrical	$99.1	$30.00	3304	$99.4	$33.00	3012	$115.3	$36.30	3176
Carpentry	$63.7	$22.00	2895	$62.1	$24.20	2566	$72.1	$26.62	2708
Masonry	$26.3	$25.00	1196	$43.5	$27.50	1582	$50.5	$30.25	1669
Painting	$21.4	$28.00	1070	$37.3	$30.80	1211	$43.2	$33.88	1275
Total	$535.7		19,305	$621.4		19,859	$720.8		20,941

To achieve the sales goals, the services offered will be targeted to the commercial, retail, institutional, and residential market areas. Most segments will be targeted for contract work, while residential stand alone homes will be targeted mostly for "as called" rather than "contract" service business. Market areas are as shown.

Service categories	Commercial* Offices	Commercial* Warehouse	Commercial* Factory	Retail* Strip mall	Retail* Enclosed mall	Retail* Stand-alone store
HVAC	c	c	c	c	c	c
Plumbing	c	c	c	c	c	c
Electrical	c	c	c	c	c	c
Carpentry	c	c	c	c	c	c
Masonry	a	c	c	c	c	c
Painting	b	c	c	c	c	c

	Institutional* Govt.	Institutional* Religious	Institutional* Medical	Residential* Condo.	Residential* Apts.	Residential* Stand-alone home
HVAC						
Plumbing	c	c	c	c	c	b
Electrical	c	c	c	c	c	a
Carpentry	c	c	c	c	c	a
Masonry	a	a	c	c	c	a
Painting	a	a	c	c	c	a
	a	a	c	c	c	a

* Sales strategy:
a = as called
b = both, as called and contract
c = contract

It can be seen that the majority of the offerings will be targeted for contract work. This supplies the most stable source of income and allows for the most efficient use of resources. However, efforts need to be continued to sell to residential stand-alone homes because it is an excellent source of referrals with high margins.

The sales plan for the multi-service users should be both broad and focused.

Broad includes:

- Development of a general purpose/information brochure.
- Development of a Web site based on the brochure.
- Get listed on local Chamber of Commerce–type Web sites.
- Periodic newspaper advertisements.
- Advertisement on local area Web sites. This will eventually replace Yellow Page advertisement.
- Continuation of Yellow Page advertisements for year 2002. Evaluate for continuing for year 2003.
- Stay personally in touch with clients. This may require hiring full-time QC people.
- Set up incentive programs to encourage staff and employees to sell as an informal part of their responsibilities.

Focused includes:

- Networking through trade groups.
- Networking through local business organizations such as
- Business referral groups.
- Chamber of Commerce.
- Rotary
- Similar organizations.
- Networking focused on getting one service into a client's property, then expanding to do other work. Consider loss leaders to get into desired properties.

All sales plan activities will become the basis of goals under the objectives and goals scenario. They will have measured start and complete dates as well as specific measurables to be sure the goals are accomplished.

The sales plan illustration is typical of what we need for a small business. It has all of the elements of a plan: targets by sales dollars and volume, a basic breakdown by segments, and how the company intends to go to market. Obviously it can be more complicated as the complexity of the business grows, as successful ones do in time. The important factor is that the plan really does spell out how much, when, and how achieved. The sales plan leads to the next step of business planning, creating the production plan.

Example of a Production Plan

The production plan defines how the company will deliver the products and/or services sold following the dictates of the sales plan. Therefore as expected, the production plan takes as one of its inputs the deliberations and final conclusions of the sales plan. It mixes this with existing or planned capacity and capability to produce a method of doing the work called for, and doing it in a manner that will generate cash flow as specified by the financial plan. (This is an iterative process with finance that will become evident as we proceed with the example.) The production plan also is a major input into the general and administrative (G&A) plan for human resources. Let's take a look at the mechanics of producing a production plan and then at an example generated for the ABC Maintenance Company.

Production Plan Development Strategy

Here in outline format are the steps we will take to prepare the production plan for ABC Maintenance Company. The primary data to start the development is the most recent sales plan.

- Determine whether the sales plan is cyclic or steady state. This determines the strategy of the production plan.
  - *Cyclic* indicates a small core of capability that can be expanded quickly to meet changing needs. This usually means relatively fewer permanent employees and lots of temporary staff.
  - *Steady state* indicates little movement in staffing levels (mostly all permanent) with gradual changes.
- Determine lead-time averages for sales to delivering services for each service category.
  - *Lead-time averages* indicates the window of visibility of loading levels. For example, if the lead time is 6 weeks, then the operations people know that orders taken now will not be completed for 6 weeks.
  - The longer the lead time, the longer the event horizon for operations loading becomes.
- Break down sales into service units and average hours/service unit. This becomes the current output labor standard
- Based on hours, calculate labor hours per service category by week/month/quarter, year based on sales plan offset by lead time.
- Group service categories with transferable skills, such as commercial plumbing similar to residential plumbing. Create job categories.
- Create matrix- job categories, hours required per week through calendar periods.
- Develop personnel H/C needs, using 40 hr/week/person/job category as availability to work.
- Compare personnel needs vs. current staffing levels.
  - Show short fall or excess capacity on calendar spreadsheet.
  - List employee needs or excesses.
- Factor in productivity improvements expected over business plan period; coordinate with Measurable Goals Program Plan.

- Develop plan to hire or layoff as required.
- Try to balance labor into other job categories via training.
- Use O/T for temporary needs for additional personnel.
- Hire new employees with sufficient lead time to make them effective when additional help is required.
- Try to grant vacation instead of layoff for short variances with needs.
- Develop training plan (see G&A plan)
  - For all skill sets
  - Quality
  - Record keeping

Production Plan

Keeping in mind the production development plan strategy, let's see how that works out for a production plan for the ABC Maintenance Company.

Example: Production Plan, ABC Maintenance Company

Service offerings are divided into 6 distinct categories. After 1 year of record keeping it has been shown that the selling price of the quoted job is made up of 52% direct labor, 48% materials, overhead, and margin. Therefore, it is possible to calculate the number of labor hours that have to be booked per year to achieve the sales goals (see sales plan).

Hours of labor per segment have been calculated to represent the sales goals. The number of personnel required per segment are based on 2080 hours per year per person. Numbers have been rounded to the nearest 1/2 person. An evaluation of the work (HVAC) indicates that the seasonal factor needs to be considered. An estimated 15% swing maximum between winter (the lowest) and summer can be factored into the headcount requirements. Anything beyond that would require additional staffing than shown. To accommodate this variation:

- The company will work overtime at the busiest times. A full work shift on Saturdays gives the company 16.6% more capacity.
- Work in all segments is not anticipated to peak at the same time. Therefore, there is an opportunity to cross-train and have additional personnel available at peak times. This will minimize the need to use overtime.

Lead time between receipt of order to delivery of service is not a factor. Work is generally performed as it becomes available. For the purpose of production planning and setting staffing levels, we can say that there is virtually no backlog; such as work on hand but unable to be performed because of lack of personnel or equipment.

Productivity has been accounted for with a 10% reduction in hours for equivalent sales dollars for each year of the plan.

Figures are shown below for years 2003, 2004, and 2005.

Service categories	2003 Service hours/yr	2003 Calc. staff required	2003 Staff plan	2004 Service hours/yr	2004 Calc. staff required	2004 Staff plan
HVAC	6143	2.95	3	6591	3.17	3.5
Plumbing	4697	2.26	2.5	4897	2.35	2.5
Electrical	3304	1.59	2	3012	1.45	1.5
Carpentry	2895	1.39	1.5	2566	1.23	1.5
Masonry	1196	0.58	1	1582	0.76	1
Painting	1070	0.51	1	1211	0.58	1
Total	19,305	9.28	11	19859	9.55	11
Average hours/year/employee = 2080						
Full time:			10			9
Part time:			2			4
Total staff			12			13

Service categories	2005 Service hours/yr	2005 Calc. staff required	2005 Staff plan
HVAC	6950	3.34	3.5
Plumbing	5163	2.48	2.5
Electrical	3176	1.53	1.5
Carpentry	2708	1.30	1.5
Masonry	1669	0.80	1
Painting	1275	0.61	1
Total	20,941	10.07	11
Average hours/year/employee = 2080			
Full time:			9
Part time:			4
Total staff			13

During the slow period, the company may have excess staff for work available. The simplest strategy would be to lay off junior seniority personnel to get down to required staffing levels. However, this comes with a price: severance pay and a debilitating effect on morale and company loyalty. While it is difficult to put a price on the latter, it would be substantial. Therefore, the strategy must be to minimize layoffs. This can be done by encouraging vacation time during the slow period and granting leaves of absence to those desiring it. Keep in mind that these same people had been working overtime just a few months ago and will be in need of mental and physical rest.

The company will develop a training plan (see G&A plan)

- For all skill sets
- Quality
- Record keeping
- Safety

The plan's focus will be on achieving the highest level of cross-training possible to improve the ability to staff all segments on an as-needed basis and to do this with minimal personnel additions.

This concludes the example of the production plan. As we can see, the major emphasis of the plan is staffing. This is also true for companies that make "hard" products. People resources are always the most critical factor for any company and must be evaluated carefully. Keep in mind, as the example shows, people issues are more than simply solving equations. Emotional factors, as well as calculations, must be considered. Note that the example plan goes to some lengths to state that a layoff is not a viable action, not because it doesn't mathematically solve a cash problem, but rather because it damages the morale of the staff. With poor morale, worker efficiency and quality will deteriorate. That is a factor that can not be calculated but certainly is not zero. It is important that business plans be more than financial as we see in this example. The business plan must put forth the best strategy for being successful, hence the morale factor in the production plan segment.

For product-producing companies, the production plan would also show the materials quantities, as broken down by the bills of materials, used for each product. These would be presented in a spreadsheet by usage and by time period, usually in monthly buckets. Cost for the materials may also be shown, but usually it's not and instead is included in the pro-forma P&L for each year in the finance plan.

Example of a General and Administrative Plan

The sales plan and production plan are very specific plans. The next section of the business plan is the catchall: the general and administrative (G&A) plan. Let's explore it.

General and Administration Plan Development Strategy

The general and administrative plan, being a pot porri of everything not included in the specific plans (sales, production, and finance plans), requires considerable patience and determination to do properly. Picture the scenario. You've finished all of the terribly detailed work to determine how the company will go to market and whether it will be fortunate enough to receive orders, and then how it will deliver on them in a timely, quality-acceptable, and profitable manner. This, you believe, is the core, except for doing the finance numbers, of the effort to produce a business plan. Wrong. Nothing will happen properly unless the people who will do the job are trained, and there are enough people, and furthermore, they are motivated to do the best job they can. This catchall of indirect activities affecting the outcome of the direct activities is the domain of the general and administrative plan. The plan is dominated by employee-relations activities. So it looks inward instead of outward as the other plan segments must. It can be thought of as "touchy-feely" instead of direct cause-and-effect relationships. And this makes it somewhat uncomfortable to most entrepreneur types, who are prone to be direct-action take-charge people. The G&A plan needs to focus on doing what has to be accomplished to staff the company with skills necessary to do the company's work. The staff also has to be melded into a cohesive team to perform at its best. We need to proactively plan how that is to be accomplished.

This plan becomes the most difficult of all the plans to develop since it deals with abstracts, not hard facts. How this can be accomplished is shown in the example of the G&A plan development shown below.

Example: A General and Administrative Plan Development Strategy

- Investigate organization needs (primarily for staff functions—labor accounted for under the production plan but shown on the organization charts)
- Use basic functions model:
  - Marketing/Sales
  - Operations
  - Finance
- Determine how each is performed in the company:
  - Identify what tasks need to be done under each of the 3 categories (Musts and Wants).
  - Create current organization structure.
  - Investigate work load per incumbent.
  - Determine if Musts are being done, and the degree of Wants being done.
  - If all Musts are not being done and some Wants are, determine if work emphasis can be shifted to cover more of the Musts.
- Create list of Must tasks that are not being done and categorize under basic functions model.
- Create desired organization chart.
- Determine what Wants are to be included in organization and add to desired organization chart.
- Make plans to add/transfer/delete staff to meet needs of the desired organization chart.
- Investigate training needs to support desired organization chart, including labor force training:
- List basic skills levels for each job.
- Match incumbents with skills levels through a skills inventory assessment.
- Identify lacking skills.
- Develop plans to obtain necessary skills (see measurable goals program plan).
- Develop pay plan for staff and hourly personnel (see measurable goals program plan) based on:
- Job rating system
- Job rate evaluation
- Job rate point system matched to $/hr or annual salary
- Compatible with geographic and industry averages

- Develop benefits plan compatible with employee needs and affordability (see measurable goals program plan).
- Health
- Retirement
- Vacation
- Education
- Develop a documents control system to retain company records that allow the user to trace orders through all phases of the business cycle:
- Create an audit trail for documents. Documents consist of any and all information used in obtaining orders and delivery of services.
- Create and/or integrate a system for the following auditable documents (see measurable goals program plan):
  - Orders control
  - Services dispatch
  - Quality audits
  - Finance interactions with customers and vendors
  - Payroll

General and Administrative Plan

The G&A plan strategy shown above contains virtually all items that need to be covered for a full human resources program for a mature company, usually with an employee base above 100 people. For companies just starting up, the key items would be an organization chart showing how the company is structured to do marketing, operations, and sales, and who does each job. The other items that should be considered for a startup company is a pay plan, a training plan, and a document control scheme.

So for the startup enterprise, ABC Maintenance Company, the G&A plan would be considerably less than a full-fledged plan as considered in the strategy. However, keep in mind that as the company grows, more elements of the full strategy would need to be added. This is done during the annual review feature of business plans whereby the participants get the opportunity to review and revise the plan at least annually based on then-current realities. Let's take a look at ABC Maintenance Company's G&A plan as it exists at year 1 of the business plan.

Example: General and Administrative Plan, ABC Maintenance Company

The G&A plan is established to provide support for sales and operations in carrying out the direct business activities of the company. The G&A plan provides for organization, staffing, compensation, and training necessary to operate the business in an efficient and coordinative fashion.

Organization

The organization for the company comprises a hierarchical style with some matrix management appended to it. The basic organization chart with current incumbents is shown in the attached organization chart (Figure 9-4).

Due to the small nature of the firm, we are capable of handling many different activities by staff carrying out multiple functions. It is recognized that finance may require a full-time person in the second year of the plan due to an anticipated increase in business volume. It is also anticipated that Operations can continue as currently structured for the entire business plan since most of the complex work of HVAC, plumbing, and electrical is already handled on a contract-by-contract basis with long-term relationships with the subcontractors.

This organization is sufficient for the volume of work expected over the business plan period. The span of control is such that the levels of management between the highest and lowest are at a maximum of 3 formal layers, but for practical purposes are in only 2 layers. The 3 layers are found in the Operations branch from the President/CEO to the Operators at the job site. It has been set up so supervisors have a prac-

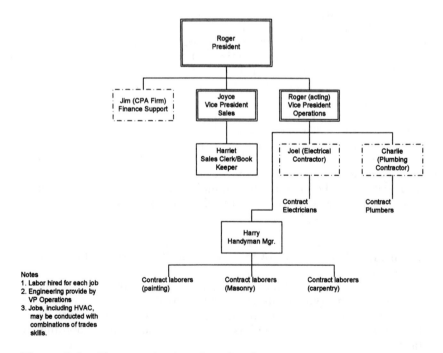

Figure 9-4. The organization chart for the ABC maintenance company.

tical span of control that allows tight governance of all jobs. It is also enhanced by extensive use of Nextel radio/telephone communications to keep up with the dynamic nature of work dispatch requirements. This basic organization structure has been in existence since the founding of the company and is working well at the current projected sales rate per the sales plan; sales are forecasted at a nominal increase of 12%. Therefore, there is no need to increase permanent staffing during the entire forecast period with the possible exception of a bookkeeper clerk. There will likely be no gains in Operator staff over the forecast period.

Compensation
A pay plan for the company is set at the mid-point of current compensation for office and trades personnel based on salary rates obtained from the Department of Labor for the South East Florida locale.

Presently the company does not have any insurance and savings plans for employees. The company will investigate the feasibility of initiating these benefits during the course of the business plan period. It is recognized that such benefits would enhance the company's ability to attract and keep the skills levels necessary to grow.

Training
An employee handbook will be developed to codify relationships between employees and the company. Major sections will include, but not be limited to:

- Hourly pay plan (as discussed above)
- Attendance and tardiness requirements
- Leave of absence
- Vacation
- Holidays
- Benefits
- Work rules
- Training and education
- Use of company property

Attendance and tardiness is a concern because it affects the company's ability to deliver its services to customers. It is essential that all employees understand the detrimental effects of poor reporting to work habits, to themselves, as well as to the company. A vacation plan already exists, as does a listing of company-paid holidays. However, both need to be formalized with equitable rules for vacation days and holiday pay earned and when they can be used. This will be accomplished early in the business planning cycle and will be incorporated into an employee manual. In addition, the related topics of benefits and work rules will be addressed.

Training and education benefits will be developed and placed in the employee handbook for reference. Training will primarily be on the job, with some classroom setting demonstrations. Education is defined as attendance in company-financed courses aimed at improving general as well as specific skills.

Document Control System

A significant portion of training for staff will be to develop a Documents Control System to retain company records. Lots of components of the needed system already exist and will be incorporated into the finished product. By having such a system, the user will be able to trace orders through all phases of the business cycle to:

- Create an audit trail for documents. Documents consist of any and all information used in obtaining orders and delivery of services.
- Create and/or integrate a system for the following auditable documents (see Measurable Goals Program Plan):
  - Orders control
  - Services dispatch
  - Quality audits
  - Finance interactions with customers and vendors
  - Payroll

For a company the size of the ABC Maintenance Company, the previous example is sufficient for a G&A plan. However, before we leave this segment of the business plan, I believe it is necessary to discuss pay plans.

For the ABC Maintenance Company it was sufficient to find out what the median compensation ranges for the handyman and sales clerk are in the entrepreneur's locale and set that as the fair wage for those 2 people. As far as he and his wife were concerned, they would make do with whatever the business could afford. And of course, their personal expenses as much as feasible would be borne by the company. This is the way of life of most small startups. However, as the company grows, this back-of-the-napkin approach to compensation planning no longer is satisfactory. To compete, the company will have to adopt insurance plans and retirement plans. They will also have to adopt a salary plan that is considered fair by the employees. When does a formal salary plan need to be considered? Probably when the number of employees is such that most employees are fully occupied with only one type of task, such as painting, carpentry, etc., because this will force definitions of job worth between different jobs.

Example of a General and Administrative Plan Pay Plan Development Strategy

Let's take a look at a typical pay plan and how it's structured, and how the differentiation between pay for different jobs is achieved.

Example: The purpose is to establish a rational and defensible method of differentiating pay levels between various jobs. We do this by objectivising an essentially subjective system. We know instinctively that some jobs are worth more than others for the success of the company. For example, a licensed electrician is going to be worth more than an electrician's assistant. That's simple and there would be few if any arguments to the contrary position. However, is the licensed electrician worth more to the company's success than the licensed plumber? Hard to tell and in truth it would probably change from time to time, depending on the scope of jobs a company undertakes; and that would be the case for our example company, ABC Maintenance.

How we do the differentiation can be complex or simple, depending on how objective we wish to paint the rational. The most common method is to create an evaluation point system to measure each job against, and then create pay grades based on the points. Let's say ABC Maintenance Company has prospered and 5 years in the future it needs to establish a pay system based on points. Let's take a look at the mechanics of creating the pay system.

A pay system for hourly personnel has been developed to achieve equitable pay for the various positions based on a job evaluation schedule, as shown below.

ABC Maintenance Company

	Pay grades	Pay progression $/hr (max. allowed per evaluation period)				
				Job rate pay level		
Grade	Point range	Start	2 weeks	6 months	12 months	anniversary-24 mo (after start date) review
A	52–100	$6.50	$7.00	$7.00	$7.00	inflation adjust.
B	101–150	$6.50	$7.00	$7.50	$8.00	inflation adjust.
C	151–200	$7.00	$7.25	$8.00	$9.00	inflation adjust.
D	201–250	$7.50	$7.50	$8.50	$10.00	inflation adjust.
E	251–300	$8.00	$8.00	$9.75	$11.50	inflation adjust.
F	301–350	$9.00	$9.00	$11.00	$13.00	inflation adjust.
G	351–400	$10.00	$10.00	$12.50	$15.00	inflation adjust.

Other allowable pay adjustments

1. $0.25/hr for each cross-trained* skill above pay grade, or out of department same pay grade.

    Max. 3 allowed.
    Skill adjustments: +1   +2   +3
    $ amount $0.25 $0.50 $0.75

---

* *Cross-training:* Supervisors nominate candidates to the president to receive cross-training. President approves and sets up a schedule for cross-training for the individual. An employee is certified as cross-trained when he or she achieves the accumulated time for training designated for the specific job and demonstrates proficiency.

2. $0.25/hr for acting as a designated leader
3. $0.50–$1.50 range for being designated a department leader (acting as an assistant supervisor)
4. Progression may start at any point: management decision based on starting skills of individual.
5. Progressions shown are for satisfactory performers as per table below. Lesser performance, poor attitudes, and attendance problems will stretch out the time to maximum job rate.

Evaluation of performance and attitude overall rating	Pay adjustment factor
0 to 6.0	0
6.1 to 7.0	60%
7.1 to 8.0	75%
8.1 to 9.0	90%
9.1 to 10	100%

ABC Maintenance Company hourly job rating system

	Relative Importance (used to develop degree pts.)	Lowest	Low	Degrees average	High	Highest
Job skills required	25%	13	25	50	75	100
Education required	10%	5	10	20	38	40
Responsibilities	15%	8	15	30	45	60
Effort required	25%	13	25	50	75	100
Working conditions	20%	10	20	40	60	80
Compete (ease of finding a similar job)	5%	3	5	10	19	20
Total Points	100%	52	100	200	312	400

How to apply table:

1. Rate each job for the degree of importance for each factor. Select a point factor where it fits best (lowest to highest). Assign the points for that rating from the table.
   a. *Job skills required:* (1) decide if skilled, semi-skilled, or unskilled. (2) Depending on the level of skills involved, select the appropriate degree and its point value from the table.
   b. *Education required:* Determine what level of academic education or practical training is required. High school graduate (or equivalent) is the midpoint degree level. Judge more or less education needed to do the job accordingly. Keep in

mind the need for administrative capabilities and English speaking.

Training period to become proficient in the job: The lowest skilled/knowledge jobs take the shortest time to become proficient. For classification purposes, lowest = 1–2 wk, low = 3–5 wk, ave. = 6–10 wk, high = 11–15 wk, highest = over 15 wk. For jobs that require both formal education and practical training, score each separately and then average the results for the amount of points awarded.

   c. *Responsibilities:* This applies to all aspects of the job. The lowest point value is assigned if it includes only the individual and not others, while the higher point value goes to jobs in which responsibility is for the entire group doing the work. The larger the group, or the consequences of inadequate work, the higher the point value assigned.

   d. *Effort required:* The degree of energy expended to do a job is considered. For jobs requiring no physical effort, the lowest point value is assigned. Jobs requiring the most physical labor manifested in requiring more than 2 rest periods in a 4-hour continuous timeframe should receive the highest point score.

   e. *Working conditions:* This is the physical comfort factor. The less comfortable the work environment is, the higher the degree point value selected. Working in a space not air-conditioned would receive a point score in the high range. Working in perilous conditions such as on roofs, scaffolding, and ladders would rate a highest point score. Conversely, working in an air-conditioned office would rate the lowest point score.

   f. *Compete:* This factor relates to the ability to keep personnel in the job. If the skill is in high demand, the incumbent can be lured away. A higher point score is assigned commensurate with the risk. If the skill is in low demand, then the point score assigned should be at the low end of the scale.

2. List reasons for all selections made in Step 1. This will help in making comparison for future job point evaluations.
3. Tabulate scores and record.

It is necessary to rate each existing job and any newly created hourly job with this point evaluation system to ensure that all jobs are compensated fairly. The staff would use the following form to evaluate jobs under their jurisdiction to rate jobs for pay purposes. Job titles for all departments are drawn from the list in the evaluation form.

Job Title	Job Skills	Education	Responsibilities	Effort Req.	Work Con.	Compete	Total	Job No.
			Points Assigned					
Construction Cleaner								
Electrician								
Class I								
Class II								
Licensed								
Plumber								
Class I								
Class II								
Licensed								
Masonry								
Class I								
Class II								
Painter								
Class I								
Handyman								
Class I								
Class II								
HVAC Tech								
Class I								

A similar job-evaluation system could also be set up for staff/management positions to ensure that the company is competitive in keeping and attracting competent management personnel. However, before using a staff job-evaluation program, job descriptions would have to be developed.

Example of a General and Administrative Plan Creation of Job Descriptions

While the creation of job descriptions is not part of the business plan, except perhaps as a "goal," it is necessary to understand how this process is done. Writing fair job descriptions that can be evaluated for compensation levels, similar to hourly jobs, is critical for maintaining a competent set of managers and staff. If people feel they are not being compensated adequately and fairly, they will leave. And surprisingly, more will leave if they feel the compensation system doesn't fairly reflect the value of the work they do. Creating job descriptions for a company, then, has 2 purposes: to ensure that all the steps of the 7 steps of the manufacturing system are assigned to individuals to accomplish, and to communicate the responsibilities a person has in carrying out his or her assigned job. When reviewing the entire set of job descriptions, it should be very evident how the company goes about doing its business. It is also a way of determining whether all 7 steps of the classical business system have been accounted for in the company's strategy, or whether some are being left to an ad hoc planning level of accomplishment. If any of the steps are in the ad hoc camp, then the efficiency of the firm will be less than optimum. Most likely the efficiency will be significantly less than optimum.

The process of creating job descriptions is as follows.

I. List the 7 steps of the manufacturing system on the right-hand side of a page.
II. List all the job titles you have in your company on the left-hand side of the same page.
III. Draw lines from the job titles to the appropriate step of the 7 steps of the manufacturing system. One job title may cover a multiple of steps. This is especially true for smaller companies.
IV. For the company to be performing in a planned mode rather than ad hoc, every step of the 7 steps of the manufacturing system should be linked to at least one job.
V. If there are blank steps of the 7 steps of the manufacturing system without corresponding job titles, then create new job titles or expand existing ones to cover the blanks.
VI. Create job descriptions as explained next. Good job descriptions contain topics A through E. Section F shows the general format the document will entail.

   A. Purpose of a job description:
      1. Define the scope of work to be performed.
      2. Create measurements for evaluating whether the scope of work is being accomplished.
      3. Establish the job within the hierarchy of the organization.
      4. Define skills required for the job.
   B. Scope of work to be performed:
      1. Classification
         a. Management
         b. Professional
         c. Clerical
         d. Value added/non–value added labor
      2. Specific tasks
         a. Mental
         b. Physical
   C. Measurements of accomplishment:
      1. Generic measurements for the hierarchy level
         a. Management
         b. Professional
         c. Clerical
         d. Value added/non–value added labor

2. Specific methods-related measurements
D. Organizational placement:
   1. Immediate superior position the job reports to
   2. Responsibility for lower-level jobs reporting to this job
   3. Evaluation responsibilities for subordinate jobs reporting to this one
E. Required Skills:
   1. Education level external to the job
      a. General education achieved through high school
      b. Specific professional education, such as an engineering degree
   2. Training—specifically related to the job
      Example: Specific tradesman qualification such as a licensed electrician
   3. Experience performing immediately subordinate jobs
      Example: To qualify as an HVAC technician, requires 2 years experience in electrical and plumbing work
F. General format of a job description:
   1. *Job statement.* A general explanation of the job explaining what the incumbent does and how he or she goes about doing the task. Includes the "Classification of the scope of the work to be performed." The last part of the job statement is an Organizational Placement sentence showing where the job fits within the company's structure.
   2. *Specific requirements.* A listing of all the tasks to be performed included in this job. Can be narrative or outline format. Recommended order is:
      a. Mental tasks—list specifics, such as maintain job logs
      b. Physical tasks—a listing of the job content, such as prep surface for paint, mix paint, spray paint walls, etc.
         i. List biological requirements, such as lift "x" pounds, etc.
         ii. List environmental considerations, such as work outdoors on ladders, etc.
   3. Evaluation measurements
      a. A statement on company policy for measuring and evaluating job incumbents
         i. How often?
         ii. How done?
      b. Specific measurements for the job
         i. List all measurements used to make out evaluations that are specific to the job; such as any specific quality measurements particular to the job; perhaps number of complaints received from customers, etc.
         ii. Generic measurements, such as attendance, care of company property, etc.

4. Education/training levels requirements
   a. Generic education—school grade achievement required, through college (grade 16).
   b. Specific skills gained through job training or trade school
   c. Experience levels obtained through lower-level jobs to qualify for this job.

Job descriptions are never done by only one person. The draft is done by the initiator, then it is carefully reviewed and compared with other job descriptions for compatibility. The review is usually done by a small committee familiar with the job and similar jobs. Also, the review committee is familiar with company policy concerning what is thought to be the prime strengths, that is, prime jobs the company has to carry out its business. In this manner "line" jobs are given more worth than support jobs. For example, the ABC Maintenance Company, some time in the not-too-distant future, may have an Employee Relations Manager as well as an HVAC Maintenance Manager. They may both report to the vice president of Operations, but since the HVAC activity is a line activity directly responsible for doing the work for customers, it would command a higher rate of pay than the Employee Relations job.

Job descriptions spell out how a company performs. It is essential that all companies have some sort of definition as to what the responsibilities are for each of its incumbents. Even small companies such as the current ABC Maintenance Company would need to have an understanding of what the work content is for each of its 4 employees is. In the latter case, the job description may be nothing more than a bullet listing of responsibilities and be quite informal. This is so because with a very small company, each person has unique skills and does just that. Whatever the case, defining job responsibilities is one task every entrepreneur needs to take on and master.

Example of a Financial Plan

We now have all the tactical and strategic pieces assembled for the business plan except for 3: a contingency scenario for what to do if all the planning would become obsolete or even slightly wrong, a listing of the programs to achieve the plans, and how the activities will be funded. The first and second items will be discussed later. Right now we'll look at the funding problem. This is put together in the financial plan.

Financial planning can be both very complex and deceptively simple. Basically the financial plan adds up all the costs of doing all of the various plans, compares it with the revenues the sales plan is forecasting, and creates a pro-forma P/L, cash flow, and balance sheet based on these plans, the results being the way money will be received and spent. Finance plans for business-planning purposes are always done after the other operations and sales planning have gone through the initial iteration because otherwise, the finance plan has no substance to it and would be irrelevant. Lots of times, people in operations and sales grumble that finance is dictating the plan. This is not so. Finance is just reporting back what the inputs from sales and operations and other support functions equate out to be. They are reporting where the shortfalls are with regard to sales and expenses that generate the desired profits. If the results of all the puts and takes are not sufficient for creating profits that the company would like to have, then it's back to the drawing board. This means "sharpening our pencils" and looking for more ways to optimize activities and/or create more sales within the same time period. This is where people mistakenly think finance is dictating the outcome of the business plan. They are not. They are simply defining the original goals and showing where all the variances are for all the departments in meeting the goals. It is up to the operating departments to agree or disagree with the feasibility of closing these variances. Basically, whatever the operating departments commit to will be accepted because they're the people who will actually do the work. Of course, once they see the results of their plans and what needs to be changed to get the desired results, most people are willing to make changes if they're doable.

The activities involved in doing the financial plan are essentially the same as described in Chapters 6 and 7, new products introduction and financial techniques. In both cases, the P/L statement became the mainstay of determining the value of a new product offering or for making up an operating budget. In the case of business planning, we are doing the exact same thing, only this time we are looking at how our company will do financially in the future based on plans we have to grow in our market segment. So the technique is the same. Rather than repeating the technique, let's look at what the results of financial planning would be for our example company, ABC Maintenance Company. I'm not going to present a full P/L for this example because we already have one for a similar startup company in Chapter 7 (Figure 7-4). Instead, we will look at the abbreviated version: the above and below the line model with ABC Maintenance Company data. It would look as shown in Figure 9-5.

The abbreviated P/L shows the company achieving its planned-for profit margins. It does so because we've contrived it to do so with cost of goods sold being the labor content of the cost to do the job. All other costs are included in the operating expense line. This includes the minimal materials

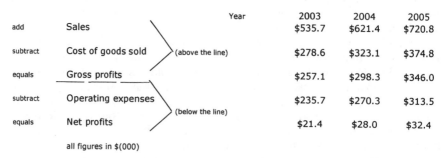

Figure 9-5. The abbreviated P/L chart for the ABC maintenance company.

this service company uses while performing maintenance activities. Notice how large the operating expense line is. This division of costs is a reasonable assumption for a small startup company.

Overriding the sales and labor to do the work is the important inclusion of the overhead expenses. This includes all the items as explained in Chapter 7 and shown in Figure 7-4. I point this out because a business plan must recognize the enormity of the indirect or overhead expenses that will have to be borne by the company before a profit is realized.

Too many young companies make the mistake of visualizing only the gross profit line and are rudely awakened by the reality of "takeaways" before bankable profits; the net profits line is achieved. The business plan must deal with these overhead expenses realistically before setting off to do business. These plans must prudently measure all costs to set appropriate selling levels. The entrepreneur then must be objectively honest and answer the question: "Is our product (or service) appealing enough to achieve sales at the prices the P/L arithmetic requires?" If the answer is anything but an affirmative yes, then the entrepreneur must think very carefully about continuing. This is the cold hard truth that the financial plan must illuminate.

Example of Programs To Support Strategies (Goals and Supporting Projects)

Plans are very necessary for setting directions. However, no plan is complete without defining what the measurable achievement points are and then determining an implementation scenario. So programs need to have goals that are measurable (see Chapters 3 and 4 for a discussion of objectives, goals, and project relationships). Once goals are set, we can define action steps (projects) to achieve them.

The sales, production, G&A, and financial plans all set directions for the company. ABC Maintenance Company, as small as it is, now has defined plans. We have gone from an informal, almost seat-of-the-pants type of direction setting to one where an integrated logic prevails. Now we have directions for all of the segments of the emerging company and all of those directions are compatible. Without the business plan focus, operations, sales, and marketing could easily stray and find they are working at cross-purposes, and many times finding out only when one segment is "burned" by the actions of another. Imagine a scenario where sales is striving to sell only HVAC cleaning services, to the point that the only work the company has is within this segment. This has caused an overload. The company cannot deliver the HVAC services to its clients on time, while at the same time the carpentry, masonry, and painting resources are idle. How could this have happened? Well, if the sales department is unaware of the capacity of the various operations subdepartments and their sales plan has not been used as the start of the production plan workup, then it is easy to stray. The business plan makes this more difficult to happen. It requires each department to integrate with the others so each is using common data to create the company's plan from the viewpoint of their specific responsibilities.

Example: If there are 6143 hours of sales ($217.3 K) forecasted by the sales plan for HVAC in 2003, then the corresponding production plan needs to have 6143 hours of HVAC capacity at roughly the same time the sales are to happen. The production plan would recognize this need and establish the mechanisms for having in place or hiring employees and training them to do HVAC. At the same time, the finance plan would look at the revenue generated by the 6143 hours of HVAC maintenance activities and compare it with the costs to deliver, and in this case, perhaps consider adding and training an additional person to become proficient in the company's HVAC services methodology.

For example, if the production plan shows an existing capacity for 6000 hours of HVAC maintenance and that the capability of doing 6143 hours requires an additional person, who if trained properly can work 2000 hours annually, the question becomes how much does it cost to add 2000 hours capacity and is it worth it because we need only 143 hours more? The financial plan would calculate the additional costs and ensuing profits. Perhaps it would show that the company should strive for either 6000 hour sales or 8000 hour sales because the interim level, 6143 hours, can't possibly yield the profit desired. This would be true because adding 2000 hours of capacity and using only 7.15% of it doesn't cover overhead costs. The answer to the question would be for sales to define whether it is practical to get more sales to fill the capacity. If yes, they would increase the sales plan to perhaps 7500 hours (taking into account the training time for the new operator to be fully productive). If no, they would back off to 6000 hours. In both cases it yields plans for production where human resources are being used at the higher capacity level, even though at 7500 hours there would be a training cost associated with gaining the additional capacity. The finance planning activity would recognize both scenarios, and if the 7500-hour version was still profitable (even though perhaps not as much, percentage-wise, as the 6000-hour plan), it would report such to both operations and sales. Then all 3 would discuss the short- and long-term implications of both versions of HVAC work capacity. However, whatever they choose would be doable and acceptable and integrated throughout the company.

With an integrated measurable plan in place, we would need to convert the data to goals and subsequent project. We would do that as discussed in Chapters 3 and 4. It is a straightforward process, but quite often it is confused and leads to goals being put into the business plan that are out of synch with the results that have been agreed to with the sales, production, G&A, and finance plans. This happens because the goals and their subsequent projects are not vetted against these plans. They are, instead, approximations of what the plans really call for. The way to prevent this from happening is to do the following.

1. List the end results that the various plans call for.
2. Write a goal with the end results being the measurable point, both magnitude and time cycle.

3. Check that the goal supports at least one of the ongoing objectives. If it doesn't, the plan itself is not compatible with the vision and mission statement of the company.
4. There should be at lease one goal for each objective. However, a goal may support more than one objective.
5. Check to ensure that plans, goals, and projects support all of the objectives. If they do, we have an integrated business plan. If they do not, we have an inconsistency somewhere that needs to be found and corrected.

Let's use the HVAC work as an example of checking and setting goals. Our sales plan now calls for 7500 hours of HVAC work and the production plan calls for adding 1 person to payroll to make that happen. Goals could possibly be constructed as follows: Sales: Sell 7500 hours of HVAC work for revenues of $225 K, evenly divided by calendar quarters.

Operations: Add 1 additional HVAC operator during the 1st quarter of 2003 and train to yield first-year results of 1500 billable hours.

Finance: Achieve $7.5 K in additional revenue in 2003 by adding 1 additional HVAC operator.

We would now check whether the objectives that the company has decided are pertinent to its vision and mission statement to ensure that these goals support at least one of them. Those objectives were outlined at the beginning of the business plan. Here are the objectives that these goals, derived from the functional plans, support:

Operations-oriented

- Improve productivity to levels that are compatible with maintaining a positive cash flow.
- Provide a positive work experience for employees.

Finance-oriented

- Maintain a positive cash flow to negate the need to use line of credit financing.
- Improve profit margins by reducing break-even costs to be at the rate of best in industry standards.

Sales-oriented

- Improve market share to become a top-tier maintenance provider in the geographic area we service.
- Create pricing structures to stimulate sales and increase profitability.

We can readily see that the specific goals do support these broad-based statements of intent. Remember, it is not necessary to prove, to the degree of a geometric theorem proof, that compatibility exists between the goals and objectives. It is only necessary to agree that by doing the projects to achieve the goals, we are taking a step along the asymptotic path of reaching the objectives.

The only objectives these goals do not directly support would be the mission-oriented objectives:

- Achieve 6 sigma quality standards in all departments.
- Maintain service innovation leadership to provide best value for customers.
- Introduce new technologies to become a "world class" maintenance and related services provider as recognized by the industry.

These objectives would be supported by any of the programs necessary to improve performance, primarily in the second and third years of the plan. In both of those years, the operating expenses become a lesser percentage of sales revenue because efficiency needs to be improved to maintain a competitive position compared with competitors. In that case, every reasonable methodology would need to be looked at to see how efficiencies can be improved. These would lead to direct matches of objectives to plan-derived goals. However, we should not leave the impression that the goals previously mentioned do not inspire support for the mission-oriented objectives; they do. For example, in training the new HVAC operator, it would be foolish to not include the latest techniques in methodology. In fact, the premise for this fictitious company is that the owner/entrepreneur thought he had a better way of performing HVAC maintenance, among other things. So what these plan-derived goals do directly is support the sales, operations, and finance objectives. But they are also very compatible with the mission objectives. There is no out-of-synch relationship here. Therefore, we can state that all 5 check steps have been examined and are in compliance. The goals are suitable extensions of the plans and they do support the vision, mission, and objectives of the ABC Maintenance Company.

Example of Contingency Plans

We've reached the end of the business plan development. The only thing left to do is to evaluate the need for contingencies in case our forecasts and other scenarios are just plain wrong. What happens if the market situation and segmentation is not correct? After all, the level of business we would expect could be different from what is actually happening, and ABC Maintenance Company does not have a history of predicting sales levels. If that's the case, all the planning is for naught. Yes? Absolutely not!

By planning, we've set a baseline against which to measure every real happening. Without planning, we have no way of knowing whether we're good, bad, or mediocre. If we have a plan and the actual results miss the mark, we are creating a variance. We have facts against which to judge our original assumptions, which can then be used to correct and modify these assumptions for future planning. We have created the ability to iterate for future improvements, better planning.

Contingency planning is the creations of variance situations before they happen so we can also create reasonable corrective actions to mitigate those fictitious variances. Think of it as insurance against negative happenings (although it can be for positive happenings, too). For example, you're traveling and will be away from home for 3 days. What happens if the airline loses your luggage? Consternation! Fright! Despair! Of course, but you do things to mitigate the damage. You file a claim for the loss. You ask for and get an emergency toiletry case from the airline. You possibly buy clean undergarments for the next day, assuming the airline will not locate your luggage quickly. So you've taken actions based on understandable contingencies for that situation. But this is not insurance against the potential loss of luggage. Insurance could have been packing conservatively so you could include your needed belongings in a cabin carry-on bag; or if that's not possible, then distributing items between checked luggage and carry-on cases is an option. Contingency planning is both the plan of what to do if the airline loses your luggage and doing things in advance so that a loss of luggage would have a less negative impact on you. Now let's look at what contingency planning activities could be considered for the business plan of ABC Maintenance Company.

The ABC Maintenance Company depends primarily on sales to meet its goals. It feels that labor to deliver the services is not a major issue. It has key in-house personnel, but primarily it contracts out the work to subcontractors. Therefore, labor at worst would be an inconvenience factor of finding another subcontractor if a "regular" is no longer available. Sales, on the other hand, does affect everything the company does. If sales are not at the anticipated level, it could create a burden for the company. If they are higher than anticipated, materials expenditures will need to be increased, along with the logistics of managing the materials. Subcontractors would have to be given sufficient time to obtain additional labor. If sales are lower than expected, the opposite happens. A contraction occurs that may go so far as layoffs and canceled materials orders, including possible penalty payments.

Sales volume variances, then, is the prime cause for deviations from plan. However, other factors need to be considered for contingency planning. The first is performance. What do we do if sales are as planned but productivity is not occurring as planned? Here we are concerned with under-performance much more so than over-performance. If the staff and the subcontracted labor is not performing up to required levels, we should have a plan on how to respond. In this case, it would probably be a 3-fold response:

- Identify areas of performance variance and understand the reasons for the variance.
- Correct the variance if it can be done reasonably fast (within a month) or supplement labor to make up for the deficiency. Set in place long-term corrective actions.
- Adjust selling prices to cover increased costs.

Another common contingency would deal with the loss of key personnel. Here a preventive strategy may be called for, such as cross-training or making sure salaries are set just above the median for the geographic area.

As we can see, you may want to have contingencies for many factors. ABC Maintenance Company focuses on lack of sales but also would include the other common occurrences that could happen. How complex does a contingency plan need to be? Maybe nothing more than a sentence or two for each topic in bullet form, as shown previously for the lack of achieving the anticipated productivity. The key is being prepared and knowing when to react. Here are a few rules of thumb that ABC Maintenance Company would use and are general enough to be valid for virtually any company.

- Set levels for deployment to initiate contingency planning, for example, plan to track more than 25% off projections by evaluation period (usually monthly).
- Establish methods of observation to set off deployment.
- Establish built-in actions for variances caused by
  - Sales-received levels
  - Hours to perform services
  - Sales-related material costs
  - Loss of key personnel
- Establish contingency actions (which may be multiple steps) and have measurements to evaluate their effectiveness.
- Establish set points where contingency actions are stopped and the regular plan is resumed.

These guidelines are essentially the core of the contingency plan. They tell when to start and stop contingency planning. If you think this discussion of contingency planning for the ABC Maintenance Company is a bit touchy-feely and not precise, you're correct. Contingency planning can never be hard and fast. It must be flexible and meet the current situation as it occurs. The only thing that remains relatively fixed is the points at which contingency planning ought to be considered. Notice I didn't say "implemented," just "considered." For example, if the ABC Maintenance Company is shy on sales this month, contingencies should be considered. However, if we know a major sale will likely close within weeks, it makes no sense to implement contingency cutbacks, only to have to rescind the action a few weeks later.

Contingency planning must remain flexible. The best advice I can give is to understand what is really important to react to, and have an idea how to do that. For ABC Maintenance Company, the keys are sales volume followed by operator performance. Both items affect cash flow and therefore need to be acted upon when they're not at a level desired. For our example company, reacting means getting more sales and fixing poor performance. Simple direction, but inevitably it will require hard work to fix if it occurs. The point is that contingency planning will get the corrective actions going sooner rather than later.

SUMMARY

The business plan is the culmination of the structuring of an enterprise. It is a way of expressing the philosophy of a company, as well as defining how it will proceed with doing business. Each business plan is different. Not only is it different for different companies, but different for the same company as time goes by. A business plan for a startup company is going to be different than that for the same company later on in its development because the business situation is different. However, the premise of how we construct the business plan is always the same.

The sequence shown in Figure 9-3 is always the same. First, the reason for being must be stated. This is the vision, mission statement, and objectives. Then we examine where the company is with respect to its markets and the geo-political situation it finds itself in. With these constraints understood, we then construct the sales plan, production plan, general and administrative plan, and finance plan, which are all interwoven. With completed plans in place, we now devise the programs to bring those plans to fruition. And finally we add in the contingencies for how we will react if our plans do not come together as designed and we find we're faced with different realities than forecasted. So we end the business plan on a note of uncertainty but with a fall-back position if needed.

It's only fitting that a business plan should end in uncertainty because the world of business, especially an entrepreneurial startup, is laced with hope rather than fait accompli, and probabilities rather than certainties. A business plan is our best way of visualizing what we need to do to give us the best chance of being successful. If we look at it that way and understand that it is our best-guess benchmark, we will at least be marching in a direction that is probably our best path toward business success. Remember, without a plan, any road traveled is as good as another and only happenstance will get us to a desirable location.

Chapter 10
Toward World-Class Performance*

An entrepreneur's dream is to found a company that will become an icon of visibility to industry, commerce, and the general public, something like the general recognition of General Electric or Microsoft or Boeing as famous (some may say infamous) and successful companies. Of course that rarely happens; however, a startup company can become world class in its performance and reap generous profits for its owners. This doesn't happen in a vacuum. It comes about because the company's personnel understand world-class performance and how to achieve it. This book has been a treatise on world-class performance because an entrepreneur is forced to be at that level if she expects her company to be successful. I have outlined an approach of doing business based on the world-class model. Now I will cap off this entrepreneur's instruction guide by summarizing that performance ethos and hopefully pulling it all together in a manner that makes sense for the entrepreneur. I will return again to the 7 steps of the manufacturing system, which is the best model I know of for consciously approaching world-class status. Think of it as a model to focus on while implementing the techniques espoused in this book.

INTRODUCTION

Even the largest companies have finite resources and must use them properly. So it's understandable that the entrepreneur starting her own company needs to also obey this rule of business survival. We must do our best to minimize outflow of resources to maximize sustainable profits by finding best practices for managing our businesses. This is obvious. The trick is how to do it. The methodology of doing so exists and can be learned, as I've demonstrated in this book.

Throughout the earlier chapters I have alluded, sometimes outwardly, mostly subtly, that the only way to find best practices is to first understand the science of business, that is, to understand the process of obtaining orders through actual delivery of products or services to customers. The process of how this is done is known as the 7 steps of the manufacturing system, which I've referenced many times. (Don't be fooled—the 7 steps apply to any organization that has a 'deliverable' for its customers, not only

---

* Note: This chapter is based in part on the author's paper "Making it in the competitive world" printed in *Mechanical Engineering* magazine, May 1998, published by ASME International.

to producers of physical products.) In Chapter 4, I asked you to take my word for it that the sequence of manufacturing a product or service has to follow these 7 steps. In this chapter, I'm going to show you why. I want to focus on the 7 steps and show how faithfully understanding each step and then executing them well is the only path to world-class performance. The entrepreneur who understands the 7 steps has a significantly better chance of creating a survivable company than those who do not. We will explore the science of business by developing an understanding of the 7 steps beyond the application for organizing a company that I discussed in Chapter 4. By doing so, I hope to show you how to measure your company against "best practices" so you can plan to achieve optimal results. I will explain ways to evaluate how well an entrepreneur's startup company does 7 steps. We'll then explore ways of comparing current methods with the accepted best practices by doing a gap analysis, and finally, how to develop plans to close the gap between the ideal and a company's specific situation.

BASIC TRUISMS

Business leaders should keep 3 axioms in mind:

- All companies comply with the 7 steps. Those that do so consciously enjoy higher profitability.
- The closer a company comes to achieving best practices in complying with the 7 steps, the higher its profitability will be.
- With the ability to emulate best practices of world-class performers readily available, you cannot afford the status quo of the do-nothing option and still survive.

Small companies can be world class. In fact, small companies have an advantage because they can focus faster on tasks than large companies. So emulating best practices is a tool the entrepreneur needs to learn how to use and use it effectively.

UNDERSTANDING THE 7 STEPS OF THE MANUFACTURING SYSTEM

Best practices rely on implementing the dictates of the 7 steps of the manufacturing system (see Figure 7-9 for a listing of the 7 steps). So a logical place to start our discussion of how to approach world-class performance is via a thorough understanding of those steps.

An In-depth Discovery Discussion of the 7 Steps of the Manufacturing System

With respect to the entrepreneur's startup company, what does "making it in the competitive world" mean? It depends on your perspective. From an employee's view, he or she would expect a decent salary and some degree of job security. A supplier to the entrepreneur's company would say the company has "arrived" if he get paid consistently on time for goods or services provided. The customer's view always depends on receiving products and/or services that meet quality requirements and are delivered on time for his or her needs. People who have invested money in the startup company will say the company has "arrived" when they see their equity increasing and the dividends representing a satisfactory rate of return. All stakeholders (employees, suppliers, customers, investors) have common shared views of what constitutes "making it":

- They do not feel they've invested time, effort, or money in a losing cause.
- Their finite efforts and expenditures are yielding the expected returns.

So how do companies "make it in the competitive world" when the ability to succeed is buffeted by many intangible and uncontrollable variables? By the luck of the draw? Can we say business success is only slightly more predictable than the spin of a roulette wheel? All this represents a fatalistic approach: You react to immediate change as best as you can. You take what you can get and hope the pluses outweigh the minuses. Or, is success a result of a planned proactive attack? Here we would consciously try to do the essential core operations of a business correctly. We would take actions to minimize surprises thus blunting the effects of outside influences. The latter is the correct choice. Companies that understand the entirety of their essential core operations and take charge of their destiny will succeed far more often than those that do not.

The 7 steps of the manufacturing system are the entirety of the core operations of any business. The questions the entrepreneur must ask about the 7 steps are:

- Do I know what they are?
- Am I controlling them?

By the end of this chapter, I hope I've put it all together for the entrepreneur and his or her answers will be: "Yes and Yes." The first 9 chapters have focused on techniques and reasons for applying them. They represent the "how to do it" aspects of being a successful entrepreneur with a new startup business. The focus is different to a degree in this last chapter. Here my purpose is to sum up and explain why I recommend those techniques as being essential for business success, not so much from the viewpoint of each specific technique, but from the viewpoint of entire process (call it a system). The 7 steps offer opportunities to analyze a company's essential core operations, see where there is room to improve, and know where improvement efforts need to be, all in a synergistic compatible approach.

A Description of Each of the 7 Steps and How They are Interlinked with one Another

I have explained previously that a process called the 7 steps of the manufacturing system is a series of events that all companies who produce goods and services adhere to for each and every product they produce. Only the very best, the world-class producers, are consciously aware they are doing these same events over and over again. This series of events is the business system. We describe it in 7 discrete but interlinked steps. It is not a computer system. It is not software, although it forms the road map for developing integrated business software. The business system is a logical approach to solving the problems of business, that is, how to deliver goods and services on time and at a profit; it is as old as the industrial revolution itself. To reiterate: Those companies that are aware that the 7 steps exist and consciously try to make sure they perform all of them to the best of their ability, are the successful companies. Those companies that fail do not try to be in compliance and more than likely don't even know the 7 steps exist.

Why the Step Order Is the Way It Is and Why It Must Remain That Way

The steps follow a definite sequence and are mainly done in sequential order. We can see that there is definite linkage between all the steps. And it should be "intuitively obvious" that the predecessor step must be done before the successor step. But is it really so obvious? How many manufacturing companies try to make a product before the design is finalized and then are baffled as to why costs are so prohibitively high? This leads to operations Rule 1.

> Rule 1. Never attempt a successor step until the predecessor step is "really" complete.

To do otherwise is akin to taking off on your leg of a relay race without first receiving the baton.

The other "intuitively obvious" observation is that the content of what happens at each step needs to relate to the same set of facts.

Example: Many times in industry we find that salespeople have not fully transferred customer requirements to design engineering and manufacturing so the company blithely goes about designing and building a product that the customer "almost" wants. This leads to my operations Rule 2.

> Rule 2. Make sure the data being acted on at each step of the manufacturing system is consistent and identical between predecessor and successor steps.

Confusion reigns supreme when Rule 2 is violated. It is like salespeople saying the customer wants green tomatoes. In response, the engineering team concocts a way to make green fried tomatoes because it's more elegant to do so. But it turns out that all the customer really wanted was tomatoes that weren't overripe.

What Rules 1 and 2 show us is that we should do everything in business with a well-thought-out logic process. We should make sure that the entire team is aware of what that logic is. This way, we obtain an optimized focused effort by the entire company team. It stands to reason that if everyone knows what play we're running, we have a better chance of success than if only the quarterback and runningback are allowed to know the details of the play.

A DISCUSSION OF EACH OF THE 7 STEPS

Let's look at the particulars of each step of the manufacturing system. What are the optimum approaches so your company can be world class and truly be "making it"? By investigating the nature of each step, we can find the path to world-class performance. But before we start, we should understand that the 7 steps of the manufacturing system applies equally to services as well as physical goods. A company that contracts to mow your lawn and trim your shrubs must go through the exact same logic as one that is contracting to build your patio. World-class companies fulfill the intent of each step for everything they do, regardless of what the deliverable is.

Step 1: Obtain Product Specification

This is the sales and product-design phase of producing a product. Sales people have to determine what the customer wants and then transfer the information to the design team to conceptualize how to create it.

This is a very iterative step whereby sales must be very careful to fully understand the true needs of the customer. Sometimes it's easy because your business is a ready-to-serve business and the customer comes to you. Sometimes this is difficult because customers do not know what they want until the new product is a reality.

It is the latter where world-class companies differentiate themselves from the also-rans. Oftentimes we think we see products created, and then a market created for them. Or so it seems. In reality, what is happening is that marketing and sales are discerning what the customer wants by focusing on future desires. For example: modern production technology has allowed us to mass produce products, thus lowering their costs. This in turn makes products more readily available for the masses. Now people are enjoying products that weren't previously within their means and they want more. They want products to be individualized at mass–production costs. Knowing that this is the trend, astute marketers, teaming with design engineers, are stretching the technological envelope. They're saying, "See, we can give you individualized designer jeans at the same price as the massed-produced variety were last year." They know they can do this because they've bought "smart" sewing machines using the latest chip technology. They are obtaining product specification in an abstract manner and are using so-called soft science, psychology, to do so. This is obtaining product specification generically speaking by tapping the unconscious desires of the body public marketplace and discerning what their company should produce (always in line with their company's capabilities) to obtain product specification.

How do the best companies ensure that product specifications meet their factory's capabilities? At a minimum, they employ some form of a concurrent engineering team. The team is formed from all functions of the company to usher products from conception to distribution. The best companies do one more thing: In addition to simply discussing products with their customers, they expand to a more definitive analysis by employing a pragmatic Quality Functional Deployment (QFD), as discussed in Chapter 2. If your company doesn't have QFD capability, it is not in the category of "making it." It is spinning the roulette wheel on every job it takes on.

Step 2: Design a Method for Producing the Product, Including the Design and Purchase of Equipment and/or Processes to Produce, If Required

This step used to be considered the manufacturing engineering step. In actuality it is that plus a continuation of the design phase of Step 1. This may sound like a contradiction of terms, but it's not. Ever since the advent of producibility engineering in the early 1980s and its successor, concurrent engineering, these methods have become the integrated way to manufacture. We have truly recognized that design includes 3 phases, and they are dependent on each other.

As a reminder, the phases of design are as follows:

- *First phase—the concept design phase.* Product specification is related to test for compatibility with the laws of science (which it must do).
- *Second phase—the producibility design phase.* The design is tested to see if it is technically and economically feasible to produce in the intended factory. (Yours.)
- *Third phase—the manufacturing facilities design phase.* The jigs and fixtures and tooling are designed to be compatible with the proposed concept design. This is the portion of design where we figure out how to create a reality of the concept in a manner that satisfies the customer and all of our stakeholders.

Everybody must do the concept phase. Where world-class companies excel is that they put as much effort into the second and third stages as they do the first and it pays off. This is where the concept design is tested for its robustness. If it isn't, robust, there will be many failures in manufacturing, which means low product yield and high manufacturing losses, all of which probably spells doom for the economic viability of the product.

The emerging entrepreneur must engage effectively in all 3 phases of design If not, his or her company is not "making it." Not to require that all 3 phases of design be employed is the easiest way to go bankrupt.

Example: In a very difficult manufacturing situation, semiconductor chip making, we see strict adherence to the practice of all 3 phases of design because these companies know that to do so is a prerequisite for success. They've learned that there is no choice but to do equally effective design work in all portions of design to survive. Perhaps the severity of their environment causes them to be very precise. But the world-class performers would have done it anyway because they are determined to keep their costs down and profits high. If your company doesn't have this attitude, it is not world class.

Simply put, we must have the very best designed factory procedures to produce the very best concept design. Companies that put all their creative talent into concept design and then treat creating methods for producing as an afterthought, or worse yet, put lesser talent on that task, will fail. Perhaps not tomorrow but sooner than you think.

Step 3: Schedule to Produce

This is the coordination step and if not done well, it will spell doom for the company. How many companies do you know of that have great designs, great equipment and facilities, but still can't deliver on time? How often does the lowest cost quote producer fail to meet production due dates? Probably more often than you'd like to admit. Scheduling is as critical to world-class company performance as having the products customers want to buy.

The design step output is used to create a coordinated production schedule for all parts, subassemblies, and assemblies related to a company's products. We do this by creating a workstation route for where work will be done and in what sequence, all derived from an engineering bill of materials (BOM). The BOM is really part of the concept design that shows what the product will be made of and what the sequence order of fabrication will be. With the route and BOM as a guide, the company can construct a coordinated sequence to ensure that the proper parts are done on time to meet all the assembly needs. The last component is to factor in the time to do each step of the sequence. Times for each step come from some aspect of

- Scientific time standards
- Stopwatch time standards
- Estimates

Many companies spend enormous sums in perfecting designs, but get themselves into company–killing conflicts by not being able to schedule their factories. This doesn't happen to world-class companies because they've learned 2 very important facts about competition. In order to sell, they need to

- Know how to make their products, and
- Know how long it should take to do so.

By knowing the method and how long it should take, they can schedule their factory. By scheduling their factory they know what promise dates to give their customers and they have a very good prognosis of meeting the schedule. This makes them reliable vendors with the ability to generate additional orders, as long as their product meets customer needs.

There is only one way for companies to schedule effectively. The schedules need to be dynamic, not static. To do this they need to have a Manufacturing Resources Planning system (MRP II) of some type, capable of fast recasting of schedules as events happen. This has to be integrated with a Computer–Aided Process Planning system (CAPP). These systems must all be driven by cycle times derived from a consistent definition of times to accomplish jobs, preferably a scientific time standards system. These 3 items sometimes are marketed under different terminologies. A common one is "Enterprises Resources Planning" system.

All world-class scheduling systems have one thing in common. They're driven by a common integrated information data system. We call that system computer integrated manufacturing (CIM). CIM-based scheduling systems need to be embraced wholeheartedly; there is no choice. To say "No, I'll pass," is tantamount to committing corporate suicide. CIM with its subsets, one of which is MRP II, is an absolute necessity for the world-class company. No other way of performing integrated schedules exists in a manner that you could say is even a reasonable alternative.

What about Just In Time (JIT)? Isn't that an alternative to MRP II and perhaps better? No. JIT is not a scheduling system. JIT is a philosophy borrowing heavily from traditional industrial engineering theory that says simply: eliminate waste. In its most popularized form, that's taken to mean the elimination of the waste of excessive inventory on hand. This means eliminate waste by going to a *pull* instead of a *push* production control philosophy. The entrepreneur must endorse the precepts of JIT to eliminate waste, and in fact we should all understand that world-class companies do this in an intuitive manner.

The entrepreneur may think, "My business is a service business; I don't need a fancy CIM integrated scheduling system." This is absolutely wrong. First of all, integrated scheduling systems need be no more complicated that running an Excel program or D-base III.

Second, service businesses do need MRP II or something similar because they do make something in the broadest sense of the definition. They provide a value–added function, otherwise they wouldn't be compensated for it. Most likely it is a time–dependent function in that the customer wants the service at a particular time. Therefore, the need to schedule is there. If the company is successful, it will need to juggle demands from many customers at the same time. Remember, in any business, and particularly service-oriented ones, there is a strong need to satisfy customers based on the customers' time frames or risk of losing them to competitors. This means they must schedule and comply with the Two Knows (how to make the product and how much time it should take). Virtually everything a hard goods manufacturer has to schedule for, a service provider does too. Even workstations are similar. The service firm's workstations do not make chips or transform materials, but they do create the basis of the service. It may be a design workstation or even a word processor, but the fact is it only does one job at a time. Therefore, the throughput of work must be scheduled. World-class service firms need an MRP II–type scheduling algorithm. They cannot compete effectively without one.

Step 4: Purchase Raw Materials in Accordance with the Schedule

Coordination of purchasing with the other steps, once more, is the key to success that companies "making it" employ. To purchase effectively, we need to know precisely what it is that has to be purchased and when the materials will be required. World-class companies take advantage of the integrated nature of CIM to use the same database information from design and scheduling to create purchase orders for materials.

The modern manufacturing resources planning system, MRP II, evolved from its progenitor materials requirements planning, MRP, where "make or buy" decisions were made for every item on the engineering BOM. Those that were "buy" were time-phased via MRP and ordered accordingly.

By integrating make or buy via the MRP II system, world-class companies create seamless integration of internal "make" items with external purchased items. Materials arrive at workstations when needed in accordance with the routing instructions. The integrated nature of the scheduling algorithm makes it possible to purchase materials with enough lead time to ensure on-time delivery vs. actual need. Also–ran companies continuously struggle because vendor-supplied items are not coordinated tightly with needs; hence, their ability to deliver on time is jeopardized.

World-class companies create special relationships with their vendors to gain competitive edges in purchasing. The main one is creating a supply chain with their vendors, as we discussed previously. They treat vendors as an extension of their own in–house workstations. However, before they engage a vendor, they make sure they are qualified to meet their high quality standards, the same as required for internal operations. They go so far as to assist vendors in upgrading their quality and production-management skills as required so they do not become weak links in the supply chain. They also, as policy, strive to maintain long-term relationships with specific sets of vendors through long-term contracts for services and supplies. By doing all of this, they are steps ahead of their competitors in guaranteeing high-quality materials and services, as well as fair and reasonable prices.

Step 5: Produce in the Factory

This is the transformation phase of raw materials to finished product, commonly called the value-added step. This is where work is accomplished that directly affects the customer's receivable entity. So we can see that it applies equally to service firms as well as goods producers.

World-class companies make no distinction between external and internal workstations for exercising management control. In fact, the only differentiations they make (as we discussed previously) are that internal workstations are considered to be "owned" while external are "rented." They integrate workstation activities on the basis of their master schedule output from MRP II and make no distinctions as to where the workstation is located. They allow, and in fact insist, that vendors have the same scheduling information that their in-house workstations receive.

World-class companies mitigate differential in labor costs by maintaining tight control on how work is handled at workstations. They employ "short-interval scheduling" techniques that factor in workstation methods, time standards, maintenance criteria, suitability of materials, and operator training. They factor these with accomplishment expectations over short periods, usually no longer than half of a work shift.

They also vigorously investigate failures of any kind for root cause and set immediate corrective actions. And they use the workstation operator as a member of their production team, not as a cipher to be controlled or as a mere nuisance to be put up with. This is much the same as a test pilot being a member of the technical team developing a new aircraft and not just a human guinea pig in the cockpit. This philosophy ensures the best performance on the shop floor and gives a company the highest probability of success. The entrepreneur must insist that his or her company embrace this philosophy. If he or she doesn't, the company won't "make it" and will be constantly squandering resources by chasing low labor cost around the globe. For an extensive discussion on the topic of "short-interval scheduling," I recommend you read the pertinent chapters in my book *Fundamentals of Shop Operations Management: Work Station Dynamics,* published by ASME Press and SME (available from both sources).

Step 6: Monitor Results for Technical Compliance and Cost Control

Measuring is a method of obtaining feedback on how well a company is implementing its plans and, as well, a check on how effective the plans are to start with. In many ways, this step monitors the effectiveness of doing the "Two Knows": know how to make the product and know how long it takes to do so. The first "know" is technical compliance with the plan and is the domain of quality assurance. The second "know" refers to efficiency and effectiveness of accomplishing the plan and generally is within the domain of financial measurements. This is so because for the most part "time is money," as the saying goes, and how we achieve the plan affects costs. So we monitor results in both venues: compliance for what the product or service is suppose to do and compliance with the costs compared with what we had planned it to cost. Let's see how both of those are impacted via Step 6.

World-class companies consider the quality-assurance phase to be an ongoing process of constant vigilance and the seeking of continuous improvements. Quality assurance encompasses the product or service being provided in accordance with the plan. The plan includes technical, schedule, and cost goals, all of which can be evaluated from a quality viewpoint. Is our performance at least as good as our plan laid out? That's a question that measurements of quality performance can help answer.

The best companies work on a philosophy of Total Quality Management (TQM). Continuous improvement is ongoing, as exemplified symbolically by the TQM triangle (see Figure 5-3), as we discussed in Chapter 5.

Remember the concept of the triangle. The customer is at the apex spinning off data about the validity of the work received. This information goes down the leg of the triangle to the left corner as data to be processed. The data goes along the base of the triangle becoming an improved process. Process improvements commence at the right corner and back up to the apex for customer judgment. Then the process starts over again.

There are many approaches on how to monitor and control processes. By far the most accurate is the Statistical Process Control (SPC) technique, especially when tied into the CAPP system for developing process-monitoring steps as part of the methods plan. In fact, many world-class companies consider SPC to be part of the integrated CIM approach whereby SPC action steps are included in the MRP II scheduling algorithm, as well as the CAPP system.

Most recently we've seen world-class companies employ the "Six Sigma" method to gain even further competitive advantage by approaching the very difficult goal of "zero defects." Six Sigma is a philosophy of merging SPC, including the investigative portions of TQM, with mistake-proofing (Poka-Yoke), and team-based management concepts. The message here is that companies that are "making it" employ some form of SPC usually tied into their CIM based MRP II system. If the entrepreneur doesn't factor in the need to do so, then his on her company will have a very difficult time traveling the road to achieve world-class status.

Let's talk a bit about ISO 9000 registration as a factor in performing Step 6. ISO 9000 registration by itself signifies little in regards to adequacy of a quality system vis-à-vis the needs of a company. It states only that a company has a control system that it follows faithfully and has records to prove it. It does not judge whether or not that system is adequate for the company's needs. Employing a TQM philosophy along with tools such as SPC tailored to the needs is what makes a viable, effective quality system. On the other hand, achieving ISO 9000 registration forces the company to follow a path to recognize what its needs are such that the company is aware of what a total quality system is for them, what it needs to embrace to get there, and what is required to maintain it. It instills a systems discipline within the company that makes it more likely it will do the right things that will ensure a quality product is produced. So on the whole, ISO 9000 certification is a good thing. It is especially beneficial while engaged in doing the steps necessary to get there because it will cause the company to learn about itself and correct flaws it has, which until that time has had no driving force to fix. From the viewpoint of the startup company, reaching certification is not a prime goal just yet. The important thing would be to learn from the ISO 9000 process and set up its quality system to mimic that format. Then as it gets operations experience, it can decide a few years in the future to become certified, for the same reasons as mentioned before.

Let's move to financial compliance control. There are many approaches on how to monitor and control costs. By far, the most effective is a use of the traditional financial measurements coupled with responsibility and accountability given to operations. World-class companies use their finance departments to:

- Assist in setting policy
- Gather data
- Teach interpretation of data
- Review and comment on adequacy of corrective actions for senior management
- Find ways to raise external operating funds

World class companies use their operations departments to:

- Assist in setting policy.
- Manage operations budgets
- Interpret data
- Recommend corrective actions
- Carry out approved corrective actions

Notice, the onus for management and the responsibility for applying funds correctly is on operations, not finance. This may be a surprise to many who have been led to believe that finance manages the money. They don't. They keep score and advise as requested. Companies that are "making it" require their operations managers to be very active and take a leadership role in managing the company's funds. This is a position the entrepreneur must espouse to be successful.

In addition to just managing the funds here and now, companies that are "making it" do annual business plans with inputs from all departments. They set financial objectives and goals based on sales and operations realities. They establish budgets in a participative manner for all departments and for major projects. They require continuous measurements for financial and productivity results. They do monthly reviews of operations. They get to root causes for all variances above or below set threshold levels and take corrective actions to correct all deviations greater than threshold variance limits. They do monthly rolling forecasts and compare to budget with explanations for greater-than-threshold variances. And they look at variances as opportunities for improvement and go after them vigorously. The emerging entrepreneur needs to emulate these traits. She must recognize the time it will take to do all these things and make sure it becomes part of her and her colleagues management routines.

Step 7: Ship the Completed Product to the Customer

The job is not done until the product purchased from your company is delivered to the customer on time, complete, and at the expected quality level.

Making deliveries properly is a skill the entrepreneur has to master from the first job on into infinity. Just as vendors are part of your supply chain, you are part of your customers' supply chains. The world-class company's goal is to never be the weak link in any supply chain.

Shipping completed goods to customers means just that. You must use a valid system to track materials through your operation to make sure the actions are performed correctly and that they are done on time to your customers' needs. You in fact must ensure shipment of all completed products to customers and not continually ship only 90%. Proper shipping and tracking of shipments are traits of world-class performers.

Also-ran companies do very well in being 90% companies. They can get the first 90% of the customer's order out on time but the remainder is always backlogged. 90% is sufficient in school but not in the real world. Would you be satisfied with an automobile with the trim missing? Or how about the spare tire not being in the well? Of course not. World-class companies always find ways to ship 100% on time, and the entrepreneur must always do this if her or his company is to succeed.

Companies that are "making it" use integrated scheduling and control systems to track each and every aspect of production, then kit goods systematically and only release for shipment when it is all there. They use MRP II as their tool and control mechanism to make it work. Their shipping and warehousing people are trained in distribution controls and are held responsible for inventory control. In fact, their inventory control records are always 99% accurate at a minimum. They know that whatever happens beforehand, the job can be completed only by the customer taking delivery of the product and agreeing that all is in order and as expected. The entrepreneur must be able to make these claims on behalf of her or his company in order to be considered on the path to becoming a world-class performer.

WHY THE 7 STEPS SEQUENCE APPLIES EVEN THOUGH A COMPANY MAY BE IGNORANT OF ITS EXISTENCE

With this explanation of the 7 steps of the manufacturing system showing what the budding entrepreneur must do to become a world-class performer as background, let's look at the 7 steps from another perspective. What if we were blithely ignorant of their existence? Could we still be successful? Maybe, but there would be lots of "do overs" and waste along the way because the 7 steps are the natural order of business and doing them is not optional. Understanding that you must do them is an opportunity to be efficient, effective, and thus more profitable. Let's see why ignorance leads to less success and conversely why knowledge leads to optimal success.

As you can see, the thread of information cascading from Step 1 through Step 7 is very evident. The 7 steps are the only natural sequence of events that an order can take to become a shipment reality. They have to happen that way.

Whether we like it or not, we cannot start the process before we have an order to make something. This is Step 1. We then have to design some way of doing it in the factory. This is Step 2. To make something, we need to start it through its process. In other words, find a time to do it: develop a schedule. This is Step 3. Most likely we wouldn't have all the materials to do the job on hand so we have to buy a certain portion of the materials before we can make something. This is Step 4. Making something in the factory or delivering a service requires plans, tools, materials, and the availability of people who know how to do it. This is Step 5. A prudent manufacturer or supplier of services is going to want to check the work to make sure it's as the customer wants it. If it isn't and goes undetected, and hence is not corrected, customers will not give the company repeat business. This is Step 6. Also the prudent businessperson wants to know if he or she is making money doing the job for his or her customers. The businessperson wants to know if his or her planning and all other parameters were done adequately and if costs are in control. This is also Step 6. Once all the work is done, the company needs to send it to the customer and make sure it performs as required. If not, the customer will not be satisfied and the relationship is placed in jeopardy. This is Step 7.

Is it possible to do the sequence out of order? Not really. How can something be made without plans to do so or without an understanding as to what has to be done? Can we design a process without a product to make? This can be done only in generic terms and therefore nowhere near optimum. Can we schedule a factory without knowing what we're going to make, even if we know what the process may be? You can try but it's nothing but an academic exercise in "could be's." Can we start making something when we don't have the needed materials? Yes, you can start, but the delays for wait time will be very costly and therefore far from optimal performance for the client or the company. Can we have people working on a product without complete plans of what to do? Lots of companies have tried this and have gone broke in the process. Is it possible to check progress for incomplete jobs? Yes, but it doesn't validate the total job for cost or quality. Can we ship products that aren't completed? You can try if you want to alienate the customer.

It is physically possible to do the steps out of sequence, but when this happens, we find that most of the out-of-sequence work turns out to be waste and must be repeated. The result is the company will stumble upon the sequence where there is no iterations, hence no tear downs and re-starts. This turns out to be the 7-steps sequence. Now, think about the waste and frustration a company goes through every day when they're ignorant that there is a logical 7-step sequence and that preceding steps need to be done before succeeding steps in order to minimize this confusion. And now I'm talking about the surviving companies, not those that have failed and learned the lesson too late. The companies that learned and survived have only lost profits as a payment for learning the hard way. Those that didn't survive paid the ultimate cost. The entrepreneur doesn't want to be in that position. The lesson of the story is that ignorance is not bliss. Pay attention to performing each of the 7 steps as optimally as possible, and by doing so, the startup company has a significantly better chance of "making it." The world-class companies document this 7-step sequence and make sure all their people understand the order in which work needs to be done.

World-class companies also exude communications excellence based on the natural cascading of information from Step 1 through Step 7. Not only do they consciously subscribe to the tenets of the 7 steps of the manufacturing system but they've integrated the information flow within an all-encompassing CIM system. By doing so, they can respond to opportunities in a dynamic fashion. They can make changes to schedule, designs, SPC checking parameters, contents of shipments, and virtually any other demand made by their customers many orders of magnitude faster than those companies not aware of the 7 step of the manufacturing system. They can do this because even though the customer may change a parameter, the company knows which step of the 7 steps is first affected. Therefore, they know what needs to be redone and can do the teardown much more precisely with minimum waste. They know precisely where to start the recycle caused by the change. Companies that aren't aware of the 7 steps tend to stumble and take several corrective iterations before the change is properly dealt with. Companies that are working with a knowledge of " the order of things," the 7 steps, are the companies that are "making it" and will continue to in the future.

WHY UNDERSTANDING THE 7 STEPS AND THE LINKAGES BETWEEN THEM IS NECESSARY FOR DEVELOPING "COMMUNICATIONS EXCELLENCE"

Communications excellence means all functions getting the same information with the same interpretation. To be successful in the business world requires being able to cope with the inflow of information that represents the dynamics of change. Change is a constant. Status quo, a static situation, is a special case of change. It is 7 phase that will not remain the same for very long. That means communications about change have to be fast and accurate; otherwise, mistakes are made. So being able to handle change is necessary for business success.

Information regarding a change is necessary for a successful execution of the requirements of a change. And it has to be given to all components of the organization, usually at the same time. Since the 7 steps of the manufacturing system entails all activities involved in making a product or delivering a service, getting that information correctly to each step is required for success. Information requirements for all 7 steps are essentially the same:

- What is it?
- How is it suppose to work?
- Who has to do the work?
- When does the work have to be done?
- Why does it have to be done in a certain way?

Each of the organization's functions responsible for one or more of the 7 steps asks these questions pertaining to what it does. But the information source must be common to all.

The fact that the 7 steps are naturally linked makes it relatively easy to ensure that the information flow will be correct. This is because the information needed to do a successor step will come from a predecessor step. If a company recognizes that there is a linkage between steps and consciously tries to follow the steps in sequential order, the probability for the information to be correct is much higher than it would be if the steps are attempted out of order. Therefore abiding by the 7 steps in the organization of doing work makes it more likely that communications excellence can be achieved.

WHY THE 7 STEPS OF THE MANUFACTURING SYSTEM ARE UNIVERSAL, NO MATTER WHAT THE PRODUCT IS

The 7 steps apply to any enterprise that has an end product or service it provides to others. We've seen in principle how it works and how it is necessary for a company to optimize its performance. Now let's dig a little deeper to look at specific manufacturing and service company applications. We'll find that there are virtually no differences in approach. Let's look at the 7 steps and review the universality of the inputs.

Step 1. Obtain Product Specification

A specification defines the exact nature of problem to be solved. This tells us what we have to do. This information is gathered in 2 distinct ways: externally from customers, and from the creativity of internal sources that are creating products to sell. The nature of the specification will depend on the field of business the company engages in.

Examples:
- Electronic circuit board design obtained from a client looking to source manufacturing.
- Economic data for a business venture desiring to develop land for an office park.
- Contract facts pertaining to a pending breach-of-contract lawsuit
- Wing aileron design for a hypersonic transport to be built in the company's prototyping facility.

We see in the examples a whole array of data, some data for solving technical problems, other data for business problems, and even an example referring to a legal problem. The point is that they're all product specifications for solving a particular problem.

Step 2. Design a Method for Producing the Product, Including the Design and Purchase of Equipment and/or Processes to Produce, if Required

After the problem is defined with the specification, we cascade down to the planning step. This step creates a methodology of solving the problem. A problem can be technical, nontechnical, or a combination of both. Creating a method is a process of applying logic to how to solve the specific problem, such as the specification to be met.

Examples:

Technical—creating a method for painting a house
Nontechnical—getting an ironclad contract signed by the client to paint the house
Combination Technical/Nontechnical—creating a method for painting the client's house in a manner that guarantees the client will pay for the paint before it is applied (for example, perhaps strip old paint but do nothing more until client pays for new paint).

Step 3. Schedule to Produce

This is putting the job into its queue and knowing when it will start and finish with respect to other jobs committed to by the company and the client's need dates. Any type of commitment to do anything will require a time to start and a time it should be done.

Examples:

- Roast a turkey in the oven.
- Buy a new automobile.
- Build an addition to a home.
- Manufacture a robot.

To do these successfully, we need to comply with the basic tenets of manufacturing, which apply to all products and services:

Know How to Make Your Product or Deliver Your Service.

This is the methods activity—the sequence of events necessary to perform your task. And these activities must be accounted for in the schedule.

Know How Long it Should Take to Do So.

This is the time sequence activity that allows us to calculate the amount of time and calendar sequence for the job. Both "knows" together give us the necessary information to structure a schedule. We see there is no differentiation between assembly-line producers, discrete parts manufacturers, and deliverers of services.

Step 4. Purchase Raw Materials in Accordance with the Schedule

All jobs need materials of some sort to be completed successfully. This is obvious for product companies, but also true for service companies. Only in fiction are there "thinkers" who never need to document or transfer information to clients in a manner that the client can record. Those documents are made up of materials, perhaps paper and multicolored printing inks. Some jobs share materials from job to job to some degree.

> **Example:** A CPA buys a notebook computer to use at clients offices—a shared material to do the job. He provides data to clients from the computer via a "floppy disk"—a specific material for the specific job.

Step 5. Produce in the Factory

Production is a concept of making something out of raw materials. However, the term *raw materials* doesn't refer only to physical entities. It can also be ideas that grow into productive action plans. For service companies, the ratio of physical to ideas is heavily slanted to ideas. And as we would expect, for product companies the ratio of physical to ideas is heavily slanted to physical. In either case, work is carried out in the factory, an office, or the shop floor, or perhaps the customer's specific locale.

Step 6. Monitor Results for Technical Compliance and Cost Control

Regardless of product or service, we will emulate "world-class" companies and always check our work to make sure it is done correctly and at the cost levels projected.

Step 7. Ship the Completed Product to the Customer

A job isn't complete until title, information, and hard goods, if any, are transferred to the customer. Service jobs are completed when the contract to perform the service is completed and the results are available to the client for his or her use. Product hard goods jobs are completed when the contract to make the product is completed and the results are available to the client for his or her use. So there is no difference between a manufacturer's or a service provider's requirements for closing out a job.

A Demonstration of Compliance for a Service Company and a Physical Product Company

In the following example, we compare a service company to a manufacturing company.

Example:

| **Computer Co. X** | **Insurance Co. Y** |
| Product: personal computer | Product: life insurance policy |

1. Obtain product specification.

| Engineering specification | Insurance coverage range |
| Based on customer needs | Based on customer needs |

2. Design a method for producing the product, including the design and purchase of equipment and/or processes to produce, if required.

Select process plan and equipment. Develop new if required.

Design method of obtaining basic insurance based on probabilities of benefit payouts vs. premiums collected and investment returns. Also ability to secure capital.

3. Schedule to produce.

Use MRP II to load the factory in accordance with customer need dates.

Calculate the time to investigate and prepare insurance plans for customers, compare to actuarial tables and underwriter rules. Use as cycle time to schedule development of policies for clients.

4. Purchase raw materials in accordance with the schedule.

Input materials in accordance with the BOM and inventory control balancing criteria. Use MRP II to forecast needs and negotiate buy contracts.

Attract capital investment and underwriter syndication. Use policy base to attract investment funding.

5. Produce in the factory.

Make personal computers in accordance with the MRP II schedule.

Invest capital and premiums collected to cover benefits payout reserves and profits. Write policies per the schedule.

6. *Monitor results for technical compliance and cost control.*

Compare actual production of personal computers to schedule and cost projections. Compare performance to design specs. Compare customer desires with results	Compare actual production of writing insurance policies to schedule, and cost projections of raising capital to cover policy reserves for policies against plans. Compare customer satisfaction with policy protection clauses vs. needs.

7. *Ship completed product to the customer.*

Package, transport, and deliver to the customer on time.	Package, deliver, and explain benefits of the policy to the customer on time.

WHEN A STEP IS NOT READILY APPARENT: WHY THAT'S DANGEROUS AND HOW TO FIND THE MISSING STEP(S)

Missing steps or doing work out of order will inevitably cause higher costs, poorer quality, and missing schedule ship dates. All this can be mitigated by attention to detail and understanding the proper sequence. For example, skipping the quality step, Step 6, can result in undiscovered quality problems getting to the customer. A step is more likely to not be readily apparent when there is a blurring of activities.

> **Example:** During manufacture, intermediate quality checks are made during the fabrication process, such as measurements taken before a final finished cut is taken on a cabinet door. The measurement is a quality check, part of Step 6 but occurs while still doing Step 5. If care is not taken, it can be overlooked because of the intermingling, which is necessary for this process.

How do you determine whether all steps of the manufacturing system are accounted for and are being done in the correct sequence? For all operations steps, do the following:

- List all methods steps in sequence of occurrence.
- Describe the key activity occurring during that method step.
- Determine what step or steps of the 7 steps the key activity describes.
- List the methods steps in the order that they currently occur.
    - If the process is being performed correctly, there ought to be a key activity every step in sequential order.
    - If a method is out of sequence, then the 7 step natural order is not being followed and extra cost and time are occurring.
- Reorder the occurrence to correct if the method doesn't agree with the 7 step sequence

Examples:

a. Sequence *1,3,4,5,2,6,7*. In this sequence, work is being done before the method for producing is set. This typically occurs when the person(s) doing the work think they understand what has to be done and ignore the need to become familiar with the requirements (assembly operations often have this problem—usually there will be a need for a tear down and redo).
b. Sequence *1,2,3,4,5,6,5,6,5,6,7*. In this sequence, work is being done correctly but intermediate inspections are required during the process. Technically this is an error. However, often a process is not robust enough to allow planning to be totally complete before the process begins. Skilled machining and the need for intermediate data as a result of intermediate work are examples of causes. The theoretical solution is to develop a more robust process where all information can be had before starting. Often the theoretical solution is unacceptable because the development would take too long and cost too much. Then the iteration is allowed. This is a judgment call.

If a method doesn't yield all 7 steps, then the investigation is aimed at finding out why. Determine whether there are any discontinuities in information as described by the method. This is a sure identification of a missing step.

Examples:

a. Sequence 1,3,4,5,6,7. Again a case of the missing the manufacturing engineering step. We go from a design to make in the factory without specific instructions. This results in a lack of information, which is usually critical for specification compliance.
b. Sequence 1,2,3,5,6,7. In this sequence, materials are not matched with the design or manufacturing instructions, and this lack of information could result in wrong materials being used and a failed product.

A 7-step checklist and flow charts are useful diagnostic aids to prevent these types of problems.

WHY THE FACTS (DATA) NEED TO BE THE SAME AND CASCADE FROM STEP TO STEP, AND WHAT HAPPENS WHEN THIS RULE IS DISREGARDED

When we make a list of methods, from solicitation of an order to delivery to the client, we are working on a set of data that spells out the specification. If that specification is allowed to drift because the data associated with it is not transmitted accurately and completely from step to step, then the ultimate results will not represent what the company set out to do.

When data drifts from step to step, we end up with an out-of-control process. This in turn means that the company is at risk of producing a product or delivering a service that does not meet specification. With data drift we end up not knowing what is correct data and what's erroneous, and all the data, regardless of its validity, becomes difficult to track or control because the variables become too numerous. This in turn means problems with customers possibly not receiving what they ordered. Simply because of confusion, the company would have no way of knowing what's acceptable or not.

Let's look at a simple example of data drift, using a machine shop process as an example.

Example:

Step 1. A nano product has a part with a diameter of 0.001 ± 0.0002 in. This is the data to be transmitted from step to step. The requirements are documented in the BOM and in the technical specifications.

Step 2. Manufacturing engineering thinks they can hold the tolerance.

Step 3. Master scheduling allows time in the scheduling sequence to make the part within the required tolerance.

Step 4. Materials are requisitioned for the parts based on the data.

Step 5. The shop doesn't think it can hold the tolerance. They modify the data to a diameter of 0.001 ± 0.0005 in. via instructions to the operator before attempting to make the part. This would be a manageable problem if the results were reported back to Step 1. It would require a redesign and negotiations with the customer since it is a change that may or may not be suitable. The shop sets out to make parts as they think they can but "drift" from original data, which is a serious mistake if done without the concurrence of the other steps.

Step 6. QC finds the product to be out of control, with high deviations from mean beyond what the process would expect at a diameter of 0.001 ± 0.0002 in. What is the problem? The choice being made by the shop? Or is the original process to support a diameter of 0.001 ± 0.0002 in. not within the shop's capability? Because we violated data integrity from step to step, we don't know.

Doubt in process integrity is a significant depressor of business confidence. It creates a situation whereby the company is hoping for good results instead of designing for good results. The company's process is suspect if data is not transmitted accurately from step to step. Entrepreneurs of startups must be especially mindful of this pitfall because quite often they succumb to temptation and plunge ahead without robust management systems in place.

HOW TO EVALUATE YOUR COMPANY TO DETERMINE HOW WELL YOU COMPLY WITH THE 7 STEPS OF THE MANUFACTURING SYSTEM

So far, we've seen how all production of products and services obeys the dictates of the 7 steps of the manufacturing system, whether the participants are aware of them or not. We now know we have the information to understand the process and realize that it is the natural system of providing services and products. However, we haven't learned how a company can determine how well it complies with the system. We will go through that procedure now.

Setting up the Task to Use the Questionnaire Sheets

The best way to find out if your company is complying with the dictates of the manufacturing system is to ask the people who do each step. I've had great success doing this with a questionnaire (see Appendices A and B). The questionnaire is aimed at determining how close a company is to being a "world-class" competitor. The questions are designed to elicit information about how each step is being accomplished.

The questions also try to find out if there is a continuous flow of information from one step to the next. It aims to ensure that there are no discontinuities, that the system in the particular company is complete, and that the formal system can handle all aspects of the information requirements for the job. It also confirms that no ad hoc actions are required. The questionnaire covers all 7 steps in sequential order. For the entrepreneur starting up a new company, using the questionnaire may be a problem because many of the role players may not be fully aware of all of their responsibilities. In that case, the questionnaire may be more of an instruction manual. However, for the purpose of discovering how the questionnaire and its results are used let's assume the entrepreneur's company is sufficiently along in its development that the questionnaire is an appropriate approach.

The questionnaire is completed in 2 complementary ways:

- By an independent survey by people familiar with the operations.
- By a consultant questioning people who work in the area and/or are very familiar with the processes used.

Independent surveys require that the person who is filling out the form understands the questions and is able to answer the questions objectively. To minimize the possibility of subjectiveness creeping into the results, several people would be asked to fill out the form and to do it independently. The drawback of independent surveys is that the person(s) filling out the form may not be familiar with the 7 steps and the concept of linked information flow. Therefore, their responses may be biased, to a degree, based on their level of understanding of manufacturing systems.

Conversely, consultants filling out the form can only be as accurate as the ability of the consultant to elicit total and true information. The consultant understands the concept of the 7 steps, but isn't as familiar with the actual specific operations. The consultant needs to phrase questions to be precise and not vague. The consultant needs to ask the same set of questions to several persons familiar with the operations to be sure of getting all the information and not just pieces.

A preferred way to prepare for using the questionnaire is to create teams of consultants and company staff. Consultants are expected to be fully versed in the products or services the client makes or performs and they do this by properly familiarizing themselves with the client's processes beforehand. The company members of the team should be from the areas that the specific questionnaire subset is handling. Operators are suitable for Steps 5, 6, and 7. Engineers familiar with the entire operation would be suitable for all steps. Various clerks, supervisors, specialists, and managers are suitable for any step of which they have firsthand knowledge. Remember, different people have different viewpoints. So it's okay to go about asking and answering the same question more than once with different people.

Gathering Data

Using Independent Surveys

Hand out the questionnaire to those responsible for filling out portions of it or the whole questionnaire. Agree to a submission date—usually a maximum of 3 days. Brief the survey respondents on the nature of the questions. Sometimes it will be necessary to give a brief tutorial on the 7 steps of the manufacturing system. Explain to the respondents that they are required to fill out the questionnaire for those areas constituting their work assignments and that they are allowed to fill out any other portion of the questionnaire they would like to; in fact, encourage them to do so. This gives more information and places their portion in proper context, and it will aid in defining problems of execution in their areas of responsibility.

Stay in contact with the respondents during the time allowed to answer the questions. This way, the questionnaire will have a better chance of getting done on time. Answer questions as they come up from different individuals as necessary, but do not influence the answers.

At the conclusion of the briefing, thank the respondents for their cooperation. Tell them they will get the results of the survey as soon as possible, or at some other designated time. Keeping them in the loop ensures further cooperation when it's time to make corrections to processes.

Using Teams

Set aside a time to collect the data that's mutually agreeable to all team members. Then at a team meeting, brief the survey team respondents on the nature of the questions. Give a brief tutorial on the 7 steps, even if it's a repeat of what has previously been explained. By repeating the message to the entire team at the same time, you are ensuring that they're all getting the same perspective of the message. After the briefing, explain that teams are required to fill out the questionnaire for those areas constituting their work assignments. As with independent respondents, they are allowed to fill out any other portion of the questionnaire they would like to; in fact, encourage them to do so. This gives more information and places their portion in proper context, and it will aid in defining problems of execution in their areas of responsibility.

While the questionnaire is being completed, consultants can answer questions as to intent but should try not to assume an answer. Let the company team member supply the answer in his or her own words. The consultant team member should volunteer to be the "scribe." This allows the consultant to elaborate on the answer by adding in perceptions and feelings observed of the company people. Sometimes a "body language" exhibit tells more than the spoken word. The goal is to arrive at the objective truth of the situation pertaining to the compliance measured against world-class standards. The consultant should collect all the data and check that all required questions have been answered by all teams.

At the conclusion of the briefing, thank the respondents for their cooperation. Tell them they will get the results of the survey as soon as possible, or at some other designated time. Keeping them in the loop ensures further cooperation when it's time to make corrections to processes.

Collating Results of the Survey and Testing for Accuracy

Results obtained from the questionnaire will be varied and in many cases contradictory. The chore is find the true data and discard the erroneous data. This is done by making a tally sheet for each question of the survey. The subheadings of the tally sheet are as follows:

- Question to be answered
- Questionnaire reply summary
- Observations
- Recommendations

Figure 10-1 shows a summary tally sheet for a question filled out by the individual responsible for the questionnaire.

The individual responsible for the questionnaire entered data in the summary sheet for each respondent to each question under the "Questionnaire Reply Summary" heading. He entered all the data as presented by respondents as close to actual quotes as possible and Paraphrased written responses as necessary. He then interpreted the results and entered them under "Observations." If results varied, he would get clarification from respondents. It is important to try to resolve differences by learning why respondents answered the way they did. Then based on experience and on the preparation done for the survey activity, he selected the most probable answer to the question. This may be as simple as selecting by majority vote, by consensus opinion, or by knowing what answers are obviously wrong. My experience has been that by doing interviews of key personnel beforehand, I tend to learn what the correct answers for the particular company are likely to be.

What I've just described is a collating and test for accuracy procedure. It is necessary to do so and not take each reply as a simple truth that can be used in finding the real performance-level status. Very rarely will respondents lie, but quite often they simply do not know an answer and respond with their

> 2.a. Does the design of the product take into account the ability of the factory to make the product with ordinary due diligence for a satisfactory yield and cost?
>
> *Questionnaire Reply Summary:*
> - I believe so, yes.
> - Sometimes.
> - Ideally Estimating should include an evaluation of engineering and ability—sometimes this is not being done.
> - Not always. While this is the goal, sometimes custom work must be redesigned on the floor to meet unforeseen design challenges.
> - Usually.
> - Yes (three replies)
> - Company provides all supplies.
> - N/A
>
> *Observations:*
> Many things can be done to make the manufacturing more producible. For example, pushing the use of standardized fittings and fixtures as far as possible. Also, gaining an understanding of the "true" shop capabilities and factor them into the designs presented in quotes.
>
> *Recommendations:*
> Institute a design data book that spells out company manufacturing capabilities and desired standards the company would want to incorporate in its products. This design data book becomes the standard from which information is extracted to aid in estimating and quotes. Sticking to the design data book elements in quotes ensures that the company can produce the product without resorting to extraordinary means, hence higher profit margins.

Figure 10-1. Sample summary sheet of questionnaire reply.

perception of reality without really knowing they're doing so. But by asking the same question of several different people with different perspectives and by matching these statements with known facts, it is possible to very accurately portray performance level status compared with a world-class level.

The consultant would then use the observation as the basis for a plan for improvement. He would use his own experience and knowledge of business practices to develop a recommendation suitable for the client company. This would then be put into the "Recommendations" section of the summary sheet. The final activity would be to collect all of the recommendations from all of the summary sheets and then create a report showing each step of the 7 steps, and under each step list the recommendations.

HOW TO COMPARE YOUR RESULTS AGAINST BEST PRACTICES—A GAP ANALYSIS

Being able to take action to correct deficiencies is the hallmark of a successful company. But we have to make sure we're correcting the most important deficiencies and not just the ones that are the easiest to do. Gap analysis is a tool that helps to arrange the sequence of improvement events, such that the company is exploiting the ideas that will give it the most benefit with the least expenditure.

An Explanation of the Gap Analysis Technique

A gap analysis is essentially a measurement of status from a reference point. It is often confused with "benchmarking," to which it is similar. "This is where I want to be. This is where I am. The difference is the gap I want to close." This is the essence of the technique. Any entity engaged in any activity that can define how to do a task in the most optimum manner can do a gap analysis. The task is to define the optimum end point in a manner that allows it to be measured.

> **Examples:**
> - Baseball: pitching a perfect game: 27 batters, 27 outs
> - Quality: Six Sigma performance of 3.4 defects/1,000,000 occurrences

The whole process of benchmarking is an attempt to set the optimal end points for various business processes by finding others who do some of these processes at those levels and then copying what they do to the extent of practicality. This is a form of gap analysis (both titles are often used interchangeably, even though they're not quite the same). Benchmarking differs from gap analysis in that it subjectively decides that a target company does something well and that what they do is not necessarily tied to the definitions of world-class embodied in the 7 steps of the manufacturing system. For the 7 steps, we have a set of benchmarks that are designated as world-class standards, which have been described in this book. We can summarize as shown in Figure 10-2.

The measurements from the "world class" standard are made via the Investigation Points questionnaires, as shown in Appendices A and B. The methodology is demonstrated in the earlier section, "Collating results of the survey

Seven steps of the manufacturing system	What "world-class companies do	Your co.? yes / no
Obtain product specification.	Use market studies to set new product introduction goals.	
	Use Quality Functional Deployment.	
	Use Concurrent Engineering.	
Design a method for producing the product and purchase of equipment and/or processes to produce, if required.	Use 3 phases of design: 1. Concept design 2. Producibility design 3. Manufacturing facilities design	
	Create bills of materials (BOM).	
	Develop process methods.	
Schedule to produce.	Use computer aided process planning (CAPP).	
	Use scientific time standards.	
	Use manufacturing resources planning (MRP II) systems.	
	Use just-in-time philosophy.	
Purchase raw materials in accordance with the schedule.	Use MRP II purchasing scheduling module.	
	Use supply-chain philosophy.	
	Use only qualified vendors.	
Produce in the factory.	Use MRP II shop-floor scheduling module.	
	Extend schedules to include vendor make items.	
	Use short-interval scheduling.	
Monitor results for technical compliance and cost control.	Use total quality management (TQM).	
	Use statistical process control.	
	Use "zero-defect" programs.	
	Register for ISO 9000.	
	Use operations budgets.	
	Create short-range forecasts. on set schedules.	
	Do periodic financial reviews. on set schedules.	
Ship the completed product to the customer.	Use MRP II inventory control.	
	Maintain inventory records to at least 99% accuracy.	
	Ship after 100% of product is kitted.	
	Obtain customer concurrence of completed shipment, for quality and completeness.	

Figure 10-2. Manufacturing competitiveness check list.

and testing for accuracy." The fact that the world-class standards are such, has been demonstrated by performance of large numbers of companies who do as the checklist (Figure 10-2) suggests and are very profitable.

Matching Survey Results with Best Practices Standards

The completed questionnaire gives the company a road map for where opportunities exist. Each gap is listed against the step of the 7 steps to which it corresponds. The importance of each gap should be defined with reference to the company's plans and culture. Then an attempt should be made to quantify the magnitude of each gap. Generally, the larger the magnitude, the more serious the gap is to the health of the company. Sometimes a scale of 1 to 10 is used to give point values for gaps. A complete absence of ability to do the step per the questionnaire is rated as "10." A minor variance would be given a score of "1." It is a judgment factor for scoring for all gaps in between. A good scoring technique would be to gather all of the respondents, then go through the scoring with them as a group project. The simplest procedure for this group meeting would be for the consultant (or team leader) to explain the world-class standard for each question, then have the group vote on the score. Perhaps use paper ballots to minimize going along with group think results.

A histogram of the number of gaps per step of the 7 steps is an excellent way of quantifying and visualizing which of the subsets of the 7 steps is farthest from the best practice "world class" standard. A single "step" with lots of small-magnitude gaps is a serious situation because magnitudes within each step are cumulative, and this step should be high on the priority list for corrective action. A plan, then, needs to be set in place for each gap identified. We would use project management techniques as discussed in this book to plan and launch a project to eliminate each gap.

Figure 10-3 is an example of the section of the report with the gap analysis histogram.

In the example, we see that Step 3e had 3 gaps and their magnitudes were 4, 4, and 6 with a resultant score of 14. This was the highest of all of the steps contained within Step 3: Schedule to produce. This would be the first choice for corrective action because of the point score and the fact that it was the only step that had 3 problems.

Understanding and Ranking Results of the Gap Analysis

The gap analysis shows discrepancies between current practices and "world class" standards. What to do with this information is a management decision that can have enormous ramifications for the company's future, and for a startup firm, could spell the difference between survival and disembodiment. The alternatives are:

- Set a plan in place to close all gaps.
- Work on those gaps of most interest to the company.
- Do nothing.

3. Schedule to produce	Gaps	Magnitude	Score	Histogram
3a. Does a schedule exist showing all open jobs and the required complete dates?	1	5	5	x x x x x
3b. Does the schedule show the total portal-to-portal cycle time for each job at each work station?	1	10	10	x x x x x x x x x x
3c. Is the cycle time for the schedule derived from Step 2, above? What is the link or methodology for accomplishing this task?	0		0	
3d. Is there a capacity document showing the amount of work capable of being completed in a certain time period? Is it used to create the schedule?	2	5 and 6	11	x x x x x x x x x x x
3e. Does the schedule process have an update routine that feeds back actual completions for comparison to plan? If so, is it done daily and is there a policy for revising schedules if the plan becomes too far out of synch. with actual results?	3	4,4,6	14	x x x x x x x x x x x xxx x x
3f. Is the schedule of sufficient longevity that the supervisors can plan their work load before the start of the next work period (day or shift)?	0		0	
3g. Is the schedule broken down by work stations or centers and are workers told the amount of time they have to do the assigned job?	1	3	3	x x x
3h. Are overdue jobs given top priority to complete before new jobs are started?	0		0	
3i. Are there any rules established that give permission to bypass a job in the sequence to start a succeeding job?	1	8	8	x x x x x x x x

Figure 10-3. Example; numerical data for gap analysis.

Most often the decision is a financial one, based on what the company can afford. But by any management philosophy standard, the best option is the first option. However, practicality must reign. A plan can be as "laid back" or aggressive as the company may tolerate. Keep in mind that if all gaps are scheduled to be closed, there is a chance it will happen. If only certain gaps are scheduled to be closed, it is virtually certain that those gaps not on the schedule will never be closed. Any open gap is a threat to the optimum profitability of a company and in some cases could lead to the demise of the company. The point is that deciding to ignore gaps is a risky gamble that should be avoided if at all possible. Once a decision is made to close gaps, the following should be considered when determining the sequence for closing them:

- Select what to do first based on a business plan.
- Rank order of project sequence in accordance with ability to make positive change(s).

A Method for Implementing Improvement Programs to Close the Gap, Including Measurements

Many methods are available for initiating improvements. I believe the following sequence is the most straightforward and guarantees the institutional changes and cultural changes necessary to make the changes permanent.

First: Set the stage to close the gaps. I do this by establishing a participatory management style, or at least the structure of one, recognizing that this in itself can be a culture shock to a company. But it is incredibly important that you work to make this culture change stick. Being able to elicit good ideas from all sources is necessary for changes to happen. We must ensure that ideas for change are accepted and that the people involved accept ownership for their success.

Change needs to be accepted by all members of the company. The participatory style gives all employees the greatest ability to take part in decisions and have their inputs considered. By being part of the decision-making process, employees more readily accept change. It makes communicating reasons for change easier. But even so, the reasons for change must be communicated in detail, not superficially but in as much detail as employees want. Use all means available to get the message effectively communicated. Use feedback mechanisms such as surveys to ensure that the reason for change is understood.

We must set the proper background for change. We need to effectively communicate the 7 steps of the manufacturing system theory and why it is the natural way of doing business. This can be done by establishing training courses to teach awareness of the 7 steps.

DEFINING THE OPPORTUNITIES FOR IMPROVEMENTS

The gap analysis shows us what needs to be improved, but not how. By understanding what the gap is, however, we can define how to close it. Gaps can be closed by first understanding how world-class performance is achieved and then doing a second-level gap analysis of the company's current method compared with the world-class method.

The process of doing the second-level gap analysis starts with first understanding how to do the method at a world-class level. This may require an effort in "reverse engineering," and/or may come from instruction manuals and textbooks, and/or it may come from personal education and previous experiences. Then compare the world-class method sequence vs. current-method sequences. We now can see the difference but we have to define them precisely. The differences are the areas to concentrate on to implement change to gain world-class performance.

The next event is to place value on the opportunities. For each sequence with a difference between current method and world-class method, define the value of closing the gap.

- Cost savings
- Time savings

If it appears to cost more to do the world-class sequence than benefit from it, then we haven't fully understood the benefit to be gained by implementation. Let's assume that's not a problem and proceed. Now list the opportunities in order of gains. But we're still not ready to commit. We have to do another sorting using Musts and Wants.

Using Musts/Wants Techniques for Ranking

In addition to listing opportunities by magnitude of gains, we need to list them by strategic importance to the company. A preferred way to do this is by the "Kepner Tregoe" Musts/Wants process. We covered this technique earlier in the book. Now we'll employ it with respect to the 7 steps of the manufacturing system.

- Must—needed to do to "save the company"
- Wants—nice to do to improve company profits and make the most of opportunities

Musts always take precedence over Wants. Here's the procedure:

- Search database for all Musts.
- List Musts in magnitude of savings order.
- Remainder of database content are Wants.
- Rank Wants on a score of 1 to 10.
  - 10 is the highest in need to be done, that is, just under the value of a Must.
  - 1 is the lowest in need to be done; can be delayed for a considerable period of time.
- Compare Wants ranking with value magnitude.
  - Sort by Want score and magnitude with Want score taking precedence.
    - Example: a Want score of 10 and a value of $10,000 is higher in precedence than:
      - Want score 10, value $9,000
      - Want score 9, value $15,000
- Final ranking would be:
  - Musts in order of value magnitude
  - Wants in order of score, then in order of magnitude with groups of same score.

Comparing Resources with Ranked Opportunities

Using the ranked opportunities list, define the resources necessary to achieve the opportunity. Determine whether it's a make or buy, or a combination. If it is a buy service or equipment determination then we would need to

1. Determine cost
2. Determine availability time frame.
3. Compare with company's financial resources and time needed.
4. Rank in accordance with 3 to define what is feasible.

If we can accomplish in-house with current personnel, then we would need to

1. Determine types of skills necessary.
2. Estimate hours required per skill.
3. Compare to in-house skills capability and capacity.
4. Determine a time line for each opportunity.

Selecting Opportunities to be Implemented as Projects

From the list compiled in Musts and Wants order, add the resources necessary to achieve them (buy or make or both). Then projects would be selected in order of precedence with Musts opportunities always first. The active project list will be as long as there are resources to do them. The inactive list will consist of all projects that do not have resources to accomplish them. They will be listed in the same precedence protocol as the active list. As soon as more resources becomes available of any subcategory, the inactive list is searched for projects requiring that type of resource and that project is upgraded to the active list. This process should continue at least until all the Musts projects are done.

Setting up Measurements for Accomplishment of Projects

Measurements have to represent the important factors of world-class standards. There are 2 types of measurements: Absolute, which are variations of yes or no; and values, which are scales depending on objective collection of data such as parts/million or temperature. Any measurement, whether absolute or value, must be compared against a standard. Standards are set levels of performance irregardless of how the conformance to the standard is measured. A standard is required for a measurement to have any meaning.

Example:

Measurement: There are 30,000 blades of grass in the lawn.
Meaning: Who knows, depends on a lot of other factors, such as size of lawn and density of grass standard. Since there is no standard, the measurement is irrelevant.

Standards often become the goal to achieve, but are not necessarily synonymous.

Setting up the measurement first requires defining the standard. Here are 2 types of standards. A very specific standard: Cost per 100 products produced is $35. There is no doubt that the standard cost to produce is set. If we are at $35 per 100 products, we've met the standard. If we are higher, we haven't. If we are lower, we've exceeded the standard. Here's one standard that is measurable but takes some further definitions to know if we've achieved the standard: ISO 9000 certification required. By when? For what? It doesn't say. A specification may require ISO 9001–2000 certification within 1 year of gaining the contract. That means meeting the standard is irrelevant until then. Similarly, ISO 9001–2000 is generic. How about meeting the equivalent of ISO 9002, which has no design requirements compared with the equivalent of ISO 9001, which does? Or how many plants of the company are required to be certified? From this standard there are no absolutes to measure against accept the total standard, which is yes or no.

For purposes of world-class performances, measurements have fit the need to illuminate the progress toward achieving the goal. The goal can also be an incremental step toward achieving the standard. For example, the ultimate goal may be to be a Six Sigma company (which could also be the standard). Then the interim goal for this year may be to get to a Four Sigma level. The goal can also be set to exceed the standard. The standard may be achieve a "C" grade for the course to get credit for graduation, but the goal may be to achieve an "A." So we see measurements, goals, and standards must be compatible to be of use. The measurement must be relevant to achieving the goal, which in turn must match with the standard.

Example: Measuring the number of units produced may not be relevant when the standard and goal is to produce zero-defect units at a prescribed quantity. Unless the process produces only zero defects, the measurement doesn't tell the full story. The measurement may show that only 1 bullet of 10 fired penetrated the armored vest and killed the wearer, when the standard is that none can penetrate, and the goal is zero penetration. Anything above zero is meaningless. The wearer is still dead.

Measurements for closing gap activities become part of the objectives and goals routine. They should be reported periodically, matched to the normal period of activities. The reports should define variances from standard or goal as appropriate, as well as compare the current report to previous reports to show trends. Also, it's a good idea to present ideas with which to go forward. That corrects current results. The report should also have a narrative fully describing the meaning of the measurement and trends.

SUMMARY

We've now looked at the philosophy of the 7 steps of the manufacturing system and how to operate a company cognizant of them. Let's summarize the key points.

    I. The 7 steps of the manufacturing system is the natural method of producing products or delivering services for any company of any size.

    II. Successful companies will produce in accordance with this natural method whether they know it or not.

III. Companies that do not know of the 7 steps have a higher tendency to try to skip steps or to do them out of order.
IV. When this happens, many mistakes occur.
   A. Mistakes require "do overs."
      1. To correct errors in the process.
      2. To recover from scrapped materials and incorrectly finished products.
      3. To replace incorrect products or services sent to customers.
   B. Cycle time gets extended; schedules are missed.
      1. Inability to consistently do it right the first time causes instability in the entire schedule.
      2. If it gets bad enough, the client's confidence may be degraded.
V. Being cognizant of the 7 steps creates an awareness of when a method for accomplishing a task is not complete or not in the right sequence order.
   A. Awareness minimizes "dumb" mistakes.
   B. Awareness ensures that shortcuts being taken are prudent, not misguided.
VI. Those companies that consciously follow the dictates of the 7 steps are much more likely to have an effective communications system,
   A. thus, a workable CIM system.
   B. and results at a higher profit margin.

You, the entrepreneur just starting into business, can make your company into a world-class performer. It is not an impossible task. It can be done by paying attention to detail, by learning, and by bootstrapping the company to higher levels of performance. The first step is to assess where you are right now and then set out to improve. Make sure your starting level is where it should be. Look at how you presently perform the 7 steps (or intend to). Compare it against how world-class companies perform and then create a plan to get from here to there. You may need help in creating that plan and implementing the action steps. Don't be afraid to engage help, but keep in mind that you need to want to do it. The consultants you hire can't do it for you. All they can do is coach you and probably teach some needed skills. But the work of implementing is yours.

Good luck. The profits are there for the taking, but only for those who are really "making it" in the competitive manufacturing and service provider world.

Appendices

- **Investigation Points (product company)**   Appendix A
- **Investigation Points (service company)**   Appendix B

Appendices

Appendix A

INVESTIGATION POINTS (PRODUCT COMPANY)

[Please answer all questions. Be specific. Use back of page or extra blank sheets if necessary.]

Company: _____; Date: _____
Completed by: _____

Seven Steps of the Manufacturing System

1. Obtain product specification.

    1a. What are the company's products?
    _____
    _____
    _____

    1b. Does the company have a standardized design process to design its products and does it have standard designed components that are compatible with its manufacturing capabilities?
    _____
    _____
    _____

    1c. Is there a business plan setting sales goals and how the company is going to achieve them? If so, describe main points.
    _____
    _____
    _____

    1d. Does the company match its abilities with potential customers' real and perceived needs?
    _____
    _____
    _____

    1e. How does the company sell its products?
    _____
    _____
    _____

1f. How are orders obtained? What is the customer base?

1g. What is in the quote package or worksheet used to quote a customer? Is there a standard part that applies to any customer? How are specials handled?

1h. Is there a standard form that lists questions to be asked and answered for each order?

1i. How are orders received and booked?

1j. Does the customer confirm the particulars of the order? In writing? Verbal?

2. Design a method for producing the product, including design and purchase of equipment and/or processes to produce, if required.

   2a. Does the design of the product take into account the ability of the factory to make the product with ordinary due diligence for a satisfactory yield and cost?

   2b. Does the company design actively consider designs for jigs and fixtures as part of the design of the product, either generic or for a specific customer?

2c. How is the content of each job planned for the scope of work the employees need to do at each workstation?

2d. How does the company scope the time required to do each operation required to accomplish the manufacture of the product? Do standard times exist for subsets of common tasks? If so, are they based on estimates or developed time standards?

2e. How is the scope of work translated into planned hours to accomplish, tools needed to do the job, and material types and quantities? Are yield factors considered?

2f. Is there a methods sheet provided for each job? If so, does the methods sheet provide sufficient detail for workers to operate independently and know precisely what constitutes a complete job?

2g. Is there a Bill of Materials (BOM) for each job? If so, is it indented to show the proper assembly order and relationships between parts, subassemblies, and assemblies?

3. Schedule to produce.

    3a. Does a schedule exist showing all open jobs and the required completion dates?

3b. Does the schedule show the total portal-to-portal cycle time for each job at each workstation?

3c. Is the cycle time for the schedule derived from Step 2, above? What is the link or methodology for accomplishing this task?

3d. Is there a capacity document showing the amount of work capable of being completed in a certain time period? Is it used to create the schedule?

3e. Does the schedule process have an update routine that feeds back actual completions for comparison to plan? If so, is it done daily and is there a policy for revising schedules if the plan becomes too far out of synch with actual results?

3f. Is the schedule of sufficient longevity that the supervisors can plan their work load before the start of the next work period (day or shift)?

3g. Is the schedule broken down by workstations or centers and are workers told the amount of time they have to do the assigned job?

3h. Are overdue jobs given top priority to complete before new jobs are started?

3i. Are any rules established that give permission to bypass a job in the sequence to start a succeeding job?

4. Purchase raw materials in accordance with the schedule.

   4a. Does only one person have responsibility for purchasing, regardless of how many others have delegated responsibility?

   4b. Is a maximum/minimum inventory control system or equivalent used to ensure that there is an adequate supply of materials on hand?

   4c. Are special-buy purchases segregated and allocated to their designated jobs?

   4d. Is there a reservation system based on the schedule for materials to be allocated for each job?

   4e. Are material use needs forecasted based on schedule, previous history, or forecast?

   4f. Is an effort made to find the lowest-cost supplier and the most dependable suppliers?

4g. Is an effort made to buy in bulk to get volume discount prices and is this effort coordinated with the schedule and forecast?

4h. Are lead times for delivery of materials taken into account for ordering materials and are they coordinated with the schedule and forecast?

4i. Are contracts consummated with suppliers to effect a JIT relationship based on long-term commitments to buy and supplier commitments to have on hand on an as-needed basis?

4j. Have supplies been differentiated based on shelf life parameters to ensure proper limits of bulk-buy parameters?

4k. Are reports prepared periodically (at least on a quarterly basis, preferably a combination of weekly and monthly) showing as a minimum inventory levels, material orders outstanding, jobs waiting because of late materials, vendor performance measurements, quality issues, cost values for all categories, and trends measurements for all categories?

5. Produce in the factory.

   5a. Are workers trained or given an orientation before being place on the job, and does this consist of job-specific instructions and specific safety and environmental hazards they may encounter?

5b. Is work done in accordance with a daily dispatch list?

5c. Do workers know what job to work on beforehand and know how many hours are scheduled for the job?

5d. Do workers receive instructions for each assigned job showing all requirements and methods to follow to do the job?

5e. Are workers required to report back completion of assigned jobs before receiving the authorization to proceed to the next job?

5f. Do supervisors check progress and help workers eliminate obstacles/problems that may hinder the job being completed on time?

5g. Is there a procedure to list obstacles/problems that cannot be resolved quickly for a planned resolution activity?

5h. Are obstacles/problems from 5f, above, purposely resolved and the results reported back to the worker?

5i. Are workers given the opportunity to suggest ways to do the job better and are they rewarded for their initiative?

5j. Are records maintained and reports issued (at a minimum on a monthly basis) for performance based on scheduled vs. actual cycle times?

6. Monitor results for technical compliance and cost control.

    6a. Is job performance monitored for compliance with job specifications?

    6b. Are reports issued showing noncompliance with root causes identified and corrective action requirements determined and implemented?

    6c. Are compliance trend reports issued at least monthly?

    6d. Do workers certify that they did the job in accordance with method instructions and that the results comply with job specifications?

    6e. Is the Total Quality Management (TQM) theory used to encourage continuous improvement? If so, are workers trained in the basics of TQM?

6f. Are vendors monitored for compliance with material and subcontractor quality requirements?

6g. Is work force absenteeism recorded with reasons and associated corrective actions tracked daily or at least weekly?

6h. Are work hours and costs for all active jobs booked daily and are they coordinated with the work schedules of Step 3, above?

6i. Are reports generated daily showing hours worked vs. schedule and efficiency, cost for labor expended vs. planned-for jobs completed and in progress, indirect labor costs expended vs. budget, material costs expended vs. planned, and the profit or loss (P/L) preliminary estimate for each completed job?

6j. Is a pro-forma P/L issued monthly with explanations of variances?

6k. Are capital expenditures formally approved and expected results tracked? Is there an established hurdle rate policy set for capital expenditure requests?

6l. Are costs for administration, sales, R&D, finance, nondurable tools and supplies, utilities, maintenance, vehicle operations, and general repairs tracked at least quarterly?

7. Ship the completed product to the customer.

   7a. Does the worker have a checklist that he or she uses to ensure that all aspects of the job are completed before turning it over to Shipping (or the customer)? Does the worker initial or sign it?

   _____
   _____
   _____
   _____

   7b. Does the supervisor or worker inform Shipping when the job is done and available for shipment to the customer?

   _____
   _____
   _____
   _____

   7c. Are products kitted against a bill of lading to ensure that all of the customer's order is consolidated for shipments before dispatching to the customer?

   _____
   _____
   _____
   _____

   7d. Is it a policy to ask customers to verify that the work has been done to their satisfaction? Are the customers asked to sign indicating verification or items not to their satisfaction?

   _____
   _____
   _____
   _____

   7e. Is it policy to satisfy all discrepancies with the customer "on the spot" during turnover to the customer? If so, does this take priority over all other work?

   _____
   _____
   _____
   _____

Appendix B

INVESTIGATION POINTS (SERVICE COMPANY)

[Please answer all questions. Be specific. Use back of page or extra blank sheets if necessary.]

Company: _____; Date: _____
Completed by: _____

Seven Steps of the Manufacturing System

1. Obtain product specification.

    1a. What are the company's services?
    _____
    _____
    _____
    _____

    1b. Does the company have a standardized design process to design its services and does it have standard designed components that are compatible with its delivery capabilities?
    _____
    _____
    _____
    _____

    1c. Is there a business plan setting sales goals and how the company is going to achieve them? If so, describe main points.
    _____
    _____
    _____
    _____

    1d. Does the company match its abilities with potential customers' real and perceived needs?
    _____
    _____
    _____
    _____

    1e. How does the company sell its services?
    _____
    _____
    _____
    _____

1f. How are orders obtained? What is the customer base?

1g. What is in the quote package or worksheet used to quote a customer? Is there a standard process that applies to any customer? How are specials handled?

1h. Is there a standard form that lists questions to be asked and answered for each service order?

1i. How are orders received and booked?

1j. Does the customer confirm the particulars of the order? In writing? Verbal?

2. Design a method for producing the product, including design and purchase of equipment and/or processes to produce, if required.

 2a. How is each job planned for the scope of work the employees need to do?

 2b. Do standard times exist for subsets of common tasks? If so, are they based on estimates or developed time standards?

2c. How is the scope of work translated into planned hours to accomplish, tools needed to do the job, and material types and quantities?

2d. Is there a methods sheet defining the process steps to be used and in the sequence provided for each job?

2e. Is there a set of processes defined and available to provide the service for each job?

3. Schedule to produce.

    3a. Does a schedule exist showing all open jobs and the required completion date?

    3b. Does the schedule show the total portal-to-portal cycle time for each job?

    3c. Is the cycle time for the schedule derived from Step 2, above? What is the link?

    3d. Is there a capacity document showing the amount of work capable of being completed in a certain time period? Is it used to create the schedule?

3e. Is the schedule of sufficient longevity that the supervisors can plan their work load before the start of the next work period (day or shift)?

3f. Is the schedule broken down by workers or teams and are they told the amount of time they have to do the assigned job?

3g. Are overdue jobs given top priority to complete before new jobs are started?

3h. Are any rules established that give permission to bypass a job in the sequence to start a succeeding job?

3i. How is the schedule kept current?

4. Purchase raw materials in accordance with the schedule.

   4a. Does only one person have responsibility for purchasing, regardless of how many others have delegated responsibility?

   4b. What kind of inventory-control system is used to ensure that there is an adequate supply of materials on hand to do the job?

4c. Are materials reserved for each job?

4d. Is there a reservation system based on the schedule for materials to be allocated for each job?

4e. Are material use needs forecasted based on schedule, previous history, or forecast?

4f. Is an effort made to find the lowest-cost supplier and the most dependable suppliers?

4g. Is an effort made to buy in bulk to get volume discount prices and is this effort coordinated with the schedule and forecast?

4h. Are lead times for delivery of materials taken into account for ordering materials and are they coordinated with the schedule and forecast?

4i. Are contracts consummated with suppliers to effect a JIT relationship based on long-term commitments to buy and supplier commitments to have on hand on an as-needed basis?

4j. Have supplies been differentiated based on shelf life to ensure proper limits of bulk-buy parameters?

4k. Are reports prepared periodically (at least on a quarterly basis, preferably a combination of weekly and monthly) showing as a minimum inventory levels, material orders outstanding, jobs waiting because of late materials, vendor performance measurements, quality issues, cost values for all categories, and trends measurements for all categories?

5. Produce in the factory (provide value-added work to be applied to the service).

5a. Are workers trained or given an orientation before being place on the job, and does this consist of job-specific instructions and specific safety and environmental hazards they may encounter?

5b. Is work done in accordance with a daily dispatch list?

5c. Do workers know what job to work on beforehand and know how many hours are scheduled for the job?

5d. Do workers receive instructions for each assigned job showing all requirements and methods to follow to do the job?

5e. Are workers required to report back completion of assigned jobs before receiving the authorization to proceed to the next job?

5f. Do supervisors check progress and help workers eliminate obstacles/problems that may hinder the job being completed on time?

5g. Is there a procedure to list obstacles/problems that cannot be resolved quickly for a planned resolution activity?

5h. Are obstacles/problems from 5f, above, purposely resolved and the results reported back to the worker?

5i. Are workers given the opportunity to suggest ways to do the job better and are they rewarded for their initiative?

5j. Are records maintained and reports issued (at a minimum on a monthly basis) for performance based on scheduled vs. actual cycle times?

6. Monitor results for technical compliance and cost control.

6a. Is job performance monitored for compliance with job specifications?

6b. Are reports issued showing noncompliance with root causes identified and corrective action requirements determined and implemented?

6c. Are compliance trend reports issued at least monthly?

6d. Do workers certify that they did the job in accordance with methods instructions and that the results comply with job specifications?

6e. Is the Total Quality Management (TQM) theory used to encourage continuous improvement? If so, are workers trained in the basics of TQM?

6f. Are work hours and costs for all active jobs booked daily and are they coordinated with the work schedules of Step 3, above?

6g. Are reports generated daily showing hours worked vs. schedule and efficiency, cost for labor expended vs. planned-for jobs completed and in progress, material costs expended vs. planned, and the profit or loss (P/L) preliminary estimate for each completed job?

6h. Is a pro-forma P/L issued monthly with explanations of variances?

7. Ship the completed product to the customer.

   7a. Does the worker have a checklist that he or she uses to ensure that all aspects of the job are completed before turning it over to the customer? Does the worker initial or sign it?

   7b. Does the supervisor or worker inform the customer when the job is done and available for the customer?

   7c. Is it a policy to ask customers to verify that the work has been done to their satisfaction? Are the customers asked to sign indicating verification or items not to their satisfaction?

   7d. Is it policy to satisfy all discrepancies with the customer "on the spot" during turnover to the customer? If so, does this take priority over all other work?

Glossary

Above the line: a series of accounts representing all income and expenses (called cost of goods sold) directly related to producing and selling the product or service, the result of which is the gross profit of the company.

Accounts payable: liabilities due for payment in the current period.

Accounts receivable: cash assets due to the company during the current period, such as cash for sales made.

Accruals: obligations or sales that are credited exactly when they are committed, not when the transaction (passing of cash between the parties) occurs.

Assets: everything a company owns converted to equivalent cash value plus cash on hand.

Balance sheet: a financial measurement showing a snapshot of a company's financial strength at any particular time by playing off its assets vs. its liabilities. The balance sheet shows what the company is worth if it were to be liquidated at that point in time.

Basic tenets of manufacturing: the fundamental theory of manufacturing; the 2 "Knows"—know how to make the product and know how long it should take to do so.

Below the line: a series of accounts representing the overhead costs associated with producing and selling the product, the total of which is subtracted from the gross profit to obtain the net profit (before federal and state income taxes).

Best practices: the process of doing work in a manner that is considered "world class."

Body language: portraying meaning without spoken or written words through the use of body positions and gestures.

BOM: bill of materials; a concise summary of the sequence of manufacturing along with the materials required in the required quantities.

Budgets: based on the pro-forma profit or loss (P/L) statements, are detailed breakdowns of accounts assigned to the various departments of the company for a measurement period.

Business plan: a method of achieving a set of established goals within the confines dictated by the strategic situation.

Business team: a semi-permanent assemblage of individuals from many different disciplines making up the needs to further a specific business objective. Quite often assembled to design, build, and launch a new product or service.

Capability: the equipment and skills to perform a manufacturing or service operation in a manner that meets design requirements and at the same time can produce profits.

Capacity: a measure of the ability to make products to required design standards at a rate suitable to economic viability.

Cash-flow statement: a financial measurement that shows the up-to-the-minute cash receipts and dispersements of a company. It is akin to a general ledger or the balance in a checking account.

Commercially viable: In a business sense, an idea that can be translated to a product or service produced and sold at a cost and price that is acceptable to the producer and customer.

Communications: the process of transmitting a thought from one mind to another. Or for modern CIM applications, the ability to transmit data from one computer to another.

Communications excellence: The ability to transfer information to another person and be assured that the transmission is fully received and understood. Or for modern CIM applications, the ability to transmit data from one computer to another with 100% accuracy.

Communications symbols: words, spoken or written, or data such as digital data used by a fiber-optics cable.

Computer aided process planning (CAPP): the use of computers to coordinate and calculate process planning times used for scheduling production. A subcomponent of CIM.

Computer Integrated Manufacturing (CIM): a generic term meaning achieving communications excellence via computer synergism throughout the business entity.

Computer Numerical Control (CNC): using computer hardware and software to drive and operate a machine tool or robot. The term CNC is used to denote an automated machine or robot.

Concurrent engineering: the process of a team approach toward doing the conceptual ideas, design, development, production sales, and deliveries of products or services in a parallel-step approach rather than the traditional function-series approach.

Contingency planning: the creation of variance situations before they happen so we can also create reasonable corrective actions to mitigate those fictitious variances. Corrective actions would be implemented if the fictitious variances become realities.

Core essential functions of a business organization: marketing, operations, and finance.

Cost of goods sold (COGS): comprises all of the direct costs associated only with the making of the product or producing a service. Typically includes direct labor and applied materials.

Downsizing: reducing staff and functions in a company to save operating expenses. Often done without regard to strategic implications.

E-commerce: doing business through electronics interchanges such as local area networks, the Internet, etc., using CIM technologies such as ERP.

Engineering: the application of scientific principles to solve business problems.

Entrepreneur: one who organizes, manages, and assumes the risk of a business or enterprise.

ERP: enterprise resources planning; see definition of MRP II.

Fire in the belly: an expression meaning an extremely strong desire to succeed that drives a person to overcome all obstacles to success.

Fixed expenses: operating expenses that management cannot control except in a strategic sense over a long time period, such as rents and non-breakable leases.

Gap analysis: a process of evaluating how a company performs the 7 steps of the manufacturing system compared with "world-class" standards.

Globalization: referring to competing in markets once closed and protected, now opened due to paradigm shifts in ease of communications and transportation.

Goal: a singular or multiple sets of specific measurable statements of intent that define when the desires (or portions thereof) or the objective(s) will be accomplished and bounded by time. In an industrial setting, the time frame is seldom longer than 1 year.

Gross profit: the subtraction of COGS from total sales.

Gun for hire: in a business sense, an employee hired for a specific skill who leaves once that skill is no longer required; a permanent employee positioned between a temporary worker and a consultant.

Hearing: being sensitive to acoustic energy from verbal communications, but deriving no meaning from it.

Intrapreneur: title given to a business team leader whose task is to cut through large company red tape and create new products and/or services for her or his corporation as speedily and with as much zeal as an entrepreneur working on her or his own.

Job description: a narrative explaining the responsibilities, authority, and accountability of a position within the organization of a company. Usually used to measure performance of an incumbent and qualifications for a new-hire to the company.

Job rating system: a scheme for evaluating the worth of all jobs within a company in order to create an ascending list of importance. Used for establishing a pay plan for the company.

Just in time (JIT): a popularized version of an industrial engineering practice for effective use of resources to eliminate waste.

Lean and mean: downsizing to the lowest level practical and still having a staff capable and dedicated to optimizing profits.

Liabilities: everything a company owes to the outside world.

Liquidity: The cash assets of a company at any particular time. Any assets that are not in cash are nonliquid and not usable in paying off accounts payable.

Listening: deriving meaning and comprehension from verbal communications.

Margin: the difference between a product or service selling price and cost to produce.

Mission: a clearly stated purpose showing how tasks will be organized to make the vision a reality.

Mistake-proofing (Poka-Yoke): a methodical technique in which you (1) look at a process or a design to define probabilities of doing each subset of the task without making a mistake, (2) then start with the highest probability of failure to create improved methods that will eliminate or at least minimize the probability of failure, and (3) continue this process for all task subsets until the entire process has been mistake-minimized.

MRP: (sometimes called "little mrp"): materials requirement planning. MRP is a computer program for determining the quantities of materials required to support scheduled production and the sequence desired.

MRP II: (sometimes called "big mrp") manufacturing resources planning. MRP II is an outgrowth of MRP and is an integrated scheduling and feedback system that allocates workstation time and delivery of materials in accordance with a master schedule. It also coordinates support services such as design of fixtures and jigs to support production. In its most encompassing form, it is also known as enterprise resources planning, ERP.

Net profit: the subtraction of operating expenses from gross profit.

New Product Introduction (NPI): a logical process of going from an idea to a commercially viable product based on the "scientific method."

Objective: a singular or multiple sets of broad-based generalized statements of intent that define how the mission will be implemented. Usually ongoing for as long as the organization exists.

Operating expenses: all expenses associated with producing a product or service with the exception of COGS. Sometimes referred to as overhead expenses.

Order entry: the process of converting a request for work into a fulfillment plan within the organization to produce a product or service. Can be externally oriented such as a sales order, or internal such as a maintenance request. Usually a part of an ERP system.

Organization structure, hierarchical: a rigid structure that follows a military style, from small units to large groupings, each with singular reporting sequences of subordinate to superior.

Organizational structure, matrix: a semi-fluid structure in which employees report to a functional manager for an administrative purpose and a team for work direction. Employees can be reassigned to teams at will to suit the present needs of the company.

Outsourced products: parts and/or completed products once made internally now made in vendor facilities, usually to take advantage of lower costs.

Overhead expenses: the entirety of the operating expense portion of the P/L that is prorated against products or services offered in some sort of equitable manner, then added to COGS to arrive at a selling price.

Owners' equity: the cash value of the worth of the company that the owners would receive if the company could be liquidated.

Paternalism: referring to the practice of companies to set policies to generally retain employees for their entire work careers.

Payback period: a criterion for evaluating the worth of a capital investment based on dividing the gross costs of a project by the expected annual savings or profit from the investment. This shows the time in years it takes for a project to pay for itself.

Pay plan: a logical approach on how to compensate the various positions within a company's organization. Usually based on the job-rating system.

Present-worth method: a financial analysis procedure using the time value of money concept to evaluate the worth of a capital investment.

Principles of motion economy: an industrial engineering body of knowledge based on the physics of human body motions related to energy expenditures and the times to do the motions.

Producibility engineering: The process of creating designs that can be made efficiently in the factory for which it was intended. This is done by matching process capability with design functionality requirements.

Product life cycle: The characteristic "S" curve a product goes through from conception to removal from the marketplace due to lack of desirability on the part of the consuming public.

Profit: the difference between the selling price of a product or service and the cost to produce it.

Profit or loss statement: a financial measurement shows whether the company made money or lost money over a reporting period, by stating sales for the period and subtracting all costs, both direct and indirect, that occurred over the same time period.

Pro-forma profit or loss (P/L) statement: a financial measurement identical to a profit or loss statement except that it is based on expected future happenings; used to validate projects of all types before commencing.

Project management: the art of managing a nonrepeatable multistep assignment that has a beginning and an end point with desired results that are measurable and bounded by time. It is the formalization of tasks into achievable and measurable steps in order to reach a set goal.

Projects: a specific plan with measurable steps, bounded by time, that lead to the accomplishment of a goal. One or a set of projects may be required for the accomplishment of a single goal.

Pull production system: a scheduling system based on the set required finish date where operations cycle times are subtracted from the finish date to obtain a start date. Also called backwards scheduling.

Push production system: a scheduling system based on starting operations on a set schedule to finish in accordance with cumulative operation cycle times. Also called forward scheduling.

Quality Functional Deployment (QFD): a methodology of matching customer needs with company capabilities to determine whether the company can supply those needs.

Reengineering: a euphemism for downsizing, usually done with the aid of consultants.

Return on Investment (ROI): a financial calculation procedure for evaluating the worth of an investment decision, usually for capital equipment.

Risk analysis and management: the process of defining the level of uncertainty, such as the probability that an unplanned event will happen, and then developing contingencies to deal with unplanned events.

Rose-colored glasses: as in seeing the world through rose-colored glasses; meaning being overly optimistic to the point that it is unrealistic and perhaps damaging.

Routing: also known as route. The sequence of workstations a product will traverse through from raw materials to finished product, as defined by the Bill of Materials (BOM).

Scientific method: a procedure for investigating a phenomenon that can be applied to all problem solving. Involves making an observation, creating a hypothesis of why the observation is as it is, testing the hypothesis, revise the hypothesis if results are insufficient, test the revised hypothesis, iterate the hypothesis/test cycle to reach a satisfactory conclusion.

Scientific time standard: a time standard for doing work based on the principles of motion economy and other precise data related to machine and human movements.

Seven steps of the manufacturing system: the fundamental theory of control of an enterprise providing goods and/or services.

Short-interval scheduling: a method of monitoring production to ensure that goals are attained through creating an output level agreed to between operator and supervisor and identifying and correcting impediments to achieving the goal. Usually employed in time periods of less than a shift.

Short-range forecast: a financial forecast, based on the profit or loss statement formatted budget, of expenses and credits (sales), usually for up to 3 months.

Six Sigma: a philosophy of merging SPC, including the investigative portions of TQM, with mistake proofing (Poka-Yoke) and team-based management concepts.

Statistical process control (SPC): The use of statistical techniques to evaluate the effectiveness of operations in achieving set goals for production. Commonly used to determine whether a manufacturing or office process is "under control" for producing products at the desired quality level.

Stopwatch time standard: a time standard for doing work based on repeated observations recorded and averaged via a stopwatch and applied to related machine and human movements.

Supply chain: a group of companies contractually linked to produce a product much the same as a vertically integrated company would.

Strategic planning: the process of evaluating the environment a company chooses to compete in, by identifying the constraints and creating a path for the company to traverse to meet its goals.

Total quality management (TQM): a philosophy of striving for continuous improvement by measuring results of previous actions against established goals and then correcting the process to achieve better future results.

Variable expenses: operating expenses that are deemed to be controllable by management over a relatively short term, such as purchases of stationery and supplies.

Vertically integrated company: a firm that owns all the necessary capabilities and resources for making its products from conception to delivery and service to its customers.

Virtual factory: a group of companies that make up a supply chain to manufacture a product from raw materials to delivery and customer service, whereby they are joined contractually to perform subtasks of the whole. Thus acting in the aggregate as a vertically integrated company.

Vision: a concept or formulation of a desirable end point.

Work Breakdown Structure (WBS): The subdividing of work into sufficient detail so that predecessor steps tie directly to the next step without any missed activities.

Workstation: a place where raw materials or ideas are transformed through value-added processes to products or services to meet customer needs.

World-class standards: an expression connotating the best way to perform a specific task or plan.

Selected Related Readings

Bower, Joseph L. *The Craft of the General Manager,* Harvard Business School Press, Boston, 1991.

Bradford, David L.; Cohen, Allan R. *Influence Without Authority,* John Wiley & Sons, New York, 1990.

Breyfoggle III, Forrest W. *Implementing Six Sigma, Smarter Solutions Using Statistical Methods,* John Wiley & Sons, New York, 1999.

Buzan, Tony; Dottino, Tony; Israel, Richard. *The Brain Smart Leader,* Gower Publishing Limited, Brookfield, VT, 1999.

Carlson, Richard K., *Personal Selling Strategies for Consultants and Professionals,* John Wiley & Sons, New York, 1993.

Fogg, C. Davis. *Team-Based Strategic Planning, A Complete Guide to Structuring, Facilitating and Implementing the Process,* Amacom, New York, 1994.

Delaney Jr., Robert V.; Howell, Robert A. *How to Prepare an Effective Business Plan: A Step-by-Step Guide,* Amacom, New York, 1986.

Griffith, Gary K. *The Quality Technician's Handbook,* 4th Edition, Prentice-Hall, EngleWood Cliffs, NJ, 2000.

Hargrove, Robert. *Mastering the Art of Creative Collaboration,* McGraw-Hill, New York, 1998.

Kepner, Charles H.; Tregoe, Benjamin B. *The New Rational Manager,* Princeton Research Press, Princeton, NJ, 1981.

Koenig, Daniel T. *Manufacturing Engineering; Principles for Optimization,* 2nd Edition, Taylor & Francis, Washington, DC, 1994.

Koenig, Daniel T. *Fundamentals of Shop Operations Management: Work Station Dynamics,* ASME Press & Society of Manufacturing Engineers, New York, 2000.

Lewis, James P. *The Project Manager's Desk Reference,* 2nd Edition, McGraw-Hill, New York, 2000.

Martin, Paula; Tate, Karen. *Project Management Memory Jogger,* Goal/QPC, Salem, NH, 1997.

Marrus, Stephanie K. *Building the Strategic Plan: Find, Analyze and Present the Right Information,* John Wiley & Sons, New York, 1984.

McMullen, Jr., Thomas B. *Introduction to the Theory of Constraints (TOC) Management System,* The St. Lucie Press, Boca Raton, FL, 1998.

Pfeffer, Jeffrey. *The Human Equation, Building Profits by Putting People First,* Harvard Business School Press, Boston, 1998.

Shim, Jae K.; Joel G. Siegel. *Budgeting Basic & Beyond,* Prentice-Hall, New Jersey, 1994.

Index

A
Above the line 166, 169, 173
Abstractions 105, 106
Acceptable imprecision 207
Accounting system 175
Accounts payable 165
Accounts receivable 165, 189
Action steps 41, 42, 65, 66, 68
Administrative plan 237, 242
Administrative tasks 190, 192
Application courses 21
Assessment 239, 241
Assessment development plan 240
Asset 159, 161, 163
Assumptions 102
Available precision 207

B
Balance sheet 157–161, 163, 166, 171, 183, 224, 261
Bank loan 185
Bar coding 212
Barcode 214
Barriers to effective communication 114, 116
Barriers to effective transmittal 103
Basic tenets of manufacturing 79, 149, 287
Below the line 166, 169, 173
Benchmarking 53, 202, 297
Best practices 271, 272
Bill of materials (BOM) 137, 138, 149, 150, 167, 212, 249, 277, 279, 292
Body language 100–103
Bootstrap 202, 203
Brainstorming 134
Budgets 166, 168–177, 179, 282
Business
  control system 187
  cycle 171
  owner 71
  plan 26, 74, 168, 193, 215–217, 220–229, 233, 236, 239, 243, 249, 252–254, 258, 261, 263–266, 268, 282, 300
  planning 64, 76
  product team 71
  purpose of 71
  system 274
  team 8, 16, 21, 23, 26, 78, 92–94, 96, 99, 116
  transaction 207

C
Calendar checklist 191
Calendar of standard events 190
Capacity 138, 139, 141, 142
Capacity requirements planning 88
Capital plan summary 224
Capstone design courses 16, 18, 21
Career security 7
Carpenter's rule 47
Cash
  check sheet 181–183
  flow 5, 180, 184, 189, 261
  negative 184
  reserves 36
  statement 157, 158, 163, 165, 166, 171, 183, 224
Cause-and-effect planning 45
Change 46, 48, 58, 301
  control 46
  mandated 47
Chronological list 127, 128
Client 7
CNC 139
Cohesive plan 79
Collateral 185, 187
Collection agency 189, 190
Commercial viability 127, 128, 132
Commercialize 126, 127
Common core experiences 103, 104

Common data 263
Communicating with customers 114
Communicating with potential
   customers 114
Communication
   loop 115
   process 107
   skill 19
Communications excellence 99,
   101–103, 106, 122, 286
Communicator 100–102, 105, 106
   company loyalty 39
Compensation 253
   planning 254
Competition 234
Computer-aided process planning
   system (CAPP) 278, 281
Computer-based Manufacturing
   Resources Planning or Enterprise
   Resources Planning (MRP/ERP)
   187
Computer-integrated manufacturing
   (CIM) 19, 278, 279, 281, 285
Computer Numerical Control (CNC)
   49, 52, 54, 55, 59
Concept of customer 120
Concurrent engineering 276
Consultant philosophy 8
Contingencies 45, 70
Contingency planning 45, 46, 59, 60,
   63, 65
Continuous improvement 121, 280,
   281
Cost of goods sold (COGS) 146, 149,
   150, 167, 262
   as a percentage of sales 176
Commercialization 155, 156
Commercially viable 134
Common databases 205, 206
Communications excellence 120, 207,
   285
Competition 218, 235–237
Concurrent engineering 276
Concurrent schedule 129
Contingencies 266, 267, 269

Contingency plan 216, 226, 227,
   266–268
Core essential organization 83, 86, 92
Core strengths 137
Corporation 72
Cost of goods sold (COGS) 146
Credit cards 186
Critical path diagram 44
Current liabilities 165
Curriculum 14–16, 18, 26, 32

D
Data 240
Database 76, 77, 88, 100, 210–213
Decision rule tree 62
Design engineering 134, 140
Direct labor as a percentage of sales
   176
Discontinuities 45
Dividends 72
Do good factor 73
Document control system 254
Downsized 10
Downsizing 4

E
E-commerce 195, 205, 206, 209, 211,
   213, 214
Economic variance 177
Employee handbook 253
Employment summary 223
Engineering 136, 142
   definition of 14
Enterprises resources planning
   system 88, 278
Entrepreneur 11, 12, 26–28, 35, 44,
   45, 56, 62–64, 74, 86, 89, 101,
   122, 142, 153, 162, 184–187, 209,
   228, 249, 261, 263, 276, 280–284
Entrepreneurial spirit 7
Entrepreneurship 35
Environment 46
Environmental influences 218, 234,
   235

Equity 161–163, 184, 201
Executive summary 216–218, 225
Expediting collections 187

F
Facilitator 130
Factoring 186, 187, 189
Feedback loop 108, 109
Feel good 74
Feel good motive 73
Finance 83
  function 80
  plan 223, 237, 242, 269
  task 92
Financial
  compliance control 282
  evaluation 49, 50, 51, 53
  measurements 180, 224
  plan 246, 261–264
Financials 216, 224, 226
Fixed expenses 150
Flow type production 207
Fulfillment 205–207, 209, 211
Full resources 8
Fuzzy logic 20

G
Gantt charts 44
Gap analysis 272, 297, 301
General and administration plan 222, 242, 246, 248–251, 254, 269
General ledger 165, 169, 171, 174, 175, 188
General plan 237
Generic product 196
Globalization 5
Goal
  group 37
  singular 37
Goals 66, 74, 76–78, 218, 221, 223, 224, 237, 245, 258, 263–265
Gross income as a percentage of sales 176
Gross profit 146, 147, 166–168, 263

Group dynamics 203
Group technology 211, 212

H
Hearing 109, 111
Hierarchical organization structure 1
Human capital 36
Human resources 136
  program 251
Hurdle rates 36, 50, 52

I
Implied contract 7, 8
Individual contractor 39
Integrated measurable plan 264
Intrapreneurs 94–97, 99
Intrepreneur 128
Invoice 187, 188
ISO 9000 certification 304
ISO 9000 registration 281
ISO 9001 135
Iterative process 140, 143
Inventory control 283

J
Job
  descriptions 258, 259, 262
  evaluation 255
  responsibilities 261
  security 7
  shop 86
  system 258
Judgment 49–51
Just in time (JIT) 278

K
Kepner Tregoe must/wants process 302

L
Laws of supply and demand 134
Legal recourse 189, 190

Liability 159, 161–163
Lines of credit 185
Listening 109, 110, 111

M
Make or buy 51
Making it 272, 273, 279–285, 304
Management 28
  tasks 78
Mandated 47
Manufacturing 134, 136, 142, 171
  capability 6
  engineer 137, 140
  engineering 139
  effectiveness programs (MEPS) 201, 202
  resources planning systems (MRPS) 205, 206, 208, 209, 213
  resources planning system (MRP II) 86, 88, 278, 279, 281, 282
Margin 150
Market
  analysis 228
  and facilities expansion summary 224
  assessment 218, 233–235, 237, 239
  assessment and segmentation review 243
  assessment development plan 238, 240
  (customer) segmentation 218, 233, 234, 236, 240, 262
  situation 228
Marketing 83, 131, 134, 135, 142, 169
  department 83
  tasks 92
Mass production 207
Master schedule 88
Materials 137
Materials requirements planning (MRP) 205
Measurable goals program plan 246
Measurements 303, 304
Mentoring advantage 214
Mentoring relationship 202
Migating risks 62

Mission 74, 77, 226, 266
Mission statement 77, 217, 218, 221, 228, 230–232
Mistake-proofing (Poka-yoke) 281
Multifunctional team 129

N
Natural law of business 131
Negotiating 116
Negotiating QFD process 116, 117, 119
Negotiations 115
Net income as a percentage of sales 176
Net profit 147, 150, 166, 167, 168, 198, 263
New product 4
New product introduction (NPI) 127–131, 133, 134, 139, 142, 143, 146, 152, 154–157, 173, 193, 203, 204, 262
  system 132
Non-goal-mandated changes 47
Non-goal-oriented changes 48
Nonfinancial measurements 180
Note taking 113

O
Objectives 66, 76–78, 218, 221, 223, 224, 226, 228, 229, 232, 233, 263, 265, 266, 269
Objectives and goals 66, 73, 245, 282, 304
Objectives and goals management 65
On-the-job learning 21
Ongoing objectives 220, 221
Operating expenses 167–169, 173, 262
  as a percentage of sales 176
Operations 252
  function 81
  rule 1 274
  sequencing 89
  task 83, 92
Opportunities identification 131

Order entry 205–213
Orders release 89
Organization 252
　chart 91
　structure 78, 79
Overhead 147, 168
　expenses 146, 263
Owner's equity 159

P
Paradigm shift 58
Paralysis by analysis 18
Partially satisfactory solutions 18
Participatory management 301
Partnership 72
Paternalism 3, 4, 7, 10
Pay plans 254
Payback period 52–54
Percent ratios 170
Period goals 221
Pert chart 44
Phases of design 276
Plan 68
Planning 66
　contingecy 68
Post-NPI 155
Practical QFD process 117, 119
Pragmatic risk probability 59
Present worth method 51, 53
Pro-forma P/L 143, 149, 150, 152, 166, 261
Probability and statistics 48, 49
Probability assessment 55
Probability of failure 45
Producibility engineering 276
Production plan 222, 228, 237, 242, 245–247, 264, 265, 269
Product specification 83, 136
Professional engineering 12
Profit 71, 72
　for 72
　generating 37
　generation 36
　margins 199, 201
　motive 71–73
　not for 71–73

Profit/loss (P/L) 146, 149, 162, 163, 166, 169, 176, 181, 183, 224, 229, 262, 263
　expense portion 146, 147
　income portion 147
　statement 157, 158, 169, 171
Programs 216, 223, 226
Project 263, 265, 303
　management 28, 35, 37, 39–41, 44, 66, 68–70, 96, 129, 157, 203, 224, 299
　management process 40
　management skills 39, 40
　manager 4, 154
Projects 4, 66, 76–78
Prototype 136, 137, 141, 143, 155, 156

Q
Qualified supplier 197
Quality-assurance 280
Quality functional deployment (QFD) 28, 30, 31, 33, 113, 116, 117, 119, 134, 203, 204, 243, 276
　general approach 116

R
Rate of production 138
Receiver 103, 106, 108, 109
Receptor 101–103, 105
Recipients 107, 108
Reengineering 6
Regional manufacturers association 202
Resources required 216, 223, 226
Return on investment method (ROI) 51, 52
Reverse engineering 6, 301
Review of operations 179, 180, 193
Risk
　analysis/management 41, 42, 45, 47, 65
　assessment 49, 56
　avoidance 63
　controllable 64

Risk *(continued)*
  detectability of 61
  factors 55, 58, 60
  level of 54
  minimization 63
  mitigation 63
  modifiers 61
  probability 61
  probability number 61
  severity of 59, 60
  takings 95
  transfer 63
  unacceptable 64
Risks 46, 48, 50, 53, 59, 61, 69, 70
Root causes 177, 280, 282
Routes 138
Routing 137

S

S curve 141, 142
Salary plan 254
Sales department 83
Sales plan 221, 222, 228, 237, 239, 241–243, 246, 261, 263, 269
Sales plan development strategy 242, 243
Science of business 271, 272
Scientific method 11, 121, 125, 126
Segmentation 236, 241
Segmentation development plan 239, 240
Selective listening 112
Self-evaluation 220, 234
Sender 101, 103, 108
Series schedule 129
7 steps of the manufacturing system 26, 82, 83, 85, 86, 88, 127, 130, 188, 191, 201, 202, 206, 207, 258, 259, 271–273, 275, 283, 285, 292, 294, 297, 301, 302, 304
Severity 61
Severity test 60
Shop operations 137
Short-interval scheduling 280
Short-range forecast 178–180, 191, 193

Situation analysis 220, 225, 233, 234, 236
Six sigma 281, 304
Sole proprietorship 72
Specification 135
Stakeholders 273
Standards 304
Standard repeatable tasks 193
Statistical process control (SPC) 281, 285
Steady-state operations 154
Strategic issues 220, 234
Strategic planning 64, 215, 216
Strategies 216, 226
Strengths 237
Subjective process 64
Supply chain 51, 119, 137, 195–199, 201–206, 208–210, 214, 279
  mangement 96
  relationships 198
SWOT 62, 63
Symbols 100

T

Target output rate 139
Team player 71
Technical-based
  businessman/woman 8, 10, 13
  education 11
Technical capability 197
Technical problem-solver 11
Technology capability 131, 133
Theoretical QFD process 117
Thinking out-of-the-box 140
Three-month rolling forecast 177
Three phases of customer communications 114
3 phases of design 277
Time value of money 54
Total quality management (TQM) 113, 119–121, 122, 281
  triangle 120, 281
Training 253
Transactions 195
Turnaround 142
Two knows 79, 88, 278, 280

Index • 345

U
Uncertainty 45, 48, 49, 58, 59
　rating 56
Uniform products code 212
University-sponsored business-development groups 201

V
Value-added 278, 279
Variable expense 150
Variance 49, 58, 180
Venture capitalist 128
Verbal communication 113
Verbal messages 107
Vertical integration 195, 196
Vertically integrated company 198, 199, 204
Vertically oriented company 6
Virtual corporations 195
Vision 77, 217, 218, 226, 266
Vision and mission statement 73, 74, 78, 221, 265, 269

W
Weaknesses 237
Word symbols 105
Work 81–85
Work Breakdown Structure (WBS) 41–45, 65, 66, 68
Workstations 278–280
World-class companies 275, 276, 278–285, 288
World-class competitor 293
World-class performance 271–273, 275, 277, 301, 304
World-class standard 297, 299, 303
Written messges 107

Z
Zero defects 197